Margins of Error

THE WILEY BICENTENNIAL–KNOWLEDGE FOR GENERATIONS

*E*ach generation has its unique needs and aspirations. When Charles Wiley first opened his small printing shop in lower Manhattan in 1807, it was a generation of boundless potential searching for an identity. And we were there, helping to define a new American literary tradition. Over half a century later, in the midst of the Second Industrial Revolution, it was a generation focused on building the future. Once again, we were there, supplying the critical scientific, technical, and engineering knowledge that helped frame the world. Throughout the 20th Century, and into the new millennium, nations began to reach out beyond their own borders and a new international community was born. Wiley was there, expanding its operations around the world to enable a global exchange of ideas, opinions, and know-how.

For 200 years, Wiley has been an integral part of each generation's journey, enabling the flow of information and understanding necessary to meet their needs and fulfill their aspirations. Today, bold new technologies are changing the way we live and learn. Wiley will be there, providing you the must-have knowledge you need to imagine new worlds, new possibilities, and new opportunities.

Generations come and go, but you can always count on Wiley to provide you the knowledge you need, when and where you need it!

WILLIAM J. PESCE
PRESIDENT AND CHIEF EXECUTIVE OFFICER

PETER BOOTH WILEY
CHAIRMAN OF THE BOARD

Margins of Error
A Study of Reliability in Survey Measurement

Duane F. Alwin
McCourtney Professor of Sociology and Demography
Pennsylvania State University
University Park, PA

Emeritus Senior Research Scientist
Survey Research Center
Institute for Social Research
University of Michigan
Ann Arbor, MI

BICENTENNIAL
BICENTENNIAL
1807
WILEY
2007
BICENTENNIAL
BICENTENNIAL

WILEY-INTERSCIENCE
A John Wiley & Sons, Inc., Publication

Copyright © 2007 by John Wiley & Sons, Inc. All rights reserved.

Published by John Wiley & Sons, Inc., Hoboken, New Jersey.
Published simultaneously in Canada.

For general information on our other products and services or for technical support, please contact our
Customer Care Department within the United States at (800) 762-2974, outside the United States at
(317) 572-3993 or fax (317) 572-4002.

Wiley also publishes its books in a variety of electronic formats. Some content that appears in print may
not be available in electronic format. For information about Wiley products, visit our web site at
www.wiley.com.

Wiley Bicentennial Logo: Richard J. Pacifico

Library of Congress Cataloging-in-Publication Data:

Alwin, Duane F. (Duane Francis), 1944–
    Margins of error : a study of reliability in survey
measurement / Duane F. Alwin.
        p. cm.
    Includes bibliographical references and index.
    ISBN 978-0-470-08148-8 (cloth : acid-free paper)
1. Surveys.   2. Error analysis (Mathematics)   I. Title.
    HA31.2.A5185 2007
    001.4'33—dc22   2006053191

10 9 8 7 6 5 4 3 2 1

Dedicated to

Edgar F. Borgatta

and to the memory of

T. Anne Cleary
David J. Jackson
Charles F. Cannell

Contents

Preface   **xi**

Acknowledgments   **xiii**

Foreword   **xv**

1. Measurement Errors in Surveys   **1**

1.1   Why Study Survey Measurement Error?   2
1.2   Survey Errors   3
1.3   Survey Measurement Errors   6
1.4   Standards of Measurement   8
1.5   Reliability of Measurement   9
1.6   The Need for Further Research   10
1.7   The Plan of this Book   11

2. Sources of Survey Measurement Error   **15**

2.1   The Ubiquity of Measurement Errors   16
2.2   Sources of Measurement Error in Survey Reports   20
2.3   Consequences of Measurement Error   32

3. Reliability Theory for Survey Measures   **35**

3.1   Key Notation   36
3.2   Basic Concepts of Classical Reliability Theory   36
3.3   Nonrandom Measurement Error   41
3.4   The Common-Factor Model Representation of CTST   42
3.5   Scaling of Variables   43
3.6   Designs for Reliability Estimation   45

3.7   Validity and Measurement Error   46
3.8   Reliability Models for Composite Scores   51
3.9   Dealing with Nonrandom or Systematic Error   53
3.10   Sampling Considerations   55
3.11   Conclusions   57

4. Reliability Methods for Multiple Measures   59

4.1   Multiple Measures versus Multiple Indicators   61
4.2   Multitrait-Multimethod Approaches   67
4.3   Common-Factor Models of the MTMM Design   71
4.4   Classical True-Score Representation of the MTMM Model   77
4.5   The Growing Body of MTMM Studies   79
4.6   An Example   83
4.7   Critique of the MTMM Approach   91
4.8   Where Are We?   93

5. Longitudinal Methods for Reliability Estimation   95

5.1   The Test-Retest Method   96
5.2   Solutions to the Problem   101
5.3   Estimating Reliability Using the Quasi-Markov Simplex Model   104
5.4   Contributions of the Longitudinal Approach   110
5.5   Components of the Survey Response   114
5.6   Where to from Here?   116

6. Using Longitudinal Data to Estimate Reliability Parameters   117

6.1   Rationale for the Present Study   118
6.2   Samples and Data   119
6.3   Domains of Measurement   122
6.4   Statistical Estimation Strategies   127
6.5   Comparison of Methods of Reliability Estimation   130
6.6   The Problem of Attrition   135
6.7   Which Reliability Estimates?   146
6.8   Conclusions   147

7. The Source and Content of Survey Questions   149

7.1   Source of Information   150
7.2   Proxy Reports   152
7.3   Content of Questions   153
7.4   Summary and Conclusions   162

8. Survey Question Context   **165**

    8.1   The Architecture of Survey Questionnaires   167
    8.2   Questions in Series versus Questions in Batteries   171
    8.3   Location in the Questionnaire   172
    8.4   Unit Length and Position in Series and Batteries   175
    8.5   Length of Introductions to Series and Batteries   177
    8.6   Conclusions   179

9. Formal Properties of Survey Questions   **181**

    9.1   Question Form   183
    9.2   Types of Closed-Form Questions   185
    9.3   Number of Response Categories   191
    9.4   Unipolar versus Bipolar Scales   195
    9.5   Don't Know Options   196
    9.6   Verbal Labeling of Response Categories   200
    9.7   Survey Question Length   202
    9.8   Conclusions   210

10. Attributes of Respondents   **213**

    10.1   Reliability as a Population Parameter   214
    10.2   Respondent Attributes and Measurement Error   215
    10.3   Age and Reliability of Measurement   218
    10.4   Schooling and Reliability of Measurement   221
    10.5   Controlling for Schooling Differences   223
    10.6   Generational Differences in Reliability   227
    10.7   MTMM Results by Age and Education   228
    10.8   Statistical Estimation of Components of Variation   231
    10.9   Tests of Hypotheses about Group Differences   254
    10.10 Conclusions   261

11. Reliability Estimation for Categorical Latent Variables   **263**

    11.1   Background and Rationale   264
    11.2   The Latent Class Model for Multiple Indicators   265
    11.3   The Latent Class Model for Multiple Measures   272
    11.4   The Latent Markov Model   277
    11.5   Conclusions   286

12. Final Thoughts and Future Directions   **289**

    12.1   Reliability as an Object of Study   290
    12.2   Why Study the Reliability of Survey Measurement?   291

12.3 The Longitudinal Approach   299
12.4 Assembling Knowledge of Survey Measurement Reliability   301
12.5 Compensating for Measurement Error using Composite Variables   308
12.6 Conclusions   315

Appendix   Reliability of Survey Measures Used in the Present Study   **327**

References   **367**

Index   **383**

Preface

Among social scientists, almost everyone agrees that without valid measurement, there may be little for social science to contribute in the way of scientific knowledge. A corollary to this principle is that *reliability of measurement* (as distinct from validity) is a necessary, although not sufficient, condition for valid measurement. The logical conclusion of this line of thinking is that whatever else our measures may aspire to tell us, it is essential that they are reliable; otherwise they will be of limited value. There is a mathematical proof of this assertion (see Chapters 3 and 12), but the logic that underlies these ideas is normally accepted without formal proof: If our measures are *unreliable*, they cannot be trusted to detect patterns and relationships among variables of interest. Thus, reliability of measurement is a sine qua non of any empirical science.

To be somewhat more concrete, there are several reasons to be concerned with the existence and consequences of errors in social measurement. First and foremost, if we are aware of the processes that generate measurement error, we can potentially understand the nature of our results. A presumptive alternative interpretation for any research result is that there are methodological errors in the data collection, and we must rule out such methodological artifacts as explanatory variables whenever we draw inferences about differences in patterns and processes. Second, if we know about the nature and extent of measurement errors, we may (in theory) get them under better control. In the second chapter of this book, I "deconstruct" the data gathering process in survey research into *six major elements* of the response process—question adequacy, comprehension, accessibility, retrieval, motivation, and communication—and argue that discerning how measurement errors result from these components helps researchers reduce errors at the point where they are most likely to occur. Third, measurement errors affect our statistical inferences. Measurement unreliability inflates estimates of population variance in variables of interest and, in turn, biases estimates of standard errors of population means and other quantities of interest, inflating confidence intervals. Statistical analyses that ignore unreliability of variables underestimate the strength and significance of the statistical association between those variables. This underestimation not only makes the results of such analyses more conservative from a scientific perspective; it also increases the probability of type II error and the consequent rejection of correct, scientifically productive hypotheses about the phenomena of interest. Even in the simplest regression models, measurement unreliability in predictor variables generally biases regression coefficients downward, making it more difficult to reject the null hypothesis; and unreliability in both dependent and independent variables attenuates estimates of statistical associations. With appropriate measurement designs, it is possible to isolate some types of errors statistically and control for them in the analysis of data.

Over the past two decades (i.e., since the mid-1980s), we have seen a burgeoning research literature on survey methodology, focusing especially on problems of measurement. Indeed, volumes have been written about measurement errors. We probably have a better than ever understanding about the sources of measurement errors, particularly those involving cognitive processes and the effects of question wording. But little effort has been undertaken to quantify the extent of unreliability in the types of measures typically used in population surveys to help us assess the extent of its biasing effects. To be blunt, our knowledge about the nature and extent of measurement errors in surveys is meager, and our level of understanding of the factors linked to the design of questions and questionnaires that contribute to their presence is insufficient. Errors of measurement—the general class of phenomena of which unreliability is a particular type—are a bit like what Mark Twain reportedly said about the weather: "Everybody talks about the subject, but nobody does anything about it."

In this book, I argue that considering the presence and extent of measurement errors in survey data will ultimately lead to improvements in data collection and analysis. A key purpose of studies of measurement errors is to identify which types of questions, questionnaires, and interviewer practices produce the most valid and reliable data. In the chapters that follow, I consider ways in which the extent of measurement errors can be detected and estimated in research in order to better understand their consequences. The major vehicle for achieving these purposes involves a study of nearly 500 survey measures obtained in surveys conducted at the University of Michigan over the past two or three decades. Assembling information on reliability from these data sources can help improve knowledge about the strengths and weaknesses of survey data. The results of this research should be relevant to the general tasks of uncovering the sources of survey measurement error and improving survey data collection through the application of this knowledge.

Although information about the level of reporting reliability in the standard survey interview is lacking, a small and growing cadre of investigators is addressing this issue. Given the substantial social and economic resources invested each year in data collection to satisfy social and scientific information needs, questions concerning the quality of survey data are strongly justified. Without accurate and consistent measurement, the statistical tabulation and quantitative analysis of survey data hardly makes sense. Knowledge has only recently been cumulating regarding the factors linked to the quality of measurement, and I hope this study will contribute to this body of work.

Acknowledgments

There are many people to whom credit for the success of this project needs to be acknowledged. I am indebted to my daughters—Heidi, Abby, and Becky—and their families, for their love and support. My wife, Linda Wray, a social scientist in her own right, understands how important this project has been to me and has given me an unending amount of encouragement. My deepest gratitude goes to Linda—none of this would have been possible without your help, encouragement, and love (not to mention the copyediting). And more than anything, Linda, your confidence in me made all the difference.

The contributions of my research assistants over the years to this project have been indispensable. I can honestly say that this book would have been finished much earlier if it were not for my research assistants—Ryan McCammon, Dave Klingel, Tim Manning, Frank Mierzwa, Halimah Hassan, and Jacob Felson—but it would not have been as good. The efforts of Dave and Ryan in particular, who insisted that I "get it right" (to the extent that is ever possible) slowed me down. I recall many occasions on which they insisted that we not leave as many "stones unturned," which often encouraged me to take the analysis further, or to consider an alternative set of estimates, or even in some cases actually recode the data, in order to make the study as good as it could be. Ryan McCammon has played an extraordinarily important role over the past few years in helping me bring this project to completion. His technical knowledge, his expert advice in statistical matters, and his understanding of measurement issues have helped forge our collaboration on these and other projects. Pauline Mitchell, administrative assistant and word-processing expert par excellence, made it possible to express our results in numeric and graphic form. Pauline is responsible for the more than 120 tables, figures, and charts presented here, and I acknowledge her dedicated assistance. I also wish to acknowledge Pauline, Ryan, Linda, and Jake for various copyediting tasks that required them to read portions of earlier drafts. Matthew Williams of Columbus, Ohio provided stellar assistance in producing the camera-ready text for this book. Last but not least, the assistance of John Wiley's senior editor, Steve Quigley, and his staff, is gratefully acknowledged.

Some chapters presented here build on other previously published writings. Part of Chapter 3 is related to my recent entry titled "Reliability" in the *Encyclopedia of*

Social Measurement, edited by K. Kempf-Leonard and others (2004). Portions of this work are reprinted with permission of Elsevier Ltd, UK. Parts of Chapter 4 are related to my earlier publications dealing with the multitrait-multimethod approach, specifically my 1974 chapter, "Approaches to the interpretation of relationships in the multitrait-multimethod matrix" in *Sociological Methodology 1973–74*, edited by H.L. Costner (San Francisco: Jossey-Bass), and my 1997 paper, "Feeling thermometers vs. seven-point scales: Which are better?," which appeared in *Sociological Methods and Research* (vol. 25, pp. 318–340). And finally, the material presented in Chapter 10 is an extension of a chapter, "Aging and errors of measurement: Implications for the study of life-span development," in *Cognition, Aging, and Self-Reports*, edited by N. Schwarz, D. Park, B. Knäuper, and S. Sudman (Philadelphia: Psychology Press, 1999).

In 1979, I took a job at the Survey Research Center of the Institute for Social Research at the University of Michigan, where I worked for nearly 25 years. This move was motivated in part by a desire to learn more about survey research, and it paid off. At Michigan I learned how surveys were actually conducted and how measurement errors were created—by participating in surveys I conducted, as well as those carried out by others (more than a dozen surveys in all). Also, I learned how questionnaires were designed, interviewers were trained, pretesting of questionnaires was carried out, and the data were actually gathered in the field. Although it is not possible to credit all the sources of my education at Michigan, I wish to acknowledge particularly the support over the years of Jon Krosnick, Charles Cannell, Leslie Kish, Arland Thornton, Bob Groves, Jim House, Regula Herzog, Norbert Schwarz, and Bill Rodgers. Paul Beatty, whose dissertation at Michigan (Beatty, 2003) I was fortunate to supervise, taught me a great deal about the importance of cognitive factors in responses to surveys.

Over the years this project received support from the National Science Foundation and the National Institute on Aging. During 1992–1995 the project was supported by an NIA grant, "Aging and errors of measurement" (R01-AG09747), and this led to a current NIA-funded project, "Aging and the reliability of measurement" (R01-AG020673). These projects were instrumental in the work reported in Chapters 10 and 11. In between these two projects, I received support from the NSF for the project, "The reliability of survey data" (SES-9710403), which provided the overall rationale for assembling a database containing estimates of reliability and question characteristics.

Finally, I also acknowledge the support of the Tracy Winfree and Ted H. McCourtney Professorship, the Population Research Institute, and of the College of the Liberal Arts at Pennsylvania State University, support that allowed me some additional time to complete this research.

Foreword

Some projects take a long time to complete—this is true (for better or worse) of the present one. In a very real sense, the idea for this project began nearly 40 years ago, when I was a graduate student in sociology at the University of Wisconsin. In 1968, I was taking courses in reliability theory, factor analysis, and item response theory in the Department of Educational Psychology at Wisconsin (from Anne Cleary, Chester Harris and Frank Baker) in order to fulfill a Ph.D. minor requirement. At the time, I recall wondering if it would be possible to apply some of the ideas from classical psychometric theory to social science data. Large-scale survey studies were beginning to achieve greater popularity as a mainstay for social science research, and I had read several of the famous critiques about survey research. From those seeds of curiosity sown so many years ago, I now have come to understand how those "response" theories of measurement may be fruitfully applied to survey data.

I recall reading early on about the questionable role of surveys in the development of social science. The field of survey research was so under-developed in the 1950s and 1960s that Herbert Blumer (1956) could wage what seemed to many to be a credible attack on "variable analysis" in social science. Interestingly, Blumer's argument focused on the issues of *reliability* and *validity* of survey data. He believed that survey data had a high degree of reliability, but were of questionable validity. As I argue in this book, Blumer may have assumed too much about the reliability of survey data. Certainly there was little attention to the issue at the time he was writing. But, in fact, Quinn McNemar, a psychometrician, had (10 years earlier) written an important review of survey research methods in which he pointed out that survey researchers had largely ignored the problem of reliability, depending without qualms on results from single questions (McNemar, 1946). Psychometric methods had not yet made their way into survey analysis, and it was not known how to incorporate measurement errors into models for the analysis of survey data. Even the most highly regarded proponents of the quantitative analysis of survey data, Robert Merton and Paul Lazarsfeld, admitted that there was very little discussion of the art of analyzing material once it has been collected (Merton and Lazarsfeld, 1950). Later, in 1968, sociologist James Coleman observed that the investigation of response unreliability was an almost totally underdeveloped field, because of the lack of mathematical models to encompass both unreliability and change (Coleman, 1968). These arguments piqued my interest in problems of survey measurement.

I found these issues to be compelling, issues I wanted to explore further. It was my good fortune to have been accepted into an NIH training program in quantitative methodology during my graduate studies, a program initiated by Edgar Borgatta. Exposure to his work, along with that of a growing field of "sociological methodology," which included the work of James Coleman, Hubert M. Blalock, Jr., Herbert L. Costner, David R. Heise, Robert M. Hauser, and Karl Jöreskog, among others, did much to help develop an understanding of the nature of social measurement (see Blalock, 1965; Costner, 1969; Hauser and Goldberger, 1971). As one who reads the present book will discover (see Chapter 5), Dave Heise's paper on the separation of unreliability and true change in repeated measures designs and Karl Jöreskog's related work on simplex models were critical to the development of this research program (see Heise, 1969; Jöreskog, 1970).

Many of my early publications dealt with these matters (e.g., Alwin 1973, 1974), and these concerns have remained an important focus for a substantial portion of my scholarly work. It is a remarkable thing to have been driven and influenced most of my professional life by a general concern with the quality of data on which the inferences of social scientists are based. And although I have worked on a range of other topics throughout my career, a concern with measurement issues has been a keystone of my work.

I have dedicated this book to those from whom I learned the most about measurement—people whose influence I am cognizant of almost every day of my life. Ed Borgatta, now retired from the University of Washington, is without question one of the most cherished mentors I have ever had—his knowledge of measurement and his hard-nosed approach to modeling social data are attributes I hope I have passed along to my students. Anne Cleary—whose life was taken at a young age by a senseless act of terror—was an extraordinarily talented mentor who taught me just about everything I know about classical measurement theory. David Jackson, a colleague in graduate school, was my best friend. His life was taken by cancer on October 1, 2001, just a few weeks after 9/11. I still feel the grief of losing Dave, but I can say this—Dave taught me so much about measurement that I cannot think about the content of this book without thinking of what I learned from him. Charlie Cannell hired me for the job at the University of Michigan and exposed me to interviewing methodology and survey research in a way I would never have thought possible—I have only the fondest of memories of my contact with Charlie and what I learned from him.

This book took many years to complete, and the research spanned my tenure across several academic institutions. The work was influenced by many esteemed colleagues, friends, and family, and the research was supported by two federal funding agencies. I believe the time it has taken and the influence of others only strengthened the final product. To all those who read this book, I hope the underpinnings of the approach and the importance of the research agenda can have an impact on future research. To my colleagues, friends, and family who made this project possible, you have my deepest appreciation.

DUANE F. ALWIN

State College, Pennsylvania
October 2006

CHAPTER ONE

Measurement Errors in Surveys

Quality ... you know what it is, yet you don't know what it is. But that's self-contradictory. But some things are better than others, that is, they have more quality. But when you try to say what the quality is, apart from the things that have it, it all goes poof! ... But if you can't say what Quality is, how do you know what it is, or how do you know that it even exists? If no one knows what it is, then for all practical purposes it doesn't exist at all. But for all practical purposes it really does exist. ... Obviously some things are better than others ... but what's the "betterness"? ... So round and round you go, spinning mental wheels and nowhere finding any place to get traction. What the hell is Quality? What is it?

Robert M. Pirsig, *Zen and the art of motorcycle maintenance* (1974)

Measurement issues are among the most critical in scientific research because analysis and interpretation of empirical patterns and processes depend ultimately on the ability to develop high quality measures that accurately assess the phenomenon of interest. This may be more difficult in the social and behavioral sciences as the phenomena of interest are often not well specified, and even when they are, the variables of interest are often difficult to observe directly. For example, concepts like religiosity, depression, intelligence, social status, attitudes, psychological well being, functional status, and personality may be difficult to measure precisely because they largely reflect unobserved processes. Even social indicators that are more often thought to directly assess concepts of interest, e.g., variables like education, or income, or race, are not free of specification errors. Clearly, the ability to define *concepts* precisely in a conceptually valid way, the translation of these concepts into *social indicators* that have an empirical referent, and the development of survey *measures* of these indicators all bear on the extent of measurement errors. In addition, measurement problems in social science are also critically related to the nature of the communication and cognitive processes involved in gathering data from respondents (e.g., Bradburn and Danis, 1984; Cannell, Miller and Oksenberg, 1981; Krosnick, 1999; Schwarz, 1999a, 1999b; Sirken, Herrmann, Schechter, Schwarz, Tanur, and Tourangeau, 1999; Sudman, Bradburn and Schwarz, 1996; Tourangeau, 1984; Tourangeau and Rasinski, 1988; Tourangeau, Rips, and Rasinski, 2000).

With its origins in 19th-century Europe and pre-World War II American society, survey research plays an extraordinarily important role in contemporary social sciences throughout the world (Converse, 1987). Vast amounts of survey data are collected for many purposes, including governmental information, public opinion and election surveys, advertising and marketing research, as well as basic social scientific research. Some have even described survey research as the via regia for modern social science (Kaase, 1999, p. 253)—the ideal way of conducting empirical science. Many would disagree with the proposition that surveys are the *only* way to do social science, but there would be hardly any dissent from the view that survey research has become a mainstay for governmental planning, the research of large numbers of academic social scientists, and the livelihoods of growing numbers of pollsters, and marketing and advertising researchers.

1.1   WHY STUDY SURVEY MEASUREMENT ERROR?

The basic purpose of the survey method is to obtain information from a sample of persons or households on matters relevant to researcher or agency objectives. The survey interview is conceived of as a setting in which the question-answer format is used by the researcher to obtain the desired information from a respondent, whether in face-to-face interview situations, via telephone interviews, or in self-administered questionnaires. Many aspects of the information gathering process may represent sources of measurement error: aspects of survey questions; the cognitive mechanisms of information processing and retrieval; the motivational context of the setting that produces the information; and the response framework in which the information is then transmitted (see, e.g., Alwin, 1991b; Alwin, 1992; Krosnick, 1999; Krosnick and Alwin, 1987, 1988, 1989; Schaeffer, 1991b; O'Muircheartaigh, 1997).

Given the substantial social and economic resources invested each year in data collection to satisfy social and scientific information needs, questions concerning the quality of survey data are strongly justified. Without accurate and consistent measurement, the statistical tabulation and quantitative analysis of survey data hardly makes sense; yet there is a general lack of empirical information about these problems and very little available information on the reliability of measurement from large scale population surveys for standard types of survey measures. For all the talk over the past decade or more concerning measurement error (e.g., Groves, 1989, 1991; Biemer, Groves, Lyberg, Mathiowetz and Sudman, 1991; Biemer and Stokes, 1991; Lyberg, Biemer, Collins, de Leeuw, Dippo, Schwarz, and Trewin, 1997), there has been very little empirical attention to the matter. Indeed, one prescient discussion of measurement errors in surveys even stated that "we know of no study using a general population survey that has attempted to estimate the reliabilities of items of the types typically used in survey research" (Bohrnstedt, Mohler, and Müller, 1987, p. 171). Knowledge has only recently been cumulating regarding the factors linked to the quality of measurement, and we hope this study will contribute to this body of work.

Errors occur in virtually all survey measurement, regardless of content, and the factors contributing to differences in unreliability of measurement are worthy of scrutiny. It is well known that statistical analyses ignoring unreliability of measures generally provide biased estimates of the magnitude and statistical significance of the tests of mean differences and associations among variables. Although the resulting biases tend to underestimate mean differences and the strength of relationships making tests of hypotheses more conservative, they also increase the probability of type II errors and the consequent rejection of correct, scientifically valuable hypotheses about the effects of variables of interest (see Biemer and Trewin, 1997). From a statistical point of view there is hardly any justification for ignoring survey measurement errors.

1.2   SURVEY ERRORS

Terms that are often associated with assessments of survey quality, for example, the terms "bias," "reliability," and "validity," are often used in ambiguous ways. Sometimes they are used very generally to refer to the overall stability or dependability of survey results, including the extent of sampling error, nonresponse bias, instrument bias, as well as reporting accuracy. Other times they are used in a much more delimited way, to refer *only* to specific aspects of measurement error, distinguishing them from assessments of other types of survey errors. It is therefore useful to begin this discussion by clarifying how we might think about various types of survey error, how they differ from one another, and how we might arrive at a more precise definition of some of the terms frequently used to refer to levels of survey data quality involving measurement errors in particular.

In his path-breaking monograph, *Survey errors and survey costs*, Robert Groves (1989, p. vi) presents the following framework for considering *four* different types of survey errors:

> *Coverage error.* Error that results from the failure to include some population elements in the sampling frame or population lists.
>
> *Sampling error.* Error that results from the fact that a subset of the population is used to represent the population rather than the population itself.
>
> *Nonresponse error.* Error that results from the failure to obtain data from all population elements selected into the sample.
>
> *Measurement error.* Error that occurs when the recorded or observed value is different from the true value of the variable.

We consider this to be an exhaustive list, and we argue that any type of survey error can be conceptualized within this framework. The presence of any of these types of survey errors can influence the accuracy of the inferences made from the sample data, and the potential for such errors in the application of survey methods places a high priority on being able to anticipate their effects. In the worst case, errors in even one of these categories may be so great as to invalidate *any* conclusions drawn from the data. In the best case, errors are minimized through efforts aimed at their reduction and/or efforts taken to minimize their effects on the conclusions drawn, in which cases stronger inferences can be made on the basis of the data.

All of these types of *survey errors* are to some extent present in the data we collect via survey methods. It is important for users of survey data to realize that these various survey errors are *nested* in important ways (see Figure 1.1). To describe this aspect of the phenomenon, we use the metaphor of a set of interrelated structures, each inside the next, like a set of Russian *matrioshka* dolls, in which distinct levels of "nestedness" represent different "compoundings" of error (see Alwin, 1991). *Non-response errors* are nested within *sampling errors*, for example, because only those cases sampled have the opportunity to participate and provide a response to the survey and the cases representing nonresponse or missing cases, depend on which elements of the population are selected into the sample (Groves and Couper, 1998; Groves, Dillman, Eltinge, and Little, 2002). Similarly, *sampling errors* are nested within *coverage errors* because clearly the subset of the population sampled depends on the coverage of the sampling frame. Finally, measurement errors are nested within those cases that have provided a response, although typically we study processes of measurement error as if we were studying those processes operating at the population level. Inferences about measurement error can only be made with the realization that they pertain to respondents from samples of

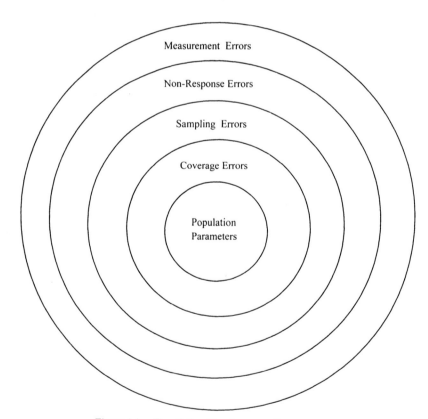

Figure 1.1. The relationship of sources of survey errors.

specified populations, and it is important to realize, thus, that our inferences regarding those processes are constrained by the levels of nestedness described here.

1.2.1   Classifying Types of Survey Error

There are a number of different ways to think about the relationship among the several types of survey error. One way is to describe their relationship through the application of classical statistical treatments of survey errors (see Hansen, Hurwitz, and Madow, 1953). This approach begins with an expression of the mean square error (MSE) for the deviation of the sample estimator ($\bar{y}$) of the mean (for a given sampling design) from the population mean (μ), that is, MSE ($\bar{y}$) = $E(\bar{y} - \mu)^2$. This results in the standard expression:

$$MSE\ (\bar{y}) = Bias^2 + Variance$$

where *Bias*^{2} refers to the square of the theoretical quantity $\bar{y} - \mu$, and *Variance* refers to the variance of the sample mean $\sigma_{\bar{y}}^2$ Within this statistical tradition of conceptualizing survey errors, *bias* is a *constant source of error* conceptualized at the sample level. *Variance*, on the other hand, represents variable errors, also conceptualized at the sample level, but this quantity is obviously influenced by the within-sample sources of response variance normally attributed to measurement error.

Following Groves' (1989) treatment of these issues, we can regroup coverage, sampling, and nonresponse errors into a category of *nonobservational errors* and also group measurement errors into a category of *observational errors*. *Observational errors* can be further subclassified according to their sources, e.g., into those that are due to interviewers, respondents, instruments, and modes of observation. Thus, Groves' fourfold classification becomes even more detailed, as seen in Table 1.1. Any treatment of survey errors in social research will benefit from the

Table 1.1.   A classification of some types of survey errors

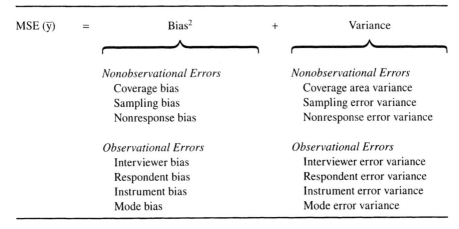

MSE ($\bar{y}$)	=	Bias^2	+	Variance
		Nonobservational Errors		*Nonobservational Errors*
		Coverage bias		Coverage area variance
		Sampling bias		Sampling error variance
		Nonresponse bias		Nonresponse error variance
		Observational Errors		*Observational Errors*
		Interviewer bias		Interviewer error variance
		Respondent bias		Respondent error variance
		Instrument bias		Instrument error variance
		Mode bias		Mode error variance

use of this classification, and further, any comparison of results across settings (e.g., across national surveys) will benefit from an understanding of the potential role of these components in the production of similarities and differences observed across settings. Ultimately, while this classification scheme is useful for pinpointing the effects of survey errors on sample estimates of means and their variances, it is also important to understand what (if anything) might be done to estimate these effects and the contributions of error sources to the understanding the results of research studies. This book focuses on one of these types of error—*survey measurement errors*—and it is hoped that the program of research summarized in subsequent chapters will improve our understanding of the effects of measurement errors on the results of surveys.

1.3   SURVEY MEASUREMENT ERRORS

Measurement represents the link between theory and the analysis of empirical data. Consequently, the relationship between measures of empirical indicators and the theoretical constructs they represent is an especially important aspect of measurement, in that ultimately the inferences drawn from the empirical data are made with respect to more abstract concepts and theories, not simply observable variables. Measurement, thus, requires the clear specification of relations between theoretic constructs and observable indicators. In addition, obtaining "measures" of these indicators involves many practical issues, including the specification of questions that operationalize the measures, and in the case of survey research the processes of gathering information from respondents and/or households.

As I indicated at the beginning of this chapter, the specification of the linkage between theory and measurement is often viewed as more difficult in the social and behavioral sciences, as the phenomena of interest are often not very well specified, and even where they are, the variables are often difficult or impossible to observe directly. The diagram in Figure 1.2 illustrates the fundamental nature of the problem of measurement. Here I have adopted a *three-ply distinction* between constructs, indicators, and measures, depicting their interrelationships. *Constructs* are the theoretical variables referred to in theoretical or policy discussions about which information is desired. *Indicators* are the empirical referents to theoretical constructs. In social surveys *measures* consist of the question or questions that are used to obtain information about the indicators. The layered nature of the distinctions of interest here can be illustrated with an example. Socioeconomic status is an example of a theoretical construct, derived from sociological theory, which can be indexed via any number of different social indicators, e.g., education, occupation, income level, property ownership. Normally, one considers such indicators as imperfect indicators of the theoretical construct, and often researchers solve this problem through the use of *multiple indicators*, combining different indicators using MIMC (multiple-indicator multiple-cause) models or common factor models for analysis (see Alwin, 1988).

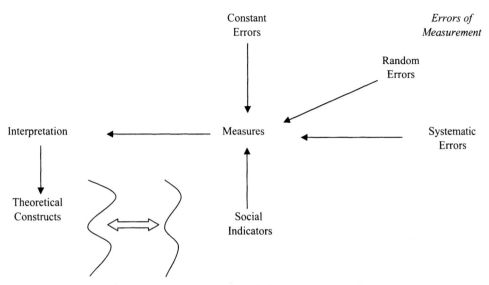

Figure 1.2.   The relationship between constructs, indicators, measures, and measurement errors.

It is important in this context not to confuse the indicators of concepts with the theoretical constructs they are thought to reflect. To do so loses sight of the purpose of measurement. In principle, it is the theoretical constructs that implicate particular indicators, not the reverse. In Kuhn's (1961, p. 190) words: "To discover quantitative regularity one must normally know what regularity one is seeking and one's instruments must be designed accordingly." The linkage between scientific theories and scientific measurement is therefore rarely traced backward, specifying constructs entirely in terms of what can be assessed at the operational level. Still, it is clearly possible for scientific data to be gathered in service of theories that are wrong or ill-conceived, and in some cases new theoretical constructs may be needed in order to account for the empirical data. The relation between the two therefore should probably be conceived of as reciprocal (as depicted in Figure 1.2).

It is also important to realize that given a particular indicator, there may be *multiple measures* that can be used to assess variation in the indicator. *It is crucial in the present context that a distinction be maintained between the idea of multiple indicators and that of multiple measures, as they refer to very different things.* In the example of the indicator, "level of education," measures may include such things as questions focusing on the number of years of schooling completed, or levels of certification, or even a test of knowledge gained from school. All such things may be legitimate measures of education, even though they may assess ostensibly different things, and it may often be the case that one is more appropriate in a given cultural context than another. The point is that within the survey context a "measure" relies on a question or a set of questions that provide the information needed to construct the indicator, and therefore "multiple" measures involve multiple replications of the measure of a given indicator.

On a practical level, once concepts and indicators are defined and agreed on, measurement is possible only if we can assume some type of equivalence across units of observation, e.g., respondents or households. As Abraham Kaplan (1964) has pointed out, the essence of measurement at an operational level lies in the principle of *standardization*, that is, the principle that units of magnitude have constancy across time and place. This, of course, implies that a regularity or stability of measures is required in order for there to be valid comparisons made across units of observation. The equivalence of some units of measurement may seem natural, as in the physical sciences—measures such as weight, mass, distance, or time—but in the social sciences most metrics are completely arbitrary, and standardization is often more an objective than a reality. However, without the ability or willingness to make such strong assumptions about *standardization* of measurement, the comparative usefulness of the information collected in surveys may be in doubt.

Efforts to measure using the survey method involve a two-way process of communication that *first* conveys what information is needed from the respondent to satisfy the objectives of the research and *second* the transmission of that information back to the researcher, usually, but not exclusively, via interviewers. The central focus of measurement in surveys, then, involves not only the specification of a *nomological network* of concepts and their linkage to "observables," but also a focus on the processes of gathering information on "measures."

A concern with the existence and consequences of errors made in this "two step" process motivates their consideration, and hence such consideration can play a role in reducing errors and better understanding the answers to survey questions. This chapter and subsequent ones focus specifically on the conceptualization and estimation of the nature and extent of *measurement error* in surveys and establish the rationale for the consideration of measurement errors when designing survey research. The diagram in Figure 1.2 specifies three types of measurement errors—*constant errors*, *random errors*, and *systematic errors*—that are of concern to researchers who have studied response errors in surveys. This and subsequent chapters will provide a more complete understanding of both the nature and sources of these types of error. We begin with a discussion of the *standards* used in the evaluation of the quality of survey measurement.

1.4   STANDARDS OF MEASUREMENT

Writing on "scales of measurement," Otis Dudley Duncan (1984a, p. 119) observed: "measurement is one of many human achievements and practices that grew up and came to be taken for granted before anyone thought to ask how and why they work." Thus, we argue that one of the biggest challenges for survey research is to figure out "how and why" survey measurement works, but also to assess when it does not work. One of the greatest potential impediments to the meaningful analysis of survey data is the existence of imperfect measurement; imperfect either in providing a *valid* correspondence of indicators to the target concept(s) of interest, or in producing *reliable* measures of those indicators.

In order to *measure* a given quantity of interest, it is necessary to (1) specify certain rules of correspondence between concepts and indicators, (2) establish the nature of the dimensionality inherent in the phenomenon, (3) choose a particular metric in which to express variation in the dimension or dimensions of the concept being measured, and (4) specify the necessary operational procedures for gathering information that will reveal the nature of differences in the phenomenon. As noted earlier, the term "measurement" implies equivalence. In *The conduct of inquiry*, Abraham Kaplan (1964, pp. 173–174) observed: "Measurement, in a word is a device for *standardization*, by which we are assured of the equivalences among objects of diverse origin. This is the sense that is uppermost in using phrases like 'a measure of grain': measurement allows us to know what quantity we are getting, and to get and give just what is called for." *Equivalence*, then, across all of the elements mentioned above, is the key to measurement. It should come as no surprise, then, that one of the major criticisms of quantitative approaches is the comparability of units and therefore of responses across respondents.

In this and subsequent chapters I discuss the issue of obtaining useful information in surveys and the problem of assessing the extent of *measurement error* and the factors that contribute to such errors. As I point out in the next chapter, errors of measurement can intrude at many different points in the gathering of survey data, from the initial *comprehension* of the question by the respondent, to the *cognitive processing* necessary to access the requested information, through to the production of a response. Clearly, the magnitude of such errors qualify the meaning one can attach to the data, and ultimately the confidence we place in survey research strategies depends intimately on the extent of measurement errors in the data.

1.5 RELIABILITY OF MEASUREMENT

Let us return to the above definition of measurement error as *the difference between the recorded or observed value and the true value of the variable* (see Groves, 1989, 1991). There have been two basic approaches to minimizing this type of error. The first is to emphasize the reduction of errors in the collection of survey data through improved techniques of questionnaire design, interviewer training and survey implementation. The second is to accept the fact that measurement errors are bound to occur, even after doing everything that is in one's power to minimize them, and to model the behavior of errors using statistical designs. The tradition in psychology of "correcting for attenuation" is an example of an approach that adjusts sample estimates of correlations based on available information about the reliabilities of the variables involved (Lord and Novick, 1968). More recently, structural equation models (or LISREL-type models) used to model response errors in surveys are another example of such an approach (see Alwin and Jackson, 1979; Bollen, 1989).

Earlier we noted that sometimes the term *reliability* is used very generally to refer to the overall stability or dependability of research results, including the absence of population specification errors, sampling error, nonresponse bias, as well as various forms of measurement errors. Here (and throughout the remainder of this book) we use the term in its more narrow *psychometric* meaning, focusing specifically on the absence of measurement errors. Even then, there are at least two different conceptions of error—random and nonrandom (or systematic) errors of measurement—that have consequences for research findings. Within the psychometric tradition the concept of reliability refers to the absence of *random error*. This conceptualization of error may be far too narrow for many research purposes, where reliability is better understood as the more general absence of measurement error. However, it is possible to address the question of reliability separately from the more general issue of measurement error and in subsequent chapters I point out the relationship between random and nonrandom components of error.

Traditionally, most attention to reliability in survey research is devoted to item analysis and scale construction [e.g., calculation of Cronbach's (1951) alpha (α)], although including *multiple indicators* using SEM models or related approaches is increasingly common (Bollen, 1989). While these procedures are invaluable and likely to reduce the impact of measurement errors on substantive inferences, they have not informed survey researchers of the nature and sources of the errors of concern. Further, these approaches generally cannot address questions of reliability of survey questions because they focus on composite scales or on common factor models of multiple indicators (rather than multiple measures). It is well known that quantities like Cronbach's α depend on factors other than the reliabilities of the component items.

While some attention has been given to this issue, we still know very little about patterns of reliability for most types of survey measures. Increasing information on survey data reliability may improve survey data collection and its analysis, and estimates of the reliability of survey measures can help researchers adjust their models. There is a large body of literature in statistics that deals with the problems of conceptualizing and estimating measurement errors (e.g., Biemer and Stokes,1991; Groves, 1991; Lyberg et al., 1997). Until fairly recently, however, little attention was paid to obtaining empirical estimates of measurement error structures (see, e.g., Alwin and Jackson, 1979; Alwin, 1989, 1992, 1997; Alwin and Krosnick, 1991b; Andrews, 1984; Bielby and Hauser, 1977; Bielby et al., 1977a, 1977b; Bound et al., 1990; Duncan et al., 1985; McClendon and Alwin, 1993; Rodgers, Andrews, and Herzog, 1992; Saris and Andrews, 1991; Saris and van Meurs, 1990; Scherpenzeel, 1995; Scherpenzeel and Saris, 1997).

1.6 THE NEED FOR FURTHER RESEARCH

Despite increasing attention to problems of measurement error in survey design and analysis, there are three basic problems that need to be addressed: (1) the lack of attention to measurement error in developing statistical modeling strategies, (2) the relative absence of good estimates of reliability to adjust for measurement error, and (3) the lack of information about how measurement error varies across subgroups of the population, for example, by age and levels of education. On the first point, many multivariate analysis techniques common in analysis of survey data—e.g., hierarchical linear models (HLM) and event history models (EHM)—have ignored explicit consideration of problems of measurement error. On the whole these approaches have not incorporated psychometric adjustments to the model (see, e.g., Bryk and Raudenbush, 1992; Tuma and Hannan, 1984; Petersen, 1993). Of course, there are exceptions in the area of event history models (e.g., Holt, McDonald and Skinner, 1991) and multilevel models (e.g., Goldstein, 1995). In fact, Goldstein devotes an entire chapter to applying estimates of reliability to multilevel models. It is important to note that rather than being a product of the HLM modeling strategy, reliability information is assumed to exist. Goldstein (1995, p. 142) states that in order for such models to adjust for measurement error, one must "assume that the variances and covariances of the measurement errors are known, or rather that suitable estimates exist."

By contrast, within the structural equation models (SEM) tradition, there has ostensibly been considerable attention to the operation of measurement errors. It is often stated that LISREL models involving multiple indicators "correct for measurement error." There are some ways in which this is true, for example, when analysts employ "multiple measures" (the same measure repeated either in the same survey or in repeated surveys) (e.g. Bielby, Hauser and Featherman, 1977a, 1977b; Bielby and Hauser, 1977; Hauser, Tsai, and Sewell, 1983; Alwin and Thornton, 1984). The same conclusion does not generalize to the case where "multiple indicators" (within the same domain, but not identical measures) are employed. There is, unfortunately, considerable confusion on this issue (see Bollen, 1989), and in subsequent chapters (see especially Chapter 3) I develop a clarification of the critical differences between "multiple measures" and "multiple indicators" approaches and their differential suitability for the estimation of reliability.

With regard to *the absence of reliability estimates*, current information is meager and unsystematic, and there are several problems associated with obtaining worthwhile estimates of measurement quality. Empirical research has not kept pace with the theoretical development of statistical models for measurement error, and so, while there are isolated studies of the behavior of measurement error, there has been no widespread adoption of a strategy to develop a database of reliability estimates. On the basis of what we know, we must conclude that regardless of how valid the indicators we use and no matter how rigorously the data are collected, *survey responses are to some extent unreliable*. More information needs to be collected on the relative accuracy of survey data of a wide variety of types (e.g., facts, attitudes, beliefs, self-appraisals) as well as potential sources of measurement error, including both respondent characteristics (e.g., age, education) and formal attributes of questions.

1.7 THE PLAN OF THIS BOOK

In this chapter I have stressed the fact that whenever measures of indicators are obtained, errors of measurement are inevitable. I have argued that one of the most basic issues for consideration in survey research is that of measurement error. This is of critical importance because measurement requires the clear specification of relations between theoretical constructs and observable indicators, as well as the specification of relations between observable indicators and potential measures. In the next chapter I "deconstruct" the data gathering process into its components in order to recognize the considerable potential for measurement error at each step in the reporting process. That chapter and subsequent ones focus specifically on the conceptualization and estimation of the nature and extent of *measurement error* in surveys and establish the rationale for the consideration of potential measurement errors when designing and conducting survey research.

There are several reasons to be concerned with the existence and consequences of errors made in survey measurement. First and foremost, *an awareness of the processes that generate measurement error* can potentially help us understand the nature of survey results. One of the presumptive alternative interpretations for any research result is always that there are methodological errors in the collection of data, and thus, it is important to rule out such methodological artifacts as explanatory variables whenever one entertains inferences about differences in patterns and processes. Second, with *knowledge of the nature and extent of measurement errors*, it is possible in theory to get them under better control. Awareness of the *six major elements* of the response process discussed in Chapter 2—question adequacy, comprehension, accessibility, retrieval, motivation, and communication—is important for researchers to understand in order to reduce errors at the point where they are likely to occur. In addition, *with appropriate measurement designs, it is possible to isolate some types of errors statistically* and therefore control for them in the analysis of data.

In the subsequent chapters I argue that the consideration of the presence and extent of measurement errors in survey data will ultimately lead to improvement in the overall collection and analysis of survey data. One of the main purposes of studies of measurement errors is to be able to identify, for example, which types of questions and which types of interviewer practices produce the most valid and reliable data. In the following I consider ways in which the extent of measurement errors can be detected and estimated in research in order to better understand their consequences. The major vehicle for achieving these purposes involves the presentation of results from an extensive National Science Foundation and National Institute of Aging-supported study of nearly 500 survey measures obtained in surveys conducted at the University of Michigan over the past several years. Assembling information on reliability from these data sources can help improve knowledge about the strengths and weaknesses of survey data. It is expected that the results of this research will be relevant to the general task of uncovering the sources of measurement error in surveys and the improvement of methods of survey data collection through the application of this knowledge.

The research addresses the following sets of questions:

- How reliable are standard types of survey measures in general use by the research community?
- Does reliability of measurement depend on the nature of the content being measured? Specifically, is factual information gathered more precisely than attitudinal and/or other subjective data? Also, do types of nonfactual questions (attitudes, beliefs and self-assessments) differ in reliability?

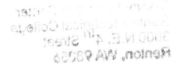

- Does reliability of measurement vary as a function of the source of the information? In the case of factual data, are proxy reports as reliable as self-reports? How reliable are interviewer observations?
- Is reliability of measurement affected by the context in which the questions are framed? Specifically, does the location of the question in the questionnaire, or the use of series or batteries of questions produce detectable differences in levels of measurement error?
- Do the formal properties of survey questions affect the quality of data that results from their use? For example, in attitude measurement, how is reliability affected by the form of the question, the length of the question, the number of response categories, the extent of verbal labeling, and other properties of the response format?
- Are measurement errors linked to attributes of the respondent population? Specifically, how are education and age related to reliability of measurement?

The present research investigates these questions within the framework of a set of working hypotheses derived from theory and research experience on the sources of measurement errors. Simply put, the analysis will focus on explaining variation in reliability due to these several factors.

While a major purpose of this book is to present the empirical results of this study, the goals of this project are more general. In addition to presenting the results of this study, we also review the major approaches to estimating measurement reliability using survey data and offer a critique of those approaches. In Chapter 3 I focus mainly on how repeated measures are used in social research to estimate the reliability of measurement for continuous latent variables. This chapter includes a rigorous definition of the basic concepts and major results involved in classical reliability theory, the major research designs for reliability estimation, methods of internal consistency reliability estimation for linear composites, and recently developed methods for estimating the reliability of single variables or items, including a brief discussion of reliability estimation where the latent variables are latent classes.

This discussion ends with a critique of the standard methods of reliability estimation in common use in survey research—internal consistency estimates—and argues that for purposes of improving survey measurement a focus on the reliability of single measures is necessary. In keeping with this critique, I then review several important developments for the examination of the reliability of single measures: the use of confirmatory factor analysis for the analysis of response errors, including the use of similar methods involving the multitrait-multimethod measurement design, reviewed in Chapter 4, and quasi-simplex models for longitudinal measurement designs, covered in Chapter 5.

Chapter 6 presents the methods used in the present study, including a description of the samples, available measures, statistical designs for reliability estimation, and the problem of attrition in longitudinal designs. The main empirical contribution of this research involves the results of a project whose aim was to assemble a database for survey questions, consisting of question-level information on reliability and question characteristics for nearly 500 variables from large-scale longitudinal surveys of national populations in the United States. The objective was to assemble information on measurement reliability from several representative longitudinal surveys, not only as an innovative effort to improve knowledge about the strengths and weaknesses of particular forms of survey measurement but also to lay the groundwork for developing a large-scale database on survey measurement reliability that can address basic issues of data quality in the social, economic, and behavioral sciences. This chapter presents these methods in four parts: (1) I describe the longitudinal data sets selected for inclusion in the present analysis, (2) I summarize the measures available in these studies and the conceptual domains represented, (3) I discuss the variety of statistical estimation strategies available for application here, and (4) the problem of panel attrition and its consequences for reliability estimation are addressed.

Using these data, there are three main empirical chapters devoted to the analysis of the contributions of various features of survey questions and questionnaires to levels of measurement unreliability, organized primarily around the topics of *question content, question context*, and the *formal properties of questions*. Chapter 7 discusses the potential effects of topic and source of survey reports on the reliability of measurement and presents results relevant to these issues. Chapter 8 discusses the *architecture of survey questionnaires* and the impact of several pertinent features of questionnaire organization on the reliability of measurement. Chapter 9 presents the basic empirical findings regarding the role of question characteristics in the reliability of measurement.

Assembling information on measurement reliability from these panel surveys will not only improve knowledge about strengths and weaknesses of particular forms of survey measurement but also lay the groundwork for developing a large-scale database on survey measurement reliability that can address basic issues of data quality across subgroups of the population. Chapter 10 presents data on the relationship of respondent characteristics—education and age—to the reliability of measurement. I partition the data by these factors and present reliability estimates for these groups. The most serious challenge to obtaining reasonable estimates of age differences in reporting reliability is the confounding of age with cohort factors in survey data. If cohort experiences were not important for the development of cognitive functioning, there would be no reason for concern. However, there are clear differences among age groups in aspects of experience that are relevant for survey response. Specifically, several studies report that educational attainment is positively related to memory performance and reliability of measurement. Since age groups differ systematically in their amount of schooling attained, cohort factors may contribute spuriously to the empirical relationship between age and measurement errors. In Chapter 11 I introduce several approaches to the study of reliability of measures of categorical latent variables. Finally, I wrap up the presentation of the results of this project by reviewing several topics where future research can profitably focus attention, turning in Chapter 12 to some neglected matters. There I also sketch out some avenues for future research on measurement errors in surveys.

CHAPTER TWO

Sources of Survey Measurement Error

Get your facts first, and then you can distort them as much as you please.
Mark Twain, quoted by Rudyard Kipling in
From sea to sea and other sketches (1899)

Reliability of survey measurement has to do with the quality of the information gathered in responses to survey questions. As I clarify in the next chapter, reliability is conceptualized in terms of the "consistency" or "repeatability" of measurement. If one could erase the respondent's memory and instantaneously repeat a particular question, a reliable measure would be one that produced the same response upon repeated measurement. There are a number of different types of measurement error—the principal ones of which are "random" and "systematic" errors—and each has a special relation to reliability of measurement. The problem in quantifying the nature of measurement error in surveys is that errors result from processes that are *unobserved*, and rarely can one directly assess the errors that occur in our data. We are, thus, forced by the nature of these circumstances to construct "models" of how measurement errors happen and then evaluate these models with respect to their ability to represent the patterns that we observe empirically. These models permit us to obtain "estimates" of reliability of measurement, which under optimal circumstances can help us understand the nature of measurement errors. The next chapter (Chapter 3) focuses on understanding how we can estimate reliability of survey measurement. In subsequent chapters I discuss the key strategies that can be used to produce estimates that are interpretable.

There are a number of misconceptions about the reliability of measurement and its utility in evaluating the quality of survey measurement. One argument that is sometimes advanced among survey methodologists is that "consistency of measurement" is of little concern, since what we desire are indications of the "validity of measurement." After absorbing the material in Chapter 3 and in subsequent chapters, one will hopefully be able to integrate the concept of reliability with other aspects of survey quality. On the basis of this understanding of the relationship between the concepts of reliability and validity, one will be able to see the truth to the claim that *reliability* is a necessary condition for validity of measurement and for scientific inference of any type. Reliability is not a sufficient condition for validity, but it is necessary, and without reliable measurement, there can be no hope of developing scientific knowledge. The obverse of this logic is that if our measures are *unreliable* they are of little use to us in detecting patterns and relationships among variables of interest. Reliability of measurement is therefore a sine qua non of any empirical science.

The researcher interested in understanding the reliability of measurement needs to have three things in order to make progress toward this goal. The first is *a model that specifies the linkage between true and observed variables*. As we noted in the previous chapter, survey methodologists define *measurement error* as error that occurs when the recorded or observed value is different from the true value of the variable (Groves, 1989). An inescapable counterpart to defining measurement error in this way is that it implies that *a definition* of "true value" exists. This *does not mean that a true value is assumed to exist*, only that the model defining "measurement error" also provides a definition of "true value." Also, as we pointed out above, the material we cover in Chapter 3 provides such a set of definitions and a specification of the linkage between true variables, observed variables and measurement errors. The second requirement for estimating reliability of measurement is a *statistical research design* that permits the estimation of the parameters of such a "measurement model," which allows an interpretation of the parameters involved that is consistent with the concept of reliability. Chapters 4 and 5 provide an introduction to two strategies for developing research designs that permit the estimation of measurement reliability for survey questions—cross-sectional designs involving the multitrait-multimethod (MTMM) approach and the quasi-simplex approach using repeated measures in longitudinal designs. Finally, researchers interested in estimating the reliability of measurement in survey data need to have *access to data gathered within the framework of such research designs for populations of interest*, and the empirical study reported in this book (see Chapters 6–10) illustrates how this can be done. Ultimately, as shown in Chapter 3, reliability is a property of specific populations, and while we tend to think about measurement quality and its components (reliability and validity) to be aspects of our measuring instruments, in reality they are population parameters. Although we estimate models for observed variables defined for individuals, we cannot estimate the errors produced by individuals—only the attributes of these processes in the aggregate. Still, understanding what these population estimates mean requires us to understand the processes that generate measurement errors (see O'Muircheartaigh, 1997; Schaeffer, 1991b; Tourangeau, Rips and Rasinski, 2000), and in order to lay the groundwork for thinking about measurement errors and eventually formulating models that depict the nature of these errors, in this chapter we discuss the sources of survey measurement errors.

2.1   THE UBIQUITY OF MEASUREMENT ERRORS

Because of their inherently *unobserved* nature, one normally does not recognize measurement errors when they occur. If we were able to identify errors at the time they occur and correct them, this would be a wonderful state of affairs, but that is extremely rare. In contrast, measurement errors typically occur unbeknownst to the investigator. For the most part, they result from inadvertent and unintentional human actions, over which we do not necessarily have control—that is, at least within the limits of our best practices as survey researchers. Some respondents or informants obviously do intentionally mislead and knowingly fabricate the information they provide, but such phenomena probably represent a very small percentage of the measurement errors of concern here. Even where respondents are "not telling the truth," they may, in fact, be unaware that they are doing so.

Some examples of what we mean by such measurement errors may be useful. One of the most common sources of error results from the fact that respondents do not understand the question or find the question ambiguous, and are unwilling to come to terms with their lack of comprehension. In our subsequent discussion we put a great deal of emphasis on the importance of writing survey questions that are simple and clear so that they are comprehensible to the respondent. Still, many errors occur because the information sought by the question is unclear to the respondent and s/he either perceives (and then answers) a different question or simply guesses about what the interviewer wants to know. Because of the pressures of social conformity respondents are often motivated to present themselves as knowledgeable, even when they are not, and *this may vary by culture*. Furthermore, survey researchers often discourage respondents from saying "Don't Know." Converse (1976–77) suggested that when respondents say they "Don't Know," this tends to be a function of a lack of information rather than ambivalence or indifference. If respondents do not have the information requested but feel pressure to provide a response, then guessing or fabricating a response will most likely produce an error. This violates the assumption often made in survey research that respondents know the answer to the question and are able to report it.

One famous example of how respondents are often willing to answer questions they do not understand or know little about is provided by Schuman and Presser (1981, pp. 148–149). In an unfiltered form of the question, they asked if respondents favored or opposed the Agricultural Trade Act of 1978 (a fictitious piece of legislation), and some 30.8% of the sample had an opinion about this. In a separate ballot they found that only 10% stated an opinion, when they explicitly offered a Don't Know alternative. It is an interesting question whether this tendency to appear knowledgeable varies by country or cultural context.

Another example of the occurrence of measurement error involves retrospective reporting where the assumption that people have access to distant events is not met. There are few surveys that do not ask respondents something about the past, and perhaps most typically about their own lives, but sometimes the gathering of such information can be alarmingly imprecise (Schwarz and Sudman, 1994). Among other things, researchers routinely question people concerning their social background or early life history, that is, what were some of the characteristics of their family of origin (e.g., father's and mother's occupations, educational levels, native origins, marital status). There are several existing tales of imperfection in the measurement of father's occupation, for example, a variable that is crucial to sociological studies of social mobility. Blau and Duncan (1967, pp. 14–15) present a somewhat disturbing case. In brief, in a small sample of men for whom they had both census records and survey data, when they compared the child's report of the father's occupation with matched census records, they found only 70% agreement (out of 173 cases). They report:

> Although 30 percent disagreement may appear high, this figure must be interpreted in light of two facts. First, the date of the census and the date at which the respondent attained age 16 could differ by as much as five years; many fathers may actually have changed occupations between the two dates. Second, reinterview studies of the reliability of reports on occupation may find disagreements on the order of 17 to 22 percent . . . even though the information is current, not retrospective.

The Blau and Duncan (1967) testimony on this point is hardly reassuring. They are claiming that the retrospective measurement of father's occupation in the United States Census is not so much worse than contemporaneous self-reports, which may be pretty poor. On the face of it, this seems highly surprising if retrospective reports of parental occupation are as reliable as self-reports, yet the conclusion has been borne out in more recent research reported by Bielby, Hauser, and Featherman (1977a, 1977b) and Hauser, Tsai, and Sewell (1983). A further interesting claim of the Blau and Duncan (1967) report is that recall actually seems to improve with the lapse of time. They report that older men were more likely than younger men to report their father's occupation reliably. This claim, however, should not be pushed too far. It is one thing to suggest that people could accurately recall dates and durations of events and transitions over the life course; it is another to suggest that reliability may increase with age (see Chapter 10).

In other cases researchers are also sometimes interested in people's perceptions of past events and social changes. In the latter type of study there is no way to measure the veridicality of perceptions, unless these reports can be validated by existing data, and it is often found that perceptions of past events are biased (see Alwin, Cohen and Newcomb, 1991). In many cases, it would be a mistake to assume that human memory can access past events and occurrences, and in general, the longer the recall period, the less reliable are retrospections (Dex, 1991). In addition, several factors affect people's abilities to recall the timing of past events. For example, one phenomenon known to exist in retrospective reports is *telescoping*, reporting events as happening more recently than they actually did. Also, more recent experiences may bias people's recollections about their earlier lives, making it difficult, if not impossible, to be certain about their validity.

We often assume that respondents have relatively easy access and can retrieve relatively recent factual material in their own lives. Since most people are employed and in the labor force, we might often assume that they can easily report information regarding their employment and pay. This may not always be the case. Bound, Brown, Duncan, and Rodgers (1990) compared company records on employment and earnings data from a single large manufacturing firm with several thousand employees to responses of a sample of those employees interviewed by telephone. They found that if the questions involve an annual reporting cycle, the measurement of earnings can be quite good, although not perfect. They find a correlation of .81 between (the log of) the company record of earnings for the previous calendar year and (the log of) the respondent report of their earnings for that year. Even though survey researchers often assume that self reports of "factual" information can be measured quite well, the fact is that even such "hard" variables as income and earnings are not perfectly measured. In this case only two-thirds of the response variance in the survey reports reflects valid variance. Other measures that Bound et al. (1990) examined performed even less well. The correlation between records and survey reports for the pay period immediately preceding the interviews was .46. Similarly a third measure of self-reported earnings from the interview, "usual earnings," correlated .46 with the average value in the records for the preceding 12 weeks. Survey reports of hours worked per week correlated in the range of .60 to .64 with the corresponding records information. Reports of hourly wages correlated even less with company records—for the most recent pay period the correlation of the survey measure of hourly earnings with the records measure is .35, which means that only about 10 percent of the variance is valid variance.

We may assume that some difficulties of measurement are universal in the sense that they represent problems that must be surmounted in all surveys, i.e., peoples' memories are poor regardless of time and place. Although one would expect to be able to obtain a higher degree of accuracy in the measurement of such things as recent employment and earnings information, some phenomena are clearly not amenable to valid and reliable retrospective measurement. For example, Kessler, Mroczek, and Belli (1994) suggested in their assessment of retrospective measurement of childhood psychopathology that while it might be possible to obtain some long-term memories of salient aspects of childhood psychiatric disorders many childhood memories are lost due to either their lower salience or active processes of repression.

Another one of the assumptions listed in the foregoing that seems to be repeatedly violated in survey practice is that the respondent is willing to put forth the effort that is needed to provide accurate information to the researcher. There is a clear recognition in the survey methodology literature that an important component in generating maximally valid data is the fostering of a commitment on the part of the respondent to report information as accurately as possible (see Cannell, Miller, and Oksenberg, 1981). It is also clear from discussions of respondent burden that high cognitive and motivational demands placed on respondents may result in a reduction in the quality of data. The potential usefulness of these developments can be seen from Krosnick and Alwin's (1989) review of possible applications of the idea that respondents maximize their utilities [see Esser (1986); also see the discussion below]. One way of phrasing the issue relies on Herbert Simon's (1957; Simon and Stedry, 1968) concepts of *satisficing* and *optimizing* behavior. The term "satisficing" refers to expenditures of the minimum amount of effort necessary to generate a satisfactory response to a survey question, in contrast to expenditures of a great deal of effort to generate a maximally valid response, or "optimizing" (see Tourangeau, 1984; Krosnick and Alwin, 1988, 1989; Alwin, 1991). We suggest that satisficing seems to be the most likely to occur when the costs of optimizing are high, which is a function of three general factors: the inherent difficulty of the task, the respondent's capacities or abilities to perform the task, and the respondent's motivation to perform the task. There is plenty of evidence in the survey methods literature that provides a basis for a "satisficing" interpretation of measurement errors: random responding (Converse, 1964, 1970, 1974), the effects of Don't Know filters (Schuman and Presser, 1981; McClendon and Alwin, 1993), the effects of offering middle alternatives (Schuman and Presser, 1981), response order effects (Krosnick and Alwin, 1987), acquiescence response bias (McClendon, 1991), and nondifferentiation in the use of rating scales (Krosnick and Alwin, 1988). The conditions under which this wide array of well-documented response errors are precisely those that are known to foster satisficing (Krosnick and Alwin, 1989; Krosnick, 1999).

To take one further example of areas in which one can expect measurement errors to occur, consider the reporting of socially undesirable or embarassing events. There has been considerable focus in the survey methods literature on underreporting problems with deviant and socially undesirable behavior (Bradburn, Rips, and Shevell, 1987; Cannell, Miller and Oksenberg, 1981). Also, on the flip side is the problem of socially desirable behaviors, e.g., church attendance in the United States or voting in national elections which tend to be over reported (see Hadaway, Marler, and Chaves, 1993; Traugott and Katosh, 1979, 1981; Presser, 1990; Presser and Traugott, 1992). Clearly what is socially desirable or undesirable in one segment of the population is not necessarily so in another, so a systematic understanding of this set of issues is necessary within the framework of a consideration of measurement errors.

All of these examples are instances in which there is a difference between the quantity or quality *recorded or observed* and the *true value* of the variable. Obviously, in order to address problems of measurement error it is important to understand what is meant by the concept of the "true" value, since error is defined in relation to it. Approaches to dealing with measurement error differ in how to conceptualize the "true" value (see Groves, 1989, 1991). To simplify matters, there are at least two different conceptions of error, based on two different conceptions of true score. "Platonic" true scores are those for which there is some notion of "truth" as is usually assumed in record-check studies (see Marquis and Marquis, 1977). Studies, for example, that compare voting records with self-reports of voting (Presser and Traugott, 1992) or those that compare actual patterns of religious behavior compared with self-reports (Hadaway et al., 1993) are implicitly using a *platonic* notion of true scores, i.e., that there is some "truth" out there that is to be discovered. This definition, however, is not generally useful because most variables we wish to measure in surveys have no "true" source against which to measure the accuracy of the survey report. "Psychometric" true scores, on the other hand, are defined in statistical terms as the expected value of a hypothetical infinite set of observations for a *fixed person* (see Lord and Novick, 1968). We return to a discussion of these matters in the following chapters dealing with the estimation of components of measurement error.

2.2   SOURCES OF MEASUREMENT ERROR IN SURVEY REPORTS

The claim that there are errors in survey reports suggests that something happens during the process of gathering data that creates this disparity between the "observed" value and the "true" value. We can perhaps begin to understand the potential errors in survey measurement if we make explicit the assumptions that are often made in the collection of survey data. These are essentially as follows: (1) that the question asked is an appropriate and relevant one, which has an answer; (2) that the question is posed in such a way that the respondent or informant understands the information requested; (3) that the respondent or informant has access to the information requested; (4) that the respondent or informant can retrieve the information from memory; (5) that respondents are motivated to make the effort to provide an accurate account of the information retrieved; and (6) that they can communicate this information into the response framework provided by the survey question. Obviously, it would be naive to assume that these assumptions are met in every case, and so to acknowledge the possibility that they are not opens the door to the conclusion that measurement errors may occur in the gathering of survey data.

As noted earlier, there are *two* fundamental strategies to dealing with measurement errors in surveys. The first is to concentrate on the aspects of the *information-gathering process* that contribute to errors and reduce them through improved data collection methods. The second is to accept the fact that measurement errors are bound to occur, even after doing everything that is in ones power to minimize them, and *to model the behavior of errors using statistical designs.* We return to the latter topic in a subsequent chapter. Here we focus on ways in which survey measurement errors can be reduced by knowing about when and how they tend to occur.

Regardless of whether one's focus is on the reduction of errors during the collection of survey data or on the modeling of errors once they have occurred, there is a common interest between these emphases in developing an accurate picture of the response process and the factors that impinge on the collection of accurate information in the survey interview. Responses to survey questions are affected by a number of factors that produce the types of errors of measurement discussed above. It is generally agreed that key sources of measurement errors are linked to aspects of survey questions, the cognitive processes of information processing and retrieval, the motivational context of the setting that produces the information, and the response framework in which the information is then transmitted (see Alwin, 1989, 1991; Alwin and Krosnick, 1991b; Bradburn and Danis, 1984; Cannell, et al., 1981; Hippler, Schwarz and Sudman, 1987; Knäuper, Belli, Hill, and Herzog, 1997; Krosnick, 1999; Schwarz, 1999a, 1999b; Schwarz and Sudman, 1994; Sirken, Herrmann, Schechter, Schwarz, Tanur, and Tourangeau, 1999; Strack and Martin, 1987; Tourangeau, 1984, 1987, 1999; Tourangeau and Rasinski, 1988).

A classic treatment of this issue by Oksenberg and Cannell (1977), for example, examines the logical flow of steps by which the individual respondent processes the information requested by a question. Their work has influenced the following discussion, and for our purposes there are essentially *six* critical elements of the response process that directly impinge on the reliability and validity of survey measurement to which we devote attention in this section (see Table 2.1). All of these factors play a role in affecting the quality of survey data, whether the question seeks information of a factual nature, or whether it asks for reports of subjective states, such as beliefs and attitudes, but they are perhaps especially problematic in the

Table 2.1. Six critical elements in the response process

1. *Question validity:* the adequacy of the question in measuring the phenomenon of interest

2. *Comprehension:* the respondent's understanding or comprehension of the question and the information it requests

3. *Accessibility:* the respondent's access to the information requested (e.g., do they have an opinion?)

4. *Retrieval:* the respondent's capacities for developing a response on the basis of the information at hand, say, from internal cognitive and affective cues regarding his/her attitude or level of approval

5. *Motivation:* the respondent's willingness to provide an accurate response

6. *Communication:* the respondent's translation of that response into the response categories provided by the survey question

measurement of subjective phenomena such as attitudes, beliefs, and self-perceptions. It is particularly important to "deconstruct" the process of questioning respondents and recording their answers within a temporal framework such as this in order to be able to identify sources of survey error.

2.2.1   Validity of the Question

The concept of *measurement validity* in its most general sense refers to the extent to which the measurement accomplishes the purpose for which it is intended. This set of considerations, thus, expresses a concern with the linkage between concepts and their indicators. For example, according to the *Standards for educational and psychological testing* (APA, 2000), we can usefully distinguish among three types of validity in this sense—content, criterion-related, and construct validity. *Content validity* refers to the extent to which a well-specified conceptual domain has been represented by available or selected measures. Too often, survey researchers do not really know what they want to measure, and they therefore beg the question of *measurement validity* from the outset. This is often seen in the way they select questions for use in surveys, namely, going to other people's surveys to see what other people have asked. This is apparently done on the assumption that the other person "knows" what they are measuring. Nothing could probably be further from the truth. Another indication of the survey researcher's opting out of the concern with valid measurement is to include several questions on the same topic, apparently on the assumption that one or more of the questions will cohere around something important. This is why factor analysis is often used after a pretest to examine whether in fact that coherence has been achieved. Generally speaking, if one knows what one wants to measure, it will take no more than a few questions to get at it.

This reflects a minimal requirement for social measurement in that the content of measurement should represent theoretical content that drives the research. *Criterion-related validity* refers to the predictive utility of a measure or set of measures—do they predict or correlate with other theoretically relevant factors or criteria? For example, the criterion-related validity of SAT scores are typically assessed in terms of their ability to predict college grades (Crouse and Trusheim, 1988). *Construct validity*, on the other hand, refers to the extent to which an indicator or set of indicators assess the theoretical or abstract concept of interest. One set of standards for evaluating *construct validity*, which is useful in the present context, was introduced by Campbell and Fiske (1959), who argued that construct validation requires evidence of both *convergent* and *discriminant* validity. Convergent validity is reflected in *high* correlations among different approaches to measuring the same trait, whereas discriminant validity is reflected by *low* correlations between measures of different traits measured by the same or similar methods. To assess these aspects of construct validity Campbell and Fiske (1959) proposed the *multitrait-multimethod matrix* (MTMM), an organized array of correlations among multiple traits simultaneously measured by multiple methods (see Chapter 3).

Despite the seeming plurality of definitions of what is meant by the concept of validity, most discussions appropriately emphasize the notion of construct validity as central to this set of concerns, namely does the set of measurement procedures produce a score that reflects the underlying construct it is intended to measure (Bohrnstedt, 1983). Validity, thus, centrally involves the question of whether the variables being measured reflect or approximate the theoretical concept of interest. However, there are rarely any direct assessments of validity in survey research. The major exceptions can be cited under the heading of *record-check* studies (Marquis, 1978), but these are very rare, and often the content of interest to social scientists cannot be validated with reference to an external criterion. Even behavior is often notoriously at odds with stated attitudes (Abelson, 1972).

Record-check validation studies, although rare, are worth emphasizing here because they illustrate the importance of obtaining estimates of the validity of survey data. Further, they illustrate how often survey researchers take the quality of their data for granted. Reviews of record-check studies reveal that considerable systematic measurement error (either over reporting or under reporting) in many cases (Marquis, 1978). *Consequently, if errors can be documented in the case of platonic true scores, one might wonder what levels of error there are when true scores are less well defined and are essentially unobservable* (see Lord and Novick, 1968).

2.2.2   Comprehension of the Question

The purpose of the survey interview is to communicate information. Language is essential to communication (Clark, 1985). Thus, it should come as no surprise that the meaning attached to the words used in survey questions, the context in which they are presented, and the comprehension of what the question is asking have important implications for the quality of the measures derived from the responses. This is a particularly difficult area for research that crosses national and cultural boundaries, because of the inherent problems of employing different languages to gather the same information. Familiarity with the content areas covered by research may be very different across cultures, and there may be very different levels of comprehension of the questions. For example, Western concepts of "justice" may be very difficult for members of Eastern cultures to appreciate, and therefore questions aimed assessments of "fairness" in social relations may be difficult for some members of Eastern cultures to understand.

Considerable research on survey methods has been stimulated by Schuman and Presser's (1981) widely cited documentation of the effects of alterations in questions and question context on response distributions, although this line of research is not about survey "measurement errors" as such. Much of this work has been characterized by "limited theorizing and many ad hoc experiments," (Hippler et al., 1987, p. 4), but it has had a role in underscoring the importance of language in shaping the meaning attached to survey responses. There is little question that the meaning and interpretation given to the question being asked has an important role in shaping the nature and quality of the data. Put simply, answers and distributions of responses depend intimately on the way the question is asked. These issues are especially critical to assessing the validity of measurement in cross-cultural survey research.

In the article by Oksenberg and Cannell (1977) mentioned above, which examines the logical flow of steps by which the individual respondent processes the information, the first step they discuss is the respondent's *comprehension* of the question. On one hand, survey researchers typically expect that respondents will know the meaning of the question and its constituent parts, and it is further assumed that all respondents will interpret these meanings in the same way. On the other hand, there are many examples from the survey methods literature wherein investigators have asked respondents to report their perceptions of the "meaning" of survey questions, which sometimes reveal substantial variation in interpretations of words and phrases (e.g., Belson, 1981; Groves, Fultz and Martin, 1992). It is important to recognize, however, that if a question is unclear in its meaning, or if parts of the question can have more than one meaning, it would hardly be possible that the question measures a single "true" value for all persons. Thus, the ability to assess measurement errors may presuppose questions and response categories that are precise in meaning.

The fascination with "question" and "context" effects on marginal distributions in surveys is understandable (see, e.g., Schuman and Presser, 1981), but it is not at all clear how far this takes us to an understanding of measurement error. To be sure, if one uses two similar questions, which have slight differences in meaning (e.g., "The government should not allow the immigration of people with AIDS into the country" vs. "The government should forbid the immigration of people with AIDS into the country") will most likely produce differences in response distributions, that is, the percentage of a given population producing an AIDS-tolerant response to the two questions. "Forbidding" something may be stronger than "not allowing" it. In psychometric terms this means that the two questions have different levels of difficulty. It in *no* sense means that the two questions do not measure the same latent attitude. Indeed, across individuals one would expect that two such questions would be highly correlated and that their latent true scores would be collinear, or nearly so. In other words, a simple true-score model might be applied to understanding their relationships and differences. But unfortunately, virtually all research that has engaged in examining the nature of "question" and "context" effects has prevented such an understanding by limiting the research to "between-subjects" designs rather than "within subjects" designs (see the discussion of this issue in Chapter 3).

2.2.3 Information Accessibility

Even if one can assume that the question is appropriate and relevant to the respondent and that his/her comprehension of what is being asked is good, it may still not be the case that the respondent can access the information. With regard to *accessibility to requested information*, there are several areas that need to be considered: (1) the respondent's access to information about the past, (2) the respondent's access to information about the present, and (3) the respondent's access to information about other people.

In one of the examples given above of typical kinds of measurement error in surveys, we mentioned that were retrospective questions about events and experiences in the past. When the respondent has *forgotten* the events and experiences in question, then we say that there is no longer access to the information. Retrieval of accessible information is another matter (see the discussion in the next section). The nature of this problem is revealed in the study of older respondents, who may be more subject to these processes because problems of memory increase with aging. There is some support for this in studies showing that older respondents are more likely to answer "don't know" to survey questions (Gergen and Back, 1966; Ferber, 1966; Francis and Busch, 1975; Rodgers and Herzog, 1987a). There is reason to expect, however, that this pattern is more prevalent for attitudes and expectations, and less so for factual content.

Access to information about the present is considerably less problematic because it is not dependent on long-term memory. There are nonetheless issues of access to information, particularly in the measurement of subjective states, such as attitudes and beliefs. Respondents who have no firm attitude or opinion on a given issue, either because they have little knowledge of the issues or have not thought about it, essentially confront an issue of access. Rugg and Cantril's classic study of question wording acknowledged that response errors were often less likely if persons "had standards of judgment resulting from stable frames of reference," as opposed to situations "where people lack reliable standards of judgment and consistent frames of reference" (Rugg and Cantril, 1944, pp. 48–49). An example of this is Converse's (1964) famous example of "nonattitudes"—respondents who do not have opinions or attitudes and therefore do not have access to any cues that will help them decide how to answer the question [see the review of this issue by Krosnick (2002)]. As noted earlier, such respondents may feel pressure to respond to survey questions because they assume interviewers want them to answer, and because of cultural norms, they believe opinionated people are held in higher esteem than the ignorant and uninformed. One interpretation is that because respondents wish to conform to these expectations and project positive self-images, they frequently concoct attitude reports, making essentially random choices from among the offered response alternatives. Critics of this viewpoint (e.g., Achen, 1975) argue that measurement errors are not the fault of respondents' random reports, but primarily due to the vagueness of the questions.

Some degree of error variation in attitude reports is likely to result from ambiguity in respondent's attitudes. It seems likely that some attitudes may be associated with univocal, unambiguous internal cues that come to mind quickly and effortlessly when a person simply thinks about the attitude object. What is often more likely to be the case involves attitudes that are associated with ambiguous or conflicting internal cues or with cues that are relatively inaccessible or come to mind only as a result of considerable cognitive effort (Fazio, Herr, and Olney, 1984). Clearly, on a given issue, some respondents are likely to have relatively consistent internal cues indicating their attitudes whereas others have highly conflicting or ambiguous cues, in which case forcing respondents to choose a single point on an attitude continuum may cause them to make such choices haphazardly, and such internal ambiguity will increase the amount of random measurement error. Some have argued that given the likelihood that such ambiguous cues may exist for a sizeable portion of the population, it might be best to filter out such respondents and ask questions only of those respondents who have firm attitudes (see McClendon and Alwin, 1993).

Gathering data by proxy in survey research is commonplace, and this is an additional issue that raises the issue of accessibility. Respondents to surveys are often asked questions about other people, including their spouse and children, and sometimes, their friends and coworkers. Because of the differences in the process of reporting about the characteristics of others is different from the more common self-report method, one would expect that the nature of measurement errors might be different for proxy vs. self-reports. Recent evidence, however, suggests that "for many behaviors and even for some attitudes, proxy reports are not significantly less accurate than self-reports" (Sudman, Bradburn and Schwarz, 1996, p. 243). This is an encouraging finding because it is seldom possible to obtain self-reports in all cases and the proxy is frequently the only source of information on the person in question, e.g., the head of household.

2.2.4 Retrieval of Information

As indicated earlier, the retrieval of information is often viewed as one of the most serious elements of the response process, given the importance of both memory and cognitive processing in formulating responses to many survey questions. Obviously, when memory fails or when retrieved information is ambiguous or imprecise, there is a tendency to create measurement errors in approximating the "true" value. There are a number of techniques developed by survey researchers to assist in jogging people's memories. Unfortunately costs are incurred in most strategies for improving the quality of the data. For example, the use of diaries may well improve the quality of retrospective data for a whole range of life history information, such as daily hassles, consumption, travel, time use, and sexual behavior. Yet such diary methods place a considerable burden on the respondent, as well as requiring substantial investment in coding and processing the often nonstandardized data (see Scott and Alwin, 1997).

One quite useful application of the technique of retrospective measurement is the *life history calendar*, an approach which employs detailed reports, often on a monthly basis, of transitions occurring across a number of life domains, for example, living arrangements, marital status, and educational status (see Freedman, Thornton, Camburn, Alwin, and Young-DeMarco, 1988). A variety of forms of life history calendars (LHC) have been used in a number of prospective panel studies. One example, reported by Freedman et al. (1988), collected monthly data on the incidence and timing of event histories within a set of interrelated domains in a sample of 900 23-year-olds, who had been interviewed once before. The sample was drawn from a population of the July 1961 birth records of first-, second-, and fourth-born white children in the Detroit metropolitan area (Wayne, Oakland, and Macomb Counties), and the mothers of this sample had been studied prospectively from the time of the birth of that child. The database for this sample of women and their children includes eight waves of personal interviews with the mothers between 1962 and 1993. In 1980, the first personal interview with the child born in 1961 was conducted, and these respondents have been reinterviewed in 1985 and 1993 using, among other measures, the LHC measurement of event histories. One of the rationales for this approach to the measurement of autobiographical information is that it is intended to improve the reliability of people's retrospective memories. The aim is to improve recall by placing the measurement of timing and duration of events in several domains (e.g., living arrangements, marriage, fertility, schooling, and work) simultaneously within the same timeframe. Thus inconsistencies in the timings of events and sequences are made apparent by the calendar itself and efforts can be taken to remove them. To our knowledge, little information about the reliability of life history calendars exists, because its assessment requires panel data. Freedman et al. (1988) give several examples of reliability assessment in their use of the LHC, including their report on employment status, which was measured on a three-category scale (full-time, part-time, no attendance/employment).

When reinterviewed in 1985 respondents appear to have a noticeable bias in reporting they had worked in 1980, when they had indicated otherwise in the 1980 interview. Data presented by Freedman et al (1988, p. 64) show that of the respondents who said in 1980 that they were not working, one third in 1985 said that they had worked. But of the respondents who reported in 1980 that they were working either full or part-time, only 7% or 8% reported in 1985 they had not worked. In this example, respondents were only 23 years old in 1980 and the period of recall was not very long. The real test is to know how well people can recall their employment status over the life course. As part of a reliability assessment of retrospective work history data in a British study on Social Change and Economic Life (SCEL), 400 Scottish couples were reinterviewed in 1986 who had originally been interviewed some 22 years earlier in 1964 (Elias, 1991). Thus it was possible to compare the person's recollections of their occupational status with what they actually said at the time. For both men and women, the agreement between their reported and recollected employment status was over 60%. However, there were major discrepancies with regard to whether a person was working or not for 20% of women and 27% of men.

Perhaps our focus on employment status has given an unduly pessimistic picture of the reliability of retrospective data, because of the ambiguities in this oversimplified categorical variable. Indeed, Freedman et al. (1988) draw attention to the fact that employment status is measured with far less reliability than reports of school attendance. Estimated reliabilities for school attendance and employment status are .90 and .62 respectively [calculated by the author from data presented by Freedman et al. (1988)]. Even more reliable was the recall of births and marriages, although, because respondents were only in their early twenties, there were relatively few such incidents. Nonetheless, enough comparison was possible to establish that the reliability of retrospective reports in these domains is extremely high.

One study that attempted to assess whether retrospective measurement of attitudes and ideological orientations was possible was based on the classic study by Theodore Newcomb, which demonstrated how women students, exposed to the rather liberal climate of Bennington College in the 1930s and 1940s, shifted their social and political attitudes in the more liberal direction. Newcomb's original study spanned the women's 4-year stay at the college, but then the same women were reinterviewed some 25 years later in 1960–61, revealing that, in general, the more liberal attitudes were retained through into middle age. In 1984, Alwin, Cohen and Newcomb (1991) conducted a nearly 50-year follow up, which gave support to a theory of political attitude development—the generational and attitude persistence model—which suggests that people are most vulnerable to historical and generational influences when they are young, and that generally, after some period of formative years, people tend to grow more rigid in their viewpoints (see Sears 1983; Alwin et al., 1991; Alwin and Krosnick, 1991a; Alwin 1994).

The Bennington College study is actually an example of a combination of a prospective attitude measurement design and a retrospective assessment of the nature of the social processes affecting the development and maintenance of attitudes over the lifespan. This example is particularly instructive for assessing the reliability of retrospective life history data because (1) it covers a very long span of time, (2) it deals with attitudes, and (3) its special value is the ability to look at what happened "objectively" to these women, as indicated by the contemporaneous attitudinal measurements, within the context of their recollections about how and why their views had changed. In 1984, the women were asked to give retrospective quantitative assessments of their attitudes at various stages in their lives, and these retrospective reports provide confirmation of the developmental processes that the prospective data revealed, that is that the women's attitudes had shifted markedly in the liberal direction as a result of their college experience, and these liberal leanings persisted over time. Thus, it appears that, at least, on issues of real import and salience retrospective attitudinal data can be quite reliable, contrary to assumptions commonly made.

Yet, the Bennington study revealed some interesting divergences between retrospective data and attitudes measured at the time, which do indicate certain consistent biases in recollections. Alwin et al. (1991, p. 133) report that among the most striking of the divergences in current and retrospective reports was the later denial of favoring Richard Nixon in the 1960 election. Of those who favored Democrats in the 1960s election, 93% had a consistent recollection in 1960. By contrast, 45% of those reporting in 1960 that they had preferred Nixon, denied that report in 1984 and claimed to have supported the Democratic candidate. It seems reasonable to assume that the 1960s report, which essentially involves no retrospection, is the more reliable and that the distortion is in the liberal direction. Whether the distortion is in order to make political orientations appear more consistent over time, or whether it merely reflects a disavowal of the disgraced president, or both, is unclear.

2.2.5 Respondent Motivation

The respondent's motivation to respond is critical to obtaining an accurate response. One recent development in the study of survey errors has been the application of principles of *rational choice* theory to the survey response process. Esser (1986) argues that survey data should be understood within the framework of the social situation that produced it, and that such an understanding will be enhanced if the investigator realizes that respondent behavior is a joint function of individual tendencies or motives to respond and the situational constraints on behavior. Empirical data, Esser (1986) argues, should be understood as the result of situation-oriented rational actions of persons. In surveys respondents are given tasks and are asked to solve them, and response behavior can only be understood if it is interpreted in this context of "problem-solving" behavior. However, the completion of survey tasks in a manner that produces reliable and valid data, that is, data free of error, depends on the respondent's "cost-benefit" analysis of various courses of action.

Thus, respondents may be viewed as "rational" actors, who perceive the elements of the situation, evaluate them against the background of their own preferences and perceived consequences—the "subjective expected utility" (SEU) values—of various action alternatives, and a certain action or response is selected according to the maximization of subjective utilities. In order to predict respondent behavior, then, one needs to understand the respondent's goal structure (e.g., providing valid data, or acquiescing to the perceived expectations of the interviewer) as well as the respondent's own SEU values associated with response alternatives. Within this framework the selection among response alternatives is made according to the interests of the respondent, and one refers to this selection as "rational choice," even if the respondent in no way corresponds to the criteria of "objective" rationality (Esser, 1986, 1993).

Some efforts have been undertaken to increase respondent motivation by providing instructions and by asking for a signed statement to provide accurate information. Cannell et al. (1981), for example, experimented with these techniques and reported that they are effective in increasing reports of typically underreported events such as physician visits and illnesses. However successful are similar efforts to improve respondent motivation, this is probably an area where there are vast differences across studies and therefore a potential source of differences in levels of measurement error.

2.2.6 Communication

Survey researchers often assume, just as they do with regard to the meaning of questions, that the categories of the response options provided by the researcher are self-evident and the respondent is well versed in how to use them. This may not routinely be the case, and difficulties with the respondent's translation of that response into the response categories provided by the survey question may create opportunities for measurement error.

One of the earliest debates among survey researchers centered on whether survey questions should be open- or closed-ended (see Converse, 1987), and although that debate has been settled largely in favor of closed-form questions, there is still some skepticism about the effects of response categories on the nature of the response (see Schuman and Presser, 1981; Schwarz, 1999b). In many areas researchers have learned that simply asking an open-ended question more often than not provides the most accurate data, e.g., in the measurement of income. Coarser categorization is then used only if the respondent refuses at the first step. However, for the bulk of sample surveys closed questions are the major approach to the transmission of information from respondent to researcher.

The meaning of the response choices may be particularly difficult if the survey question employs what are known as "vague quantifiers" (Bradburn and Miles, 1979; Schaeffer, 1991a). This can be illustrated most easily in the context of factual measurement. If a researcher is interested in determining how often people smoke cigarettes, respondents could be asked whether they smoke "constantly, frequently, sometimes, rarely, or never." However, the meanings of these response alternatives are clearly ambiguous, particularly in comparison to a question that asks respondents to report the number of cigarettes they smoked in the last 48 hours. In this latter case, the response alternatives are much clearer in their meaning. The more ambiguous the meaning of the response alternatives, the more difficulty respondents should have in mapping their internal cues onto those alternatives, and the more measurement error will result. In some cases verbal labels may enhance the quality of the responses, but in others they may simply contribute to ambiguity. As Norbert Schwarz's (1999b) research has shown, when respondents are asked to report on the frequency with which they have engaged in a specific behavior, the frequency alternatives presented in the closed-question format influence the respondent's estimates of the frequency of the behavior.

In the case of attitude measurement, for example, even given the presence of an attitude and the respondent's access to clear internal representations of that attitude, ambiguities may still remain regarding the expression of the attitude in terms of the response options provided. Error variation in attitude reports is likely to result from ambiguity offered by the survey question and the difficulty in reconciling ambiguous and conflicting judgments regarding the linkage between internal cues and the external response categories provided by the researcher. No matter how clear and unambiguous are a respondent's internal attitudinal cues, he/she must usually express those cues by selecting one of a series of response options offered by the survey question. This entails a process of mapping the internal cues on to the most appropriate response alternative. This mapping process can produce errors in several ways. For example, the respondent might find that his/her internal cues do not correspond to any of the offered response choices. For example, a respondent who feels that abortion should be legal only if the mother's life is endangered and not under any other circumstances may have difficulty in mapping that view on to the response choices of a question that simply asks whether he/she favors or opposes legalized abortion. When faced with this challenge, respondents may be forced to respond in a way that creates error.

There are several decisions that must be made in presenting response alternatives to respondents. One of the most important of these, especially in the measurement of attitudes and beliefs, is the number of response categories that should be used. Attitudes are often assumed to have both direction and intensity and researchers often attempt to obtain an attitude response on a continuum expressing both direction and intensity. Most approaches to measuring an attitude conceptualize the response scale as representing this underlying latent continuum, and the response categories are intended to map this attitude continuum. In other words, the response scale is typically intended to be (at least) bidimensional, expressing both direction and intensity. Obviously, two- and three-category scales cannot assess intensity, but they can assess direction, and in the case of the three-category scale, a "neutral" point can be identified. In all other cases, both direction and intensity can be obtained from the typical approaches to survey measurement of attitudes.

Two-category scales have higher reliabilities than many scales with greater numbers of categories (Alwin, 1992). Moreover, three category scales may actually promote less precise measurement than two-category scales, because the introduction of a middle alternative or a neutral point between positive and negative options often creates ambiguity and confusion (Schuman and Presser, 1981). Two types of ambiguity may exist. First, at the level of the latent attitude, the individual's region of neutrality may vary and in some cases may be quite wide. In other words, neutrality is not necessarily a discrete point on an attitude continuum, and there may be considerable ambiguity about the difference between neutrality and weak positive and negative attitudes. On the other hand, four- and five-category scales do not have the same problems because they provide weak positive and weak negative categories, thereby giving the respondent a better opportunity to distinguish between neutrality and weak forms of positive and negative attitudes. Second, the response scale itself may produce certain ambiguities. Even if the internal attitude cues are clear, the respondent may find it difficult to translate his or her attitude into the language of the response categories. Middle alternatives are more often chosen when they are explicitly offered than when they are not (Schuman and Presser, 1981), suggesting that the meaning of the response categories may stimulate the response. In some cases the respondent may choose the middle category because it requires less effort and may provide an option in the face of uncertainty.

2.3   CONSEQUENCES OF MEASUREMENT ERROR

One might conclude from the prior discussion that most errors of measurement in surveys are relatively minor and at worst represent the fact that in any measurement some "noise" is created that tends blur the basic "signal" or "true" nature of the phenomenon. Moreover, it might be argued much of this error is random and therefore tends to even out and not affect the kinds of aggregate statistics reported for survey samples. This view is consistent with the optimistic view that assumes that respondents are able and willing to answer the questions presented to them by survey researchers.

It is true that random errors do not affect estimates of population means, but they do inflate variance estimates and therefore affect tests of significance on those means. In other words, random "noise" may not bias the sample estimate of the population mean, but such noise adds to the response variance. For example, if the reliability of measurement for a variable is .67, this means that one-third of the observed variance is due to random errors. This has nontrivial consequences for statistical tests on sample means, since the sample variance is in the denominator of simple test statistics, for example, the t-statistic. This means that it is harder to reject the null hypothesis and tests of significance are much more conservative than would be the case under perfect measurement. In cross-national research, for example, this means that unreliability of measurement contributes to finding more similarities than differences across univariate distributions.

Random measurement errors do not bias estimates of covariances among variables in surveys, again because of the essence of the idea of randomness, but by virtue of their inflation of response variances, it is well known that they attenuate correlations among variables. In a later part of this book, after we have introduced a somewhat more formal language for discussing errors of measurement, we will show how this works. Without going into any of the details of how it is calculated, let us take the example where a true correlation between two variables is .30 and their respective reliabilities are .67, which may not be far from the typical case. In this case the observed correlation between the two variables would be .20 in that the existence of random measurement errors in the variables attenuates correlations among the variables. One can begin to see the serious impact that unreliability has on our inferences about correlations among variables when we consider the fact that if the reliabilities in this example had been .50 (rather than .67) the observed correlations would be .15 (see Chapter 12).

Given that random errors of measurement have significant consequences for inferences regarding statistical matters, especially tests of statistical significance, correlation estimates, regression coefficients, and other information based on sample estimates of response variance, it hardly seems realistic to ignore them. It is important to distinguish between random and nonrandom types of error not only because they result from different processes but also because they can have very different types of effects on quantities calculated from the sample data. Whereas random errors do not bias estimates of population means, nonrandom errors do contribute to bias as well as response variance. Further, nonrandom errors can contribute to spurious correlation among measures. In a sense, then, nonrandom or systematic errors are even worse than random errors because they bias estimates of sample means and inflate estimates of response variance and covariance. We know considerably less about how to conceptualize the influence of nonrandom errors and to separate out their effects. As will be seen in subsequent discussions, the modeling of nonrandom errors is relatively uncharted territory, although some consideration has been given to the operation of common "method" variance in the study of attitudes and other subjective variables (see Alwin, 1974; 1997; Andrews, 1984; Scherpenzeel, 1995; Scherpenzeel and Saris, 1997).

CHAPTER THREE

Reliability Theory
for Survey Measures

Although it takes a certain sophistication to recognize that the ordinary test score contains an error of measurement, few people object to this notion. An inescapable counterpart to the error of measurement does arouse objections in some quarters. . . . The concept of the true score appears to raise some philosophical problems because often the true score cannot be directly measured.

Lord and Novick, *Statistical theories of mental test scores* (1968, p. 27)

The theoretical starting point for many statistical treatments of measurement error in survey data is the body of psychometric theory known as *classical test theory* or *classical true-score theory* (CTST). As indicated in the above quote from Lord and Novick (1968) (a book that is without doubt the most comprehensive statistical treatment on the subject), the approach is a unique one. It first develops a concept of "measurement error" based on notions of propensity distributions and statistical sampling and from there develops the concept of "true score." The "true scores" in the classical model do not "exist" in reality, only in the models, so it does no good to engage in philosophical debates about the "existence" of true scores (cf. Groves, 1991). Unless the true value of a measured variable is independently known, which is rarely the case, there is no alternative other than to deal with *hypothetical* true scores. Statistical true-score models have proven useful in such cases. As we show in this chapter, the logic is straightforward, and although the earliest models for reliability estimation derived from CTST are less than optimal for evaluating the reliability of survey measures, the classical theory provides an indispensable starting point.

One thing that is common among all efforts to evaluate the reliability of measurement is the use of *repeated measures*. In this chapter and subsequent ones we consider the implications of this for the design of research that will permit assessment of reliability. One of the issues we stress is the difference between *multiple indicators* and *multiple measures*, a distinction that unfortunately has been lost on many analysts. The essence of my argument is that estimating reliability from information collected within the same interview is especially difficult, owing to the virtual impossibility of replicating questions exactly. When researchers employ similar, though not identical questions, and then examine correlation or covariance properties of the data collected, their inferences about reliability of measurement are open to critical interpretation. This results from the fact that in multiple-indicators models the components of reliable variance are more complex than in CTST models that assume multiple *univocal* measures. Multiple indicators, by contrast, contain specific components of variance, orthogonal to the quantity measured in common, and it is virtually impossible to separate these various reliable components of variance in cross-sectional data. We argue that while the use of multiple-indicators models in the context of modern structural equation models (SEM) often gives the illusion of estimating the reliability of measurement, they often miss the mark (cf. Bollen, 1989). In this chapter we provide a comprehensive foundation for understanding approaches to reliability estimation and return to a detailed discussion of these issues in Chapter 4.

In this chapter we review the most common CTST-based approaches for assessing the reliability of survey data, namely internal consistency reliability (ICR) estimates. We argue that for purposes of examining the quality of measurement for individual survey questions, the ICR methods leave much to be desired. We do not in principle object to the use of ICR methods as an approach to scale construction and item analysis, but we strenuously object to the interpretation of such *internal consistency* coefficients as estimates of reliability of survey measurement. In developing this argument we first provide a formal statement of classical true-score theory (CTST) in order to provide a basis for thinking about the reliability of survey data. We then show the relationship between classical true-score theory and the common-factor model, another classical psychometric model for understanding responses to survey questions. Using a modern structural equation (SEM) framework, permitting a more rigorous examination of some of its basic assumptions, we bring classical reliability theory into sharper relief. Through this development we present the CTST model for reliability in such a way that a more comprehensive understanding of internal consistency methods—particularly Cronbach's alpha (α) (1951) and related techniques—can be achieved and their utility for evaluating survey measures assessed.

3.1   KEY NOTATION

We follow the convention of using uppercase symbols to denote random variables and vectors of random variables; and uppercase Greek symbols to represent population matrices relating random variables. We use lowercase symbols to denote person-level scores and within-person parameters of propensity distributions. (See Table 3.1.)

3.2   BASIC CONCEPTS OF CLASSICAL RELIABILITY THEORY

On the simplest level the concept of reliability is founded on the idea of consistency of measurement. Consider a hypothetical *thought experiment* in which a measure of some quantity of interest (Y) is observed—it could be a child's height, it could be the pressure in a bicycle tire, or it could be a question inquiring about family income in a household survey. Then imagine repeating the experiment, taking a second measure

Table 3.1. Key symbols used in discussion of reliability

y_{gp}	An *observed score* for variable Y_g on person p
τ_{gp}	The *true score* of person p in measure Y_g defined as $E(y_{gp})$
ε_{pg}	The *error score* for person p in measure Y_g defined as $\varepsilon_{pg} = y_{gp} - \tau_{gp}$
S	A finite population of persons
G	The number of measures in a set of univocal measures
$E[Y_g]$	The expectation of the observed score random variable Y_g in population S
$E[T_g]$	The expectation of the true score random variable T_g in population S
$E[E_g]$	The expectation of the error score random variable E_g in population S
$VAR[Y_g]$	The variance of the observed score random variable Y_g in population S
$VAR[T_g]$	The variance of the true score random variable T_g in population S
$VAR[E_g]$	The variance of the error score random variable E_g in population S
$COV[T_gY_g]$	The covariance of the random variables T_g and Y_g in population S
$COR[T_gY_g]$	The correlation of the random variables T_g and Y_g in population S
ρ_g^2	The reliability of the gth measure
ICR_G	Internal consistency reliability for a set of G univocal measures
K	The number of sets of univocal measures
Σ_{YY}	Covariance matrix for a set of G measures in population S
Λ	The $(G \times K)$ matrix of regression coefficients relating observed measures to true scores in population S
Φ	The $(K \times K)$ matrix of covariances among latent true scores in population S
Θ^2	The $(G \times G)$ matrix of covariances among errors of measurement in population S
M	The number of distinct methods of measurement
P	The number of occasions of measurement

of Y, under the assumption that nothing has changed, that is, neither the measurement device nor the quantity being measured has changed. If across these two replications one obtains consistent results, we say that the measure of Y is reliable, and if the results are inconsistent, we say that the measure is unreliable. Of course, reliability is not a categorical variable and ultimately we seek to quantify the degree of consistency or reliability in social measurement.

Classical true-score theory (CTST) provides a *theoretical model* for formalizing the statement of this basic idea and ultimately for the estimation and quantification of reliability of measurement. We review the classical definitions of *observed score*, *true score*, and *measurement error*, as well as several results that follow from these definitions, including the definition of *reliability*. We begin with definitions of these scores for a *fixed person* (p), a member of the population (S) for which we seek to estimate the reliability of measurement of the random variable Y. Reference to these elements as persons is entirely arbitrary, as they may be organizations, workgroups, families, counties, or any other theoretically relevant unit of observation. We use the reference to "persons" because the classical theory of reliability was developed for scores defined for persons and because the application of the theory has been primarily in studies of persons as is the case here. It is important to note that throughout this and later chapters the assumption is made that there exists a finite population of persons (S) for whom the CTST model applies and that *we wish to draw inferences about the extent of measurement error in that population.*

The model assumes that Y is a *univocal* measure of the continuous latent random variable T, and that there is a set of multiple measures of the random variable $\{Y_1, Y_2, \ldots, Y_g, \ldots, Y_G\}$ that have the univocal property, that is, each measures one and only one thing, in this case T. We refer to the typical measure from such a set of univocal measures as Y_g. An *observed score* y_{gp} for a fixed person p on measure g is defined as a (within person) random variable for which a range of values for person p can be observed. In the thought experiment performed above, imagine a hypothetical infinite repetition of measurements creating a propensity distribution for person p relating a probability density function to possible values of Y_g. The *true score* τ_{gp} for person p on measure g is defined as the expected value of the observed score y_{gp}, where y_{gp} is sampled from the hypothetical propensity distribution of measure Y_g for person p. Figure 3.1 gives several examples of what such propensity distributions might look like for a Y_g measured on a continuous scale. From this we define *measurement error* for a given observation as the difference between the true score and the particular score observed for p on Y_g, i.e., $\varepsilon_{gp} = y_{gp} - \tau_{gp}$. Note that a different error score would result had we sampled a different y_{gp} from the propensity distribution of person p, and an infinite set of replications will produce a distribution for ε_{gp}. One further assumption is made about the errors of measurement in the measurement of y_{gp} and this is that ε_{gp} is uncorrelated with errors of measurement of any other Y_g. This is called the assumption of *measurement independence* (Lord and Novick, 1968, p. 44).

Several useful results follow from these simple definitions. First, the expected error score for a fixed person is zero, i.e., $E(\varepsilon_{gp}) = 0$. Second, the correlation between the true score and the error score for a fixed person is zero, i.e., $E(\varepsilon_{gp}, \tau_{gp}) = 0$. These two results follow from the fact that the true score for person p is a fixed constant. Third, the shape of the probability distributions of ε_{gp} and y_{gp} are identical and the

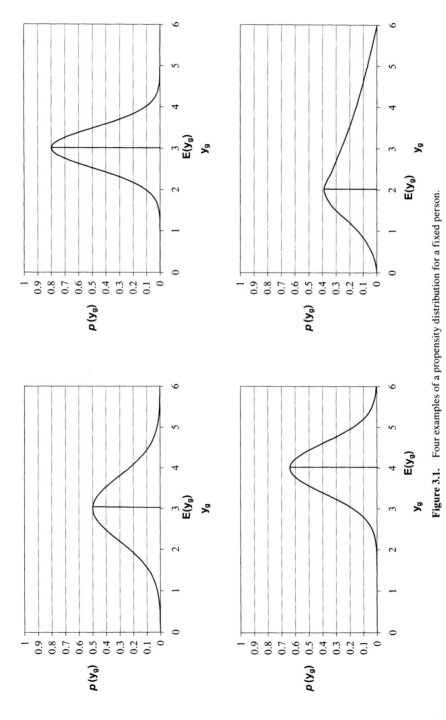

Figure 3.1. Four examples of a propensity distribution for a fixed person.

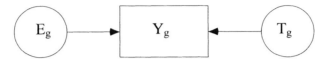

Figure 3.2.   Path diagram of the classical true-score model for a single measure.

variance of the propensity distribution for y_{gp} is equal to the variance of the error scores, i.e., $VAR(\varepsilon_{gp}) = VAR(y_{gp})$. These properties combine to define measurement error under this model as *random* error.

Given a population of persons for whom the above model holds, we can write the model $Y_g = T_g + E_g$ for the gth measure of Y, the properties of which are well known. In this model the expectation of E_g is zero, from which it follows that $E[Y_g] = E[T_g]$. Note that in contrast to ε_{gp} which is a hypothetical within-person random variable, the random variable E_g varies across persons in the population of interest. Also under this model the covariance of the true and error scores is zero, i.e., $COV[T_g, E_g] = 0$, from which it follows that the variance of the observed score equals the sum of the variance of the true score and the variance of the error score, i.e., $VAR[Y_g] = VAR[T_g] + VAR[E_g]$. The path diagram in Figure 3.2 depicts the structural relations between Y_g, T_g, and E_g.

Reliability is defined as a *population parameter*, namely, the proportion of the observed variance that is accounted for by true-score variance, which is expressed as the squared correlation between Y_g and T_g:

$$COR[Y_g, T_g]^2 = VAR[T_g] / VAR[Y_g] = (VAR[Y_g] - VAR[E_g]) / VAR[Y_g]$$

As a generic concept, then, reliability refers to the relative proportion of random error versus true variance in the measurement of Y_g, i.e., variance due to random "noise" versus variance due to "signal" to use a metaphor from telegraphy. As the proportion of error variance in $VAR[Y_g]$ declines, reliability will approach unity, and as it increases relative to $VAR[Y_g]$, reliability will approach zero. The quantity

$$COR[Y_g, Y_t] = VAR[T_g]^{1/2} / VAR[Y_g]^{1/2}$$

(the square root of reliability) is sometimes referred to as the *index of reliability* (see Lord and Novick, 1968, p. 61).

Let Y_1 and Y_2 be two measures from the set of measures defined above, such that $Y_1 = T_1 + E_1$ and $Y_2 = T_2 + E_2$. Assume further that Y_1 and Y_2 are tau-equivalent, that is, they have the same true scores, $T = T_1 = T_2$. The path diagram in Figure 3.3 illustrates the structural relations between T_1, T_2, and their errors of measurement. It follows from this set of definitions that the covariance between Y_1 and T_2 is equal to the variance of T, i.e., $COV(Y_1, Y_2) = VAR(T)$. This comes about because it is assumed from the classical assumption of the independence of errors referred to above that $COV(E_1, E_2) = 0$. With these results we can define the reliability for the two measures of the random variable of interest, Y_1 and Y_2, in the population of interest as $COV(Y_1, Y_2) / VAR[Y_1]$ and $COV[Y_1, Y_2] / VAR[Y_2]$, respectively. Such measures are referred to as

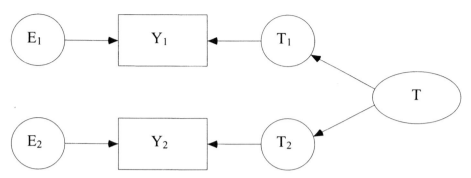

Figure 3.3. Path diagram of the classical true-score model for two tau-equivalent measures.

tau-equivalent measures. If, in addition to tau-equivalence, the error variances of the two measures are equal, i.e., $VAR[E_1] = VAR[E_2]$, this would imply equal variances, $VAR[Y_1]$ and $Var[Y_2]$, and equal reliabilities for Y_1 and Y_2. In this special case the reliability of Y_1 and Y_2 can be expressed by their correlation, $COR[Y_1,Y_2]$, since

$$COR[Y_1,Y_2] = COV[Y_1,Y_2] / (VAR[Y_1]^{1/2} VAR[Y_2]^{1/2})$$

Such measures (with tau-equivalence and equal error variances) are said to be *parallel measures*. Finally, measures are often not tau-equivalent, as in the case of different scales or metrics used to measure Y_1 and Y_2, but their true scores are linearly related (i.e., $COR[T_1,T_2] = 1.0$). When this is the case the measures are said to be *congeneric* (see Jöreskog, 1971a, 1974, 1978). Note the nested nature of the relationship between these three models: the tau-equivalent measures model is a special case of the congeneric model, and the parallel measures model is a special case of the tau-equivalence model.

3.3 NONRANDOM MEASUREMENT ERROR

Normally we think of measurement error as being more complex than the random error model developed above. In addition to random errors of measurement, there is also the possibility that ε_{pg} contains systematic (or correlated) errors (see Biemer, Groves, Lyberg, Mathiowetz, and Sudman, 1991). The relationship between random and systematic errors can be clarified if we consider the following extension of the classical true-score model: $y_{gp} = \tau^*_{gp} + \eta_{gp} + \varepsilon_{pg}$, where η_{gp} is a source of systematic error in the observed score, τ^*_{gp} is the true value, uncontaminated by systematic error, and ε_{gp} is the random error component discussed above. This model directly relates to the one given above, in that $\tau_{gp} = \tau^*_{gp} + \eta_{gp}$. The idea, then, is that the variable portion of measurement error contains two types of components, a random component, ε_{gp}, and a nonrandom, or systematic component, η_{gp}. Within the framework of this model, the goal would be to partition the variance in Y_g into those portions due to τ^*, η, and ε.

In order to do this we would formulate a model for a population of persons for whom the above model holds. We can write the model $Y_g = T_g + E_g$ for the gth

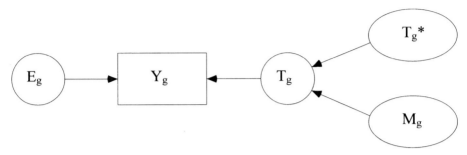

Figure 3.4. Path diagram of the classical true-score model for a single measure composed of one trait and one method.

measure of Y, where $T_g = T_g^* + M_g$. The path diagram in Figure 3.4 illustrates the structural relations between the components of Y_g. The population properties of Y_g, T_g, and E_g are the same as those given above, namely, that the expectation of E_g is zero, from which it follows that $E[Y_g] = E[T_g]$. In addition, $E[T_g] = E[T_g^*] + E[M_g]$, $VAR[T_g] = VAR[T_g^*] + VAR[M_g]$, and $COV[T_g^*, M_g] = 0$. In other words, as formulated here, in the population model the "trait" and "method" components of the true score are uncorrelated. It is frequently the case that systematic sources of error increase estimates of reliability. This is, of course, a major threat to the usefulness of classical true-score theory in assessing the quality of measurement. It is important to address the question of systematic measurement errors, but that often requires a more complicated measurement design. This can be implemented using a multitrait-multimethod measurement design along with confirmatory factor analysis, a topic to which we return in the following chapter.

3.4   THE COMMON-FACTOR MODEL REPRESENTATION OF CTST

It is a straightforward exercise to express the basic elements of CTST as a special case of the metric (unstandardized) form of the *common-factor model* and to generalize this model to the specification of K sets of congeneric measures. Consider the following common-factor model

$$Y = \Lambda T + E$$

where Y is a $(G \times 1)$ vector of observed random variables, T is a $(K \times 1)$ vector of true score random variables measured by the observed variables, E is a $(G \times 1)$ vector of error scores, and Λ is a $(G \times K)$ matrix of regression coefficients relating true and observed random variables. The covariance matrix among measures under this model can be represented as follows

$$\Sigma_{YY} = \Lambda \Phi \Lambda' + \Theta^2$$

where Σ_{YY}, Φ, and Θ^2 are covariance matrices for the Y, T, and E vectors defined above, and Λ is the coefficient matrix as defined above.

For purposes of this illustration we consider all variables to be centered about their means. Here we take the simplest case where K = 1, that is, all G variables in Y are measures of T; however, we note that the model can be written for the general case of multiple sets of congeneric measures. In the present case the model can be represented as follows:

$$
\begin{bmatrix} Y_1 \\ Y_2 \\ \vdots \\ Y_G \end{bmatrix} = \begin{bmatrix} \lambda_{1T} \\ \lambda_{2T} \\ \vdots \\ \lambda_{GT} \end{bmatrix} \times [T] + \begin{bmatrix} E_1 \\ E_2 \\ \vdots \\ E_G \end{bmatrix}
$$

For G congeneric measures of T there are 2G unknown parameters in this model (G λ_{gT} coefficients and G error variances, θ_g^2) with degrees of freedom (df) equal to .5G(G + 1) − 2G. In general this model requires a scale be fixed for T, since it is an unobserved latent random variable. Two options exist for doing this: (1) the diagonal element in Φ can be fixed at some arbitrary value, say, 1.0; or (2) one of the λ_{gT} coefficients can be fixed at some arbitrary value, say 1.0. For G measures of T the tau-equivalent model has G + 1 parameters with df = .5G (G + 1) − G + 1. For a set of G parallel measures there are two parameters to be estimated, VAR[T] and VAR[E], with df = .5G(G + 1) − 2.

Note that both the tau-equivalent measures and parallel measures form of this model invoke the assumption of tau-equivalence. This is imposed on the model by fixing all λ_{gT} coefficients to unity. In order to identify the tau-equivalent or parallel measures model, observations on two measures, Y_1 and Y_2, are sufficient to identify the model. For the congeneric model G must be ≥ 3. It should be clear that the congeneric measures model is the most general and least restrictive of these models, and the tau-equivalent and parallel measures models simply involve restrictions on this model.

What has been stated for the model in the above paragraphs can be generalized to any number of G measures of any number of K factors. The only constraint is that the assumptions of the model—univocity and random measurement error—are realistic for the measures and the population from which the data come. Although there is no way of testing whether the model is correct, when the model is overidentified the fit of the model can be evaluated using standard likelihood ratio approaches to hypothesis testing within the confirmatory factor analysis framework. There is, for example, a straightforward test for whether a single factor can account for the covariances among the G measures. Absent such confirming evidence, it is unlikely that a simple true-score model is appropriate.

3.5   SCALING OF VARIABLES

The discussion to this point assumes interval-level measurement of continuous latent variables and the use of standard Pearson-based covariance approaches to the definition of statistical associations. Observed variables measured on ordinal scales are not continuous, i.e., they do not have origins or units of measurement, and therefore should not be treated as if they are. This does not mean that where the observed variables are categorical the underlying latent variable being measured cannot be assumed to be continuous. Indeed, the tetrachoric and polychoric approaches to ordered-dichotomous and ordinal-polytomous data assume that there is an underlying continuous variable Y*, corresponding to the observed variable Y, that is normally distributed.

In Figure 3.5 we depict the contrast between interval and ordinal measurement of a continuous latent variable. The observed variable is designated as Y and the "ideal" variable corresponding to the continuous variable it measures (not to be confused with the true score) is designated as Y*. In case (a) it is assumed that there is uniform correspondence between the continuous variable, Y*, and the observed variable, that is, Y = Y*. This is what is assumed when Pearson-based product-moment correlations are used as the basis for defining the covariance structure among measures. Consensus exists that most variables having more than 15 response categories can be treated as interval scale measures. There are many examples in survey research of variables that can be considered interval measures, e.g., the measurement of income in dollars, the measurement of age in days, months, and years. In the measurement of attitudes, magnitude estimation approaches to measurement attempt to solve this problem by providing respondents with a scale that is practically continuous (e.g., Dawes and Smith, 1985; Saris, 1988), and there is some suggestion that scales with greater numbers of categories may be more reliable (Alwin, 1992, 1997).

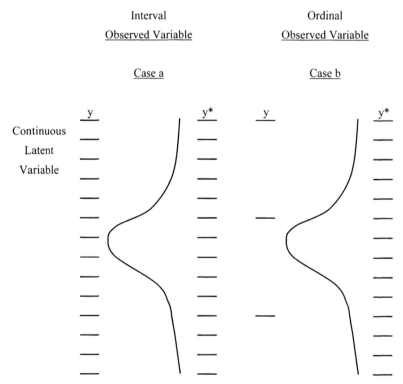

Figure 3.5. Comparison of interval and ordinal measures of continuous latent variables.

In case (b) shown in Figure 3.2 the nature of the latent variable is also continuous, as above, but the observed variable is ordinal. What distinguishes these two cases is the crudeness of measurement at the level of the observed responses. In the tradition of Karl Pearson, who invented the tetrachoric correlation coefficient, we can estimate the association between two Y* variables, given their cross-tabulation. This is the approach taken to ordinal variables in Jöreskog and Sörbom's PRELIS2 (1996a) and LISREL8 (1996b) software and Muthén and Muthén's (2001) M*plus* software, which provide estimates of tetrachoric (for dichotomous variables) and polychoric (for polytomous variables) correlations. We employ these methods in the present study and compare results obtained using the more conventional Pearson-based approach with the tetrachoric/polychoric approaches.

The use of these methods is somewhat cumbersome and labor-intensive because it requires that the estimation of the model be done in two steps. First, one estimates the polychoric or tetrachoric correlations for the observed data, which often takes a substantial amount of time, especially when the number of categories is large. The second step then estimates the parameters of the CTST model using maximum likelihood or weighted least squares. There are two basic strategies for estimating polychoric correlations. One is to estimate the polychoric/tetrachoric correlations and thresholds jointly from all the univariate and bivariate proportions in a multiwave contingency table (Lee, Poon, and Bentler, 1990). This approach is computationally very complex and not in general recommended. The other, which almost always provides the same results, is to estimate the thresholds from the univariate marginals and the polychoric correlations from the bivariate marginals (Muthén, 1984; Jöreskog, 1990). Jöreskog (1994) presents a procedure for estimating the asymptotic covariance matrix of polychoric correlations, which requires the thresholds to be equal. Bollen (1989) presents a comprehensive discussion of these methods.

Another approach to examining measurement errors, which also assumes continuous latent variables, appropriate for categorical data, is based on item response theory (IRT). Test psychologists and others have used IRT models to describe *item characteristic curves* in a battery of items. One specific form of the IRT model, namely, Rasch models, have been suggested as one approach to modeling measurement errors for sets of binary variables. Duncan's (1984b) work illustrates how the Rasch (1960, 1961, 1966a, 1966b) model can be applied to dichotomous variables measured on the same occasion. These approaches have not explicitly included parameters for describing reliability as defined here and we do not employ them here.

3.6 DESIGNS FOR RELIABILITY ESTIMATION

Generally speaking, in order to properly assess the extent of measurement error, one needs needs (1) a theoretical model that specifies the linkage between true and observed variables, (2) a research design that permits the estimation of parameters of such a model, and (3) to be able to interpret these parameters in a way that is consistent with the model linking observed, true and error components. This chapter has begun the process of developing an appropriate model for specifying the linkage between survey measures and their true scores. In this brief section we summarize the key elements of the two main types of design that exist to deal with reliability estimation, given the conceptualization of measurement error developed above.

Two general design strategies exist for estimating the reliability of measurement using repeated measures: (1) replicate (or similar) measures in the same occasion of measurement (referred to here as cross-sectional measurement), or (2) replicate measures in reinterview designs (referred to here as longitudinal measurement). The application of either design strategy is problematic and in some cases the estimation procedures require assumptions that are inappropriate given the data gathered in such designs. Estimating reliability from information collected within the same interview is especially difficult, owing to the virtual impossibility of replicating questions exactly. Researchers often employ similar, but not identical questions, and then examine correlation or covariance properties of the data collected. In other words, rather than *multiple* or *repeated measures,* investigators often use *multiple indicators* as a substitute (e.g., Bollen, 1989). However, it is risky to use covariance information from multiple indicators to estimate item reliability since items that are different contain specific components of variance, orthogonal to the quantity measured in common, and because difficulties in separating reliable components of specific variance from random error variance present significant obstacles to this estimation approach. Because of potentially biasing effects of memory and other cognitive processes, it is virtually impossible to obtain unbiased estimates of reliability from cross-sectional surveys. I return to this issue in Chapter 4.

A second approach to estimating reliability in survey data uses the multiple-wave reinterview (i.e., test-retest) or panel design. Such longitudinal designs also have problems for the purpose of estimating reliability. For example, the test-retest approach using a single reinterview must assume that there is no change in the underlying quantity being measured. With two waves of a panel study, the assumption of no change, or even perfect correlational stability, is unrealistic; and without this assumption little purchase can be made on the question of reliability in designs involving two waves. The analysis of panel data must be able to cope with the fact that people change over time, so that models for estimating reliability must take the potential for individual-level change into account. Given these requirements, techniques have been developed for estimating measurement reliability in panel designs where $P \geq 3$ in which change in the latent true score is incorporated into the model. With this approach there is no need to rely on multiple measures within a particular wave or cross section in order to estimate the measurement reliability. This approach is possible using modern structural equation models for longitudinal data, and is discussed at length in the next chapter.

3.7 VALIDITY AND MEASUREMENT ERROR

As noted in the previous chapter, the concept of *measurement validity* in its most general sense refers to the extent to which the measurement accomplishes the purpose for which it is intended. With respect to the specification of measures we emphasized the need to establish content validity of measures, referring to the extent to which a well-specified conceptual domain has been represented by available or selected measures. This set of considerations expresses a concern with the linkage between concepts and their indicators, but normally measurement at a more theoretical level emphasizes the *construct validity* of measurement as well as *content validity*. The issue raised in this context, then, asks a related, but somewhat different question: How does the true score specified, the properties of which are estimated within the framework of CTST, correspond to the conceptual variable one intended to measure? Is the true score a valid representation of the theoretical quantity of interest?

There is always some uncertainty that surrounds discussion of validity of survey measurement because the concept of validity is used in several different ways. There should be no confusion about true scores, however, as defined above. We are not measuring "truth," but rather we represent (or model) reality with a "true score" concept that is a statistical average of a hypothetical set of repetitions of measurement. Our "true scores" themselves do not "exist" in reality—only in our models. It should be clear that we are *not* entering a "true-score morass" as Groves (1991) has suggested. Classical true-score theory presents a *model* for measurement error, nothing more. The model is *neither true nor false*—it is either found to be consistent with the data, or not. The model should be evaluated with respect to its "utility," not its "truth." The utility of such a model lies in its ability to predict reality, and as long as the consensual processes of science accept the validity of a given model, it will persist. However, as a model begins to fail in its ability to describe the data, e.g., the Earth-centric view of our solar system, it will eventually fall into disrepute, even when there is strong institutional pressure in support of such a model (see Sobel, 1999).

It is a truism in CTST that the *criterion validity* of measurement cannot exceed the *index of reliability* (the square root of reliability). Criterion validity is simply defined as the correlation of the measure Y with some other variable, presumably a criterion linked to the purpose of measurement. It can be shown (Lord and Novick, 1968, pp. 61–63) that the *validity coefficient* of a measurement Y with respect to a second measurement X is defined as the absolute value of the correlation coefficient

$$COR(X,Y) = COV(X,Y) / VAR(X)^{1/2} VAR(Y)^{1/2}$$

It should be clear that the validity of measurement is an issue that cannot be settled in the abstract, but only with respect to some criterion. When the criterion involved is the "construct" that one intends to measure (a quantity that cannot be assessed empirically but only by consensus among scientists), we refer to this as *construct validity*. Within the CTST tradition, then, the main interest is in the *criterion validity* of some measurement with respect to X, and this cannot exceed the *index of reliability* (the square root of the reliability).

We can show that $COR(X,Y)$ is always less than or equal to the *index of reliability* of either X or Y. First note that the correlation between the true scores, T_X and T_Y, is equal to

$$COR(T_X, T_Y) = COR(X,Y) / \rho_X \rho_Y$$

where ρ_X and ρ_Y are defined as the indexes of reliability for X and Y, respectively, that is, $\rho_X = COR(X, T_X)$ and $\rho_Y = COR(Y, T_Y)$. From here we rewrite the above as

$$COR(T_X, T_Y) \rho_X \rho_Y = COR(X,Y)$$

Employing Chebyshev's inequality we can deduce that $COR(X,Y)$ can never exceed the *index of reliability* of either measure. Imagine in the extreme case where $COR(T_X, T_Y) = 1.0$ and $\rho_X = 1.0$—here $\rho_Y = COR(X,Y)$. So by logic, as $COR(T_X, T_Y)$ and ρ_X drop from unity, so will the correlation between X and Y, and ρ_Y is clearly seen as an upper bound on $COR(X,Y)$.

3.7.1   Record-Check Studies

As I noted in the previous chapter, there are other ways to think of true scores. "Platonic" true scores are those for which it is believed there is some notion of "truth" involved (Lord and Novick, 1968, pp. 38–39). This is usually assumed in record-check studies involving survey data (see Marquis, 1978). Studies, for example, that compare voting records with self-reports of voting (Traugott and Katosh, 1979), or those that compare actual patterns of religious behavior compared with self-reports (Hadaway, Marler, and Chaves, 1993) are implicitly using a *platonic* notion of true scores, i.e., that there is some "truth" out there that is to be discovered. This definition, however, is not generally useful because most variables we wish to measure in surveys have no "true" source against which to measure the accuracy of the survey report. "Psychometric" true scores, on the other hand, are defined in statistical terms as the expected value of a hypothetical infinite set of observations for a fixed person (see Lord and Novick, 1968).

   One thing that should be apparent is that the results obtained in record-check studies are assessments of validity, *not* reliability. We can clarify this as follows. Let Y_R be a measure of variable Y from a record source, and let Y_{SR} be a self report of the same variable. The CTST model for a finite population (S) in each case can be written as follows

$$Y_R = T_R + E_R$$
$$Y_{SR} = T_{SR} + E_{SR}$$

where these models have all the properties of classical true-score models. If we define Y_R as "truth" or at a minimum a standard or criterion against which to assess the quality of Y_{SR}, then the comparison between the expected values of their true score becomes an indicator of *measurement bias* (Marquis, 1978).

   We can define *measurement bias* as the difference in the true scores of the two approaches to measurement:

$$E[T_{SR}] - E[T_R] = E[Y_{SR}] - E[Y_R]$$

Such errors may represent a kind of constant error of measurement, as depicted in Figure 1.2 in the previous chapter, or as a kind of *differential bias* that is correlated with levels of one or the other variable, also depicted in Figure 1.2. Record-check researchers (e.g., Traugott and Katosh, 1979) typically use the term "validity" to refer to the extent of the bias, that is, the smaller the difference of means between the record source and the self-report measure, the more valid the measures.

This discussion should suffice as advance warning that the term "validity" is used in more than one way in this book, as is the case in the field of survey methods generally. As noted in the previous chapter, historically, the term *measurement validity* has a large number of different meanings. In the classical true-score tradition, for example, the *validity of a measure* with respect to a second measurement is defined as the absolute value of their correlation coefficient (Lord and Novick, 1968, pp. 61–62). If the second measure is a criterion or standard against which Y could be evaluated, as in the case of Y_{SR}, then we could use $COR[Y_{SR}, Y_R]$ as an estimate of the *criterion validity* of the self-report measure with respect to the record source.

Note that $COR[Y_{SR}, Y_R]$ is not the reliability of the self-report measure Y_{SR}. Under some circumstances, this quantity would estimate the *index of reliability*, namely, when the assumptions of parallel measures are met. At the risk of repeating the above discussion regarding the relationship between *criterion validity* and reliability, recall that the *criterion validity* of measurement cannot exceed the *index of reliability* (the square root of reliability) (Lord and Novick, 1968). Taking this discussion one step further, another way in which we may define *measurement validity* is in terms of the *construct validity* of measurement. "Validity" in this sense refers to the extent to which a measure assesses the theoretical or abstract concept that it is intended to measure. Because of several practical limitations in obtaining independent measures of the construct one intends to measure, this form of validity is difficult to establish empirically.

Often the issue of measurement validity in the *construct validity* sense rests as much on a scientific consensus as anything, rooted in practical experience and knowledge of possible measures, as much as it does on rigorous empirical demonstrations. In the present case, we would suggest that the construct validity of Y_{SR} would best be conceived of as the correlation of the true scores of Y_R and Y_{SR}, that is, $COR[T_{SR}, T_R]$, which is the true correlation among the two measures independent of measurement errors in either measure. Such an estimate would reflect the extent to which true individual differences in Y_{SR} correlate with true individual differences in Y_R independent of any *measurement bias* that may exist in the self-report measure. In the typical record-check study there is only one measure of each thing—one self-report measure and one record-report measure. It is not clear, therefore, how such an estimate of *construct validity* would be constructed in the absence of information regarding the reliability of both the record-report and self-report measures. In our assessments of validity of measurement in surveys therefore, we often settle for much less.

3.7.2   Split-Ballot Experiments

It may be of some value to cover another common approach to the study of response errors in surveys, the split-ballot randomized experiments. In such research two or more forms (or ballots) are devised that are hypothesized to evoke different responses in a given population. These may be forms of a given question, the ordering of the response options for a given question, the ordering of questions in a sequence of questions, or some other formal attribute of the question or its context. These forms are then randomly assigned to members of the sample and the effects of the different forms or methods are studied by examining differences in the marginal distributions, means, or covariance properties of the variables across experimental forms. Such designs are referred to as "between subjects" or "split sample" designs in contrast to the repeated measures approach taken in classical designs focused on reliability, which are called "within subjects" designs (Groves, 1991). Thus, these split-ballot experiments do not focus on reliability as it is defined classically, nor do they really focus on the validity of measurement. Indeed, except in a few rare instances (e.g., McClendon and Alwin, 1993), research on question wording or question context effects is not really focused on measurement errors per se, but on the cognitive processes that produce responses.

This approach may be useful for examining the constant properties of some question forms or biases linked to particular methods, but it provides little assistance in assessing components of response variation. Indeed, we suspect that many of the alternative forms of questions used across experimental forms would produce quite similar assessments of individual differences if they were both included in a within-subjects design (see Alwin, 1989). Given H experimental variations, let us represent the above CTST model for the hth experimental group as follows:

$$Y^h - E[Y^h] = T^h + E^h$$

The main focus of split-ballot experiments is on differences in the $E[Y^h]$ across the H experimental groups rather than other properties of the model. An exception to this is McClendon and Alwin's (1993) analysis of the effects of providing explicit no opinion filters on the reliability of measurement. For the typical split-ballot experiment in which there are two measures, Y_A and Y_B, we can formulate the CTST model for the measures as one of "essential" *tau-equivalence,* that is, the true scores of the two models are not equivalent, but are different only by a constant, i.e., $T_A = \kappa + T_B$. In such a case the CTST model can be stated as

$$Y_A - E[Y_A] = T_A + E_A$$
$$Y_B - E[Y_B] = T_B + E_B$$

where κ equals the difference $E[Y_A] - E[Y_B]$. In this case one would conclude that except for the constant difference in true scores, the measures are functionally equivalent. Unfortunately, there have been very few split-ballot study designs with multiple measures of social indicators of the sort implied here, and therefore these models have not been fully exploited to their ultimate potential.

3.8   RELIABILITY MODELS FOR COMPOSITE SCORES[1]

The most common approach to assessing reliability in cross-sectional surveys is through the use of multiple measures of a given concept and the estimation of the reliability of a linear composite score made up of those measures. Let Y symbolize such a linear composite defined as the sum $Y_1 + Y_2 + \cdots + Y_g + \cdots + Y_G$, i.e., $\Sigma_g(Y_g)$. Such estimates of reliability are referred to as *internal consistency* estimates of *reliability* (ICR). In this case we can formulate a reliability model for the composite, as $Y = T + E$, where T is a composite of true scores for the G measures and E is a composite of error scores. This assumes that for each measure the random error model holds, that is, $Y_g = T_g + E_g$, and thus, $T = \Sigma_g(T_g)$ and $E = \Sigma_g(E_g)$. The goal of the internal consistency approach is to obtain an estimate of $VAR(T) / VAR(Y) = [VAR(Y) - VAR(E)] / VAR(Y)$. This can be defined as a straightforward extension of the common factor model of CTST given above. The following identities result from the above development:

$$VAR(Y) = \sum_j \sum_i \Sigma_{YY}$$

$$VAR(T) = \sum_j \sum_i [\Lambda\Phi\Lambda']$$

$$VAR(E) = \sum_j \sum_i \Theta^2$$

(Note: i and j represent indices that run over the rows and columns of these matrices, where i = 1 to G and j = 1 to G.) In other words, the common factor representation of the CTST model given above for the population basically partitions the composite observed score variance into true-score and error variance. These quantities can be manipulated to form an internal consistency measure of composite reliability, as follows:

$$ICR = \frac{\sum_j \sum_i \Sigma_{YY} - \sum_j \sum_i \Theta^2}{\sum_j \sum_i \Sigma_{YY}}$$

---

[1]This discussion in this section relies on the development of similar ideas by the author in a related publication (see Alwin, 2005).

The most common estimate of internal consistency reliability is Cronbach's α (1951), computed as follows:

$$\alpha = \frac{G}{G-1}\left[1 - \frac{\sum_g VAR(Y_g)}{VAR(Y)}\right]$$

This formula is derived from the assumption of G unit-weighted (or equally weighted) *tau-equivalent* measures. The logic of the formula can be seen as follows. First rewrite $\sum_j \sum_i \Theta^2$ in the expression for ICR above as equal to $\sum_j \sum_i \Sigma_Y - \sum_j \sum_i \Sigma_T$, where Σ_Y is a diagonal matrix formed from the diagonal elements of Σ_{YY} and Σ_T is a diagonal matrix formed from the diagonal of $\Lambda\Phi\Lambda'$. Note further that under tau-equivalance $\Lambda = 1$ (a vector of 1s), so this reduces to φI, where φ is the variance of T_g and I is a $(G \times G)$ identity matrix. Note that in the population model for tau-equivalent measures all the elements in Φ are identical and equal to φ, the variance of the true score of T_g. From these definitions, we can rewrite ICR as follows:

$$ICR = \frac{\sum_j \sum_i \Sigma_{YY} - \sum_j \sum_i \Sigma_Y + \sum_j \sum_i \varphi I}{\sum_j \sum_i \Sigma_{YY}}$$

Note further that $\sum_j \sum_i \Sigma_{YY} - \sum_j \sum_i \Sigma_Y = G(G-1)\,\varphi$, and $\sum_j \sum_i \varphi I = G\,\varphi$, and thus Cronbach's α can be derived from the following identities:

$$ICR = [G(G-1)\,\varphi + G\,\varphi]/\sum_j \sum_i \Sigma_{YY}$$

$$= \left[\frac{G}{G-1}\right][G(G-1)]\,\varphi/\sum_j \sum_i \Sigma_{YY}$$

$$= \left[\frac{G}{G-1}\right]\left[\sum_j \sum_i \Sigma_{YY} - \sum_j \sum_i \Sigma_Y\right]/\sum_j \sum_i \Sigma_{YY}$$

$$= \left[\frac{G}{G-1}\right]\left[1 - \left[\sum_j \sum_i \Sigma_Y / \sum_j \sum_i \Sigma_{YY}\right]\right]$$

The final identiy is equivalent to the formula for Cronbach's α given above (see Greene and Carmines, 1979). The point of this derivation is that the ICR approach actually has a more general formulation (the congeneric measures model) for which Cronbach's α is only a special case, i.e., ICR = α when the G measures are tau-equivalent.

These methods can be generalized to the case of weighted composites, where Yw is the composite formed from the application of a vector of weights, w, to the G variables in Y, but we will not consider this case here, except to note that when the vector of weights, w, is chosen to be proportional to $\Theta^{-2}\Lambda$ such a set of weights will be optimal for maximizing ICR (Lawley and Maxwell, 1971; Maxwell, 1971).

There have been other variations to formulating ICR. Heise and Bohrnstedt (1970), for example, defined an ICR coefficient, named Ω, based on the use of U^2 in place of Θ^2 in the above formulation for ICR, where U^2 is a diagonal matrix of unique variances from an orthogonal common-factor analysis of a set of G variables without the CTST assumptions of univocity, e.g., $K > 1$. They propose partitioning Ω into its contributions from the common factors of the model, arbitrarily labeling the first factor common variance as "valid" variance and successive factor common variance as "invalid" variance. This coefficient has not seen wide use, although the logic of Heise and Bohrn-stedt's (1970) formulation has provided an interpretative framework for the consideration of reliability issues within the multitrait-multimethod design (see Chapter 4).

Although it is a very popular approach, ICR coefficients have several major shortcomings. First, ICR is an unbiased estimate of composite reliability *only* when the true-score model assumptions hold. To the extent that the model assumptions are violated, it is generally believed that ICR approaches provide a lower-bound estimate of reliability. However, at the same time there is every possibility that ICR is inflated due to correlated errors, e.g., common method variance among the items, and that some reliable variance is really "invalid" in the sense that it represents something about responses other than true-score variation, such as nonrandom sources of measurement error. ICR therefore captures systematic sources of measurement error in addition to true-score variation and in this sense *ICR coefficients cannot be unambiguously interpreted as measures of data quality.*

3.9 DEALING WITH NONRANDOM OR SYSTEMATIC ERROR

The relationship between random and systematic errors can be clarified if we consider the following extension of the classical true-score model:

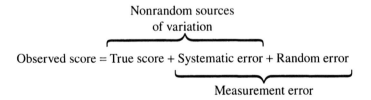

Here we have partitioned the "Measurement error" term in the preceding development into two components: systematic and random errors. In this framework the variable portion of measurement error contains two types of components, a random component and a nonrandom (or systematic) component. A focus on one or the other type of error does not negate the presence of the other, and there is plenty of evidence that both random and nonrandom errors occur with considerable regularity in most applications of the survey method. The challenge involves the design of research studies that permits the separation of systematic errors from other sources of variation in the observed measures.

Within the *classical true-score* framework, systematic sources of error cannot easily be separated from the "true score." This poses a major threat to the usefulness of classical true-score theory in assessing the quality of survey measurement, because variation in factors reflecting systematic errors are presumably reliable sources of variance. The classical framework essentially combines all sources of nonrandom variation in the observed score into the true score, leaving only random sources of error in the part identified as "measurement error." In this more complex model for measurement error, the response variance can be thought of as equaling the sum of three components: the variance of true scores in the population, the variance of the systematic error scores, plus the variance of error scores, that is, Response variance = True variance + Systematic error variance + Random error variance. On the basis of these assumptions, the *reliability of measurement* in this case is defined as the squared correlation between the systematic components and the observed scores, or put more simply, *reliability is the proportion of the response variance that is accounted for by true and systematic error scores* in the population of individuals sampled.

Is there a solution to this problem? Heise and Bohrnstedt (1970) draw a distinction between reliable variance that is *valid* versus one that is *invalid*. Estimates of reliable variance often do not make explicit the components of variation that are measured reliably, and it is clear that a given measure may index both trait and method variation (see Alwin, 1974). In some ways the use of the terms "validity" and "invalidity" to refer to components of reliability is unfortunate, and it may seem incongruous to define "validity" to be a component of reliability.

In other words, reliable measurement does not necessarily imply valid measurement. Although it is well known that the *index of reliability* [which is the square root of reliability; see Lord and Novick (1968)] does place an upper-bound on *criterion-related validity*, that is, the extent to which a measure may correlate with a theoretically defined criterion, it is illogical to reverse the implication. Such confusion about the appropriate use of the term "validity" for survey measures has led some authors (e.g., Bohrnstedt, 1983) to refer to *univocal* indicators (indicators presumed to measure one and only one thing) as "perfectly valid" because they are tied to one and only one theoretical latent variable. This is an unfortunate confusion of terms because such a latent variable, however reliably measured, may imperfectly represent the theoretical construct of interest. In the next chapter, in our discussion of the *multitrait-multimethod matrix* we employ the distinctions between *reliability, validity* and *invalidity* introduced by Heise and Bohrnstedt (1970) and in this sense refer to "validity" as a component of "reliability." By this we simply mean that a given measure reliably measures more than one thing—a trait (or valid) component and a method (or invalid) component. Within the framework of this model, the goal would be to partition the variance in the observed variable into those portions due to the true score and the two components of error. This idea can be implemented to some degree using a *multitrait-multimethod* measurement design along with confirmatory factor analytic estimates of variance components (see Alwin, 1974, 1997; Andrews,

1984; Browne, 1984; Groves, 1989; Saris and van Meurs, 1990; Saris and Andrews, 1991; Scherpenzeel, 1995; Scherpenzeel and Saris, 1997).

3.10 SAMPLING CONSIDERATIONS

The foregoing development has concentrated on the theoretical background for strategies aimed at quantifying the reliability of measurement in social research. All definitions, relationships, and results were given for a hypothetical finite population (S), and nothing has been stated up to this point about sampling. There is a tradition within sampling statistics of a concern with measurement error (Hansen, Hurvitz and Madow, 1953) that is completely consistent with the psychometric model developed above. This early literature focused on the effects of measurement errors on estimators of a population parameter, such as the mean, calculated for a random sample of the finite population. Biemer and Stokes (1991) discuss the convergences between the "statistical sampling" and "psychometric" perspectives on measurement error. Unfortunately, the more recent literature written by sampling statisticians on measurement error has not focused on sample design effects, so the nature of the design effects associated with the types of estimators employed here is relatively unexplored territory.

In the present case we have formulated sampling issues at three levels. First, in our theoretical definition of an "observed score" for a fixed person (p), we have conceived of it as involving a sampling from the propensity distribution of person a. This is consistent, not only with statistical treatments of CTST (see Lord and Novick, 1968), but also with classical definitions of measurement from the statistical sampling tradition (see Hansen, Hurwitz and Bershad, 1961, p. 33). Beyond these elemental considerations, there are two other ways in which we have considered sampling issues. The first, which we have mentioned only briefly is the idea that a set of G measures of a given indicator Y are somehow "sampled" from a universe of such measures.

Although there is some consideration of "item sampling" in the testing literature, in survey research the idea has never been developed. Survey researchers typically "pretest" measures to get an idea "if they work," although there has historically never been any systematic approach to sampling measures in the survey literature. Survey measures are typically taken as a given, and this is the manner in which we approach the "sampling" of measures in the present study. Essentially, we consider the set of measures we employ here as the population of such measures, as we define the population (see below). We assume that the quantities observed in a sample for Y_g have a particular meaning and that measures are included, if not sampled.

Also represented in Figure 3.3 is the fact that the persons assessed on the set of Y_g measures come from a finite population (S). In order to estimate reliability parameters for a given population of interest, one will need to sample the specific population using probability methods. I stress this set of considerations in order to reinforce the fact that not only is the level of reliability influenced by the properties of the measuring device and the conditions of measurement, but as a population

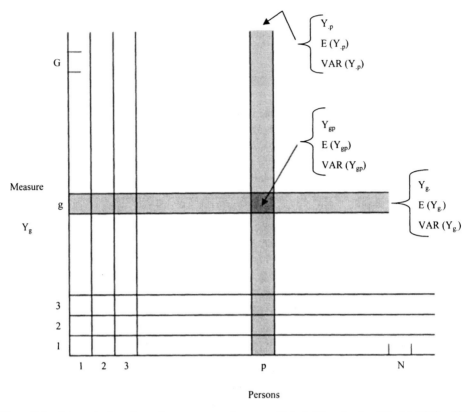

Figure 3.6. Schematization of sampling measures and persons. [Adapted from Lord and Novick (1968, p.33).]

parameter expressed by the ratio of true-score to observed score variance, it is also obviously influenced by the characteristics of the population to which the measures are applied as well.

Consistent with traditional statements of measurement issues, we have formulated the above models at the level of the population. We have assumed that the model applies to persons who are members of some finite population (S) to which we wish to generalize. The assumption we have made throughout this discussion is that the "reliability estimates" are population parameters and can be estimated only by sampling members of the population S using probability-based methods of sampling design (Hansen, Hurwitz, and Madow, 1953; Kish, 1965). Then, once measurements are taken that abide by the necessary sampling design requirements, the variance and covariance properties of the measures can be used to estimate the population parameters that represent reliability, as outlined here and in subsequent chapters. This reinforces the fact that an empirical study of reliability *cannot* focus on the individual as the unit of analysis, even though the theory is formulated at the individual level. In a study of reliability such as this, the unit of analysis is by necessity the single survey question, specific to a particular population. Thus, in subsequent chapters our orientation shifts to another level, namely, *the level of the survey question*, applied either to a sample of a broadly defined population, e.g., a national sample of households, or to a specific subgroup of that population, e.g., partitions based on age or education.

3.11 CONCLUSIONS

In this chapter we have reviewed the basic model that is a common starting point for understanding the reliability of survey data. While there are several basic features of this model that we find conceptually useful and mathematically convenient, there are at least two fundamental problems with the application of the CTST approach to survey data. The first is the assumption that the measures are *univocal*, that is, that they measure one and only one thing. The second is the assumption that the errors of the measures are independent of one another, i.e., the classical assumption of *measurement independence* (Lord and Novick, 1968, p. 44). These assumptions rule out, for example, the operation of memory in the organization of responses to survey questions within a series of questions. Obviously, respondents are fully cognizant of the answers they have given to previous questions, so there is the possibility that memory operates to distort the degree of consistency in responses. These assumptions for reliability estimation in cross-sectional studies also rule out the operation of other types of correlated errors, for example, the operation of systematic method factors. Thus, in cross-sectional data it may be impossible to assume that measurement errors are independent, since similar questions are often given in sequence or at least included in the same battery. Given the shortcomings of the ICR approaches, attention has turned to more closely examining the sources of variation at the question or item level. Two variants on the basic CTST model have been developed for this purpose: the *multitrait-multimethod measurement* design and the *quasi-simplex approach* for longitudinal data, both of which we discuss in the following chapters.

Another fundamental problem with CTST approach to reliability is that it uniformly relies on models in which the latent variable is assumed to be continuous and in which the data can be assumed to vary according to an interval scale or an approximation, e.g., one that is at least ordinal in character. These assumptions, however, are clearly problematic for categorical latent variables. Latent-class models can be used to assess the extent of measurement error in measures of categorical variables (Clogg and Manning, 1996). Several investigators have explored discrete-time Markov chain models, where the Markovian property is posited to hold at the level of the latent classes measured repeatedly (Collins, 2001; Langeheine and van de Pol, 1990). We briefly consider these models in Chapter 11, although the main empirical results we present in the bulk of this work rely on the use of continuous variable models, where it can be assumed that the latent variable measured is continuous. The categorical models, however, provide an analog to the conception of reliability involved in the structural equation modeling approaches for panel data discussed in the following chapter. We illustrate that for those event classes that do not change, reliability estimation for categorical variables is a straightforward application of SEM modeling techniques applied to latent classes. The more difficult challenge is the development of comparable models permitting change in the composition of latent classes.

CHAPTER FOUR

Reliability Methods for Multiple Measures

Both reliability and validity concepts require that agreement between measures be demonstrated . . . Reliability is the agreement between two efforts to measure the same trait through maximally similar methods. Validity is represented in the agreement between two attempts to measure the same trait through maximally different methods.

Donald T. Campbell and Donald W. Fiske (1959, p. 83)

In the previous chapter I concluded that for the purpose of evaluating the reliability of standard measurement approaches used in survey research, it is important that we move beyond estimates of internal consistency reliability (ICR) and adopt a strategy that focuses on the reliability of *individual survey measures.* As pointed out in the previous chapter, for the purposes of examining the *reliability of single survey questions* the classical ICR model is limited in four important respects: (1) it assumes the collection of multiple, or replicate, measures of the same true score, which is often problematic in practice; (2) it does not distinguish multiple sources of systematic variance, that is, it cannot separate "valid" true-score variation from "invalid" true-score variation, or from other systematic sources of variation; (3) it does not distinguish multiple sources of random variance, that is, it cannot separate "unreliable" sources of variation from random sources of "reliable" variation; and (4) it assumes the statistical independence of measurement errors across measures. The ICR approaches may be a valuable set of tools that can be used for scale construction—my point is that it may not be the best approach we have for evaluating the reliability of survey data. For these reasons, research on survey methods has increasingly turned attention to the performance of individual survey measures, and here we focus on the reliability of individual survey questions.

It is important that the reader understand that I am *not* discouraging the use of internal consistency coefficients—e.g., Cronbach's α (Cronbach, 1951) or the Heise and Bohrnstedt (1970) Ω-coefficient—in item analysis and composite score development. I simply disagree with the common practice of referring to these as "reliability coefficients," even though to suggest otherwise represents a form of heresy in many circles. Many years ago Lord and Novick (1968) showed that coefficient α was *a lower bound on reliability*. Obviously, that this coefficient is a lower-bound estimate on reliability does not make it *equal to reliability*, except in those circumstances where the assumptions are met. It may be more accurate to say that in the typical case *Cronbach's* α *is a biased estimate of reliability*, and to acknowledge this should sensitize one to the problems involved. I prefer to refer to these as "internal consistency measures," and recommend their use as a measure of "first factor saturation," i.e., as a way of establishing the degree to which a set of items "hangs together" for purposes of combining them into a scale. For reasons given in the previous chapter, I prefer to reserve the term "reliability" for more appropriate applications of the concept.

In this chapter and the next I discuss two alternative approaches that have been proposed to address some of the limitations of the ICR approach—the *multitrait-multimethod* (MTMM)/confirmatory factor analysis approach using cross-sectional survey designs and the *quasi-simplex* approach using longitudinal data. Both of these approaches involve a kind of "repeated measuring" that is increasingly viewed as a requirement for the estimation of the reliability of survey measures (Goldstein, 1995, p. 142). Both make the assumption of the independence of errors that is critical for estimation of reliability. In this and the next chapter, I review the advantages and disadvantages of these two approaches. I have employed both approaches in my own research and consider both to be valuable. Each needs to be considered within the context of its limitations, and the estimates of parameters presumed to reflect reliability need to be interpreted accordingly. The *multitrait-multimethod* design relies on the use of *multiple indicators* measured within the same interview, using different methods or different types of questions for a given concept.[1] This design represents an adaptation of the "multitrait-multimethod" (MTMM) design that was first proposed by Campbell and Fiske (1959) as an approach to establishing convergent and discriminant evidence for "construct validity" in psychology and educational research.

In the 1970s the MTMM approach was reformulated as a confirmatory factor model in which multiple common factors representing trait and method components of variation were employed (see Werts and Linn, 1970; Alwin, 1974; Jöreskog, 1974, 1978; Browne, 1984). Over the past few decades researchers have done extensive analyses of the quality of survey measures using this approach, one that has relied primarily, although not exclusively, on the analysis of cross-sectional data involving models that incorporate method variation in repeated measurement within the same survey (Alwin, 1989, 1997; Alwin and Jackson, 1979; Andrews, 1984; Andrews and Herzog, 1986; Rodgers, 1989; Rodgers, Herzog, and Andrews, 1988; Rodgers, Andrews, and Herzog, 1992; Saris and van Meurs, 1990; Saris and Andrews, 1991; Scherpenzeel, 1995; Scherpenzeel and Saris, 1997).

The MTMM approach acknowledges the fact that it is difficult, if not impossible, to obtain repeated or *replicate* measures of the same question—what I here

---

[1]As will become clear in the course of this chapter, I make a distinction between *multiple indicators* and *multiple measures*, and it is important to emphasize in this context that the MTMM approach relies on the former rather than the latter.

refer to here as *multiple measures*—in the typical survey interview. Rather, the basic MTMM design employs *multiple indicators* that systematically vary the method of measurement across different concepts. As applied to multiple survey questions measuring the same concept, the approach starts with the use of confirmatory factor analysis, implemented using structural equation models (SEM), to estimate three sources of variance for each survey question: that attributable to an underlying concept or "trait" factor, that attributable to a "method" factor, and the unsystematic residual. The residual is assumed to reflect random measurement error. The strategy is then to conduct meta-analyses of these estimates to look for characteristics of questions and of respondents that are related to the proportion of variance that is reliable and/or valid.

In the following discussion I review the basic extension of the true-score model on which the MTMM is based, review some of the findings of the extant literature employing the MTMM approach to detecting errors in survey measurement, and provide an example using this approach in order to demonstrate its potential, as well as some of its limitations. Before discussing the MTMM approaches in greater depth, however, I first review the distinction between models involving *multiple measures* and those involving *multiple indicators* in order to fully appreciate their utility. Following this discussion, I then present the common-factor model that underlies the MTMM approach and its relation to the CTST model. An appreciation of the distinction between *multiple measures* and *multiple indicators* is critical to an understanding of the difficulties of designing survey measures that satisfy the rigorous requirements of the MTMM extension of the CTST model for estimating reliability. Then, as noted, I round out the chapter with the presentation of an example and a critique of the approach as a strategy for estimating the reliability of survey measurement.

4.1 MULTIPLE MEASURES VERSUS MULTIPLE INDICATORS

The classical approach to estimating the components of response variance described in the previous chapter requires *multiple* or *replicate measures* within respondents. Two general strategies exist for estimating the reliability of single survey questions: (1) using the same or similar measures in the same interview, or (2) using replicate measures in reinterview designs. The application of either design strategy poses problems, and in some cases the estimation procedures used require assumptions that are inappropriate (see Alwin, 1989; Saris and Andrews, 1991). Estimating reliability from information collected within the same interview is especially difficult, owing to the virtual impossibility of replicating questions. In the previous chapter I made the point that it was very important to understand the difference between *multiple indicators* and *multiple measures*, and here I further clarify what I meant by that. Researchers often employ similar, though not identical questions, and then examine correlation or covariance properties of the data collected. It is risky to use such information to estimate item reliability since questions that are different contain specific components of variance, orthogonal to the quantity measured in common, and because of the difficulties in separating reliable components of specific variance from random error variance present significant obstacles to this estimation approach (see Alwin and Jackson, 1979).

4.1.1 Multiple Measures

Obviously, the design discussed in Chapter 3 requires relatively strong assumptions regarding the components of variation in the measures, notably that measurement errors are random and independent across measures. What is even more stringent is the assumption that the measures are *univocal*, that is, that each measures one and only one thing in common with other similar measures. Unfortunately, there is no way to test these assumptions with only two measures, as there are too few observed variances and covariances, i.e., the model is underidentified. The assumption of tau-equivalence can be ruled out if the measures are in distinctly different metrics. For example, Y_1 may be scaled in terms of a five-category Likert-type scale, and Y_2 may be measured on a seven- or nine-point "feeling thermometer" scale.[2] In this case while the measures may not be *tau-equivalent*, i.e., the true scores may not be equal, they may be *univocal* in the sense that they measure one and only one thing. The true scores may, for example, be perfectly correlated, that is, either may be expressed as a linear function of the other, in which case they are said to be *congeneric* (see Jöreskog, 1971), and a solution is possible for the reliability of measures where $G \geq 3$.

The *congeneric* measurement model is consistent with the causal diagram in Figure 4.1 (a). This diagram embodies the most basic assumption of *classical true-score* theory that measures are *univocal*, that is, that they each measure one and only one thing that completely accounts for their covariation and that, along with measurement error, completely accounts for the response variation. This approach requires essentially asking the same question multiple times, which is why we refer to it as involving *multiple* or *replicate measures* (we consider the multiple indicators approach below). We should point out, however, that it is extremely rare to have multiple measures of the same variable in a given survey. It is much more common to have measures for multiple indicators, that is, questions that ask about somewhat different aspects of the same things.

While this model is basic to most methods of reliability estimation, including *internal consistency* estimates of reliability considered in the previous chapter, the properties of the model very often do not hold. This is especially the case when *multiple indicators* (i.e., observed variables that are similar measures within the same domain, but not so similar as to be considered replicate measures) are used.

Let us first consider the hypothetical, but unlikely, case of using replicate measures within the same survey interview. Imagine the situation, for example, where

---

[2]Sometimes standard scores are used to avoid the problems associated with having items measured using different scales, but this does not provide a general solution to the problem.

(a) Multiple Measures

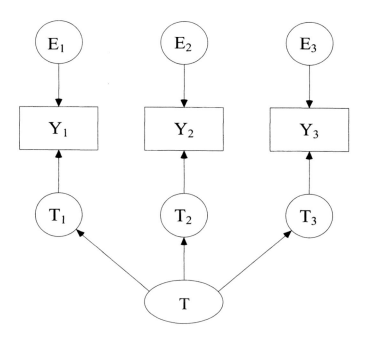

(b) Multiple Indicators

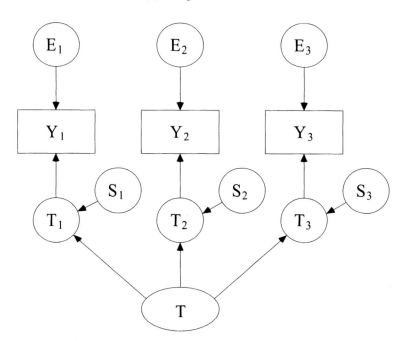

Figure 4.1. Path diagrams for the relationship between random measurement errors, observed scores, and true scores for the multiple-measures and multiple-indicators models.

three measures of the variable of interest are obtained, for which the random measurement error model holds in each case, as follows:

$$Y_1 = T_1 + E_1$$
$$Y_2 = T_2 + E_2$$
$$Y_3 = T_3 + E_3$$

In this hypothetical case, if it is possible to assume one of the measures could serve as a record or *criterion of validity* for the other, it would be possible to assess the validity of measurement, but we do not place this constraint on the present example. All we assume is that the three measures are replicates in the sense that they are virtually identical measures of the same thing. This model assumes that each measure has its own true score, and that the errors of measurement are not only uncorrelated with their respective true scores but are also independent of one another. This is called the assumption of *measurement independence* referred to earlier (Lord and Novick, 1968, p. 44).

Following the basic tenets of classical true-score theory, in the most general form of the CTST model given above—the congeneric measures model—the variances of the model are defined as follows (where $T_1 = T$, and therefore $\lambda_{1T} = 1.0$; $T_2 - \mu_2 = \lambda_{2T} T$; and $T_3 - \mu_3 = \lambda_{3T} T$):

$$VAR[Y_1] = VAR[T_1] + VAR[E_1]$$
$$VAR[Y_2] = VAR[T_2] + VAR[E_2] = \lambda^2_{2T} VAR[T_1] + VAR[E_2]$$
$$VAR[Y_3] = VAR[T_3] + VAR[E_3] = \lambda^2_{3T} VAR[T_1] + VAR[E_3]$$

Further, the covariances among the measures are written as:

$$COV[Y_1,Y_2] = \lambda_{2T} VAR[T_1]$$
$$COV[Y_1,Y_3] = \lambda_{3T} VAR[T_1]$$
$$COV[Y_2,Y_3] = \lambda_{2T} \lambda_{3T} VAR[T_1]$$

Given these definitions and the fact that the model in this simple $G = 3$ case is just-identified, there is a straightforward algebraic solution to the unknowns in these equations, as follows:

$$VAR[T_1] = COV[Y_1,Y_2] COV[Y_1,Y_3] / COV[Y_2,Y_3]$$
$$\lambda_{2T} = COV[Y_2,Y_3] / COV[Y_1,Y_3]$$
$$\lambda_{3T} = COV[Y_2,Y_3] / COV[Y_1,Y_2]$$

With these results in hand it is possible to define the reliabilities of the measures Y_1, Y_2, and Y_3 as follows:

$$\rho_1^2 = COV[Y_1,Y_2] COV[Y_1,Y_3] / COV[Y_2,Y_3] VAR[Y_1]$$
$$\rho_2^2 = COV[Y_1,Y_2] COV[Y_2,Y_3] / COV[Y_1,Y_3] VAR[Y_2]$$
$$\rho_3^2 = COV[Y_2,Y_3] COV[Y_1,Y_3] / COV[Y_1,Y_2] VAR[Y_3]$$

In other words, given the assumptions of the CTST model for multiple measures, there is a straightforward well-defined interpretation of reliability for each measure. These ideas can be generalized to the case where $G > 3$. In this case the model is overidentified, and there is no straightforward algebraic solution to the model; however, it is possible to estimate the model parameters using maximum likelihood within a confirmatory factor analysis framework (see Chapter 3).

4.1.2 Multiple-Indicators Models

These models derive from *common-factor analysis*, a psychometric tradition that predates classical true-score theory. In such models the K latent variables are called "common factors" because they represent common sources of variation in the observed variables. The common factors of this model are responsible for covariation among the variables. The unique parts of the variables, by contrast, contribute to the lack of covariation among the variables. Covariation among the variables is greater when they measure the same factors, whereas covariation is less when the unique parts of the variables dominate. Indeed, this is the essence of the common-factor model—variables correlate because they measure the same thing(s).

The common-factor model, however, draws attention not only to the common sources of variance, but to the unique parts as well (see Alwin, 2000). A variable's *uniqueness* is the complement to the common parts of the data (the *communality*), and is thought of as being composed of two independent parts, one representing *specific variation* and one representing *random measurement error* variation. Using the traditional common factor notation for this, this can be restated as follows. The variable's *communality*, denoted h_j^2, is the proportion of its total variation that is due to common sources of variation. Its *uniqueness*, denoted u_j^2, is the complement of the communality, that is, $u_j^2 = 1.0 - h_j^2$. The uniqueness is composed of specific variance, s_j^2, and random error variance, e_j^2. Specific variance is *reliable* variance, and thus the reliability of the variable is due not only to the common variance, but to specific variance as well. In common-factor analytic notation, the reliability of variable j can be expressed as $r_j^2 = h_j^2 + s_j^2$. Unfortunately, because specific variance is thought to be independent (uncorrelated with) sources of common variance, it becomes confounded with measurement error variance. Because of the presence of specific variance in most measures, it is virtually impossible to use the traditional form of the common-factor model as a basis for reliability estimation (see Alwin, 1989; Alwin and Jackson, 1979), although this is precisely the approach advocated in the work of Heise and Bohrnstedt (1970), and followed by the MTMM /confirmatory factor analytic approach to reliability estimation we discuss below (see Andrews, 1984).

The problem here—which applies to the case of multiple indicators—is that the common-factor model typically does not permit the partitioning of u_j^2 into its components, s_j^2 and e_j^2. In the absence of specific variance (what we here refer to as the "multiple measures" model), classical reliability models may be viewed as a special case of the common-factor model, but in general it is risky to assume that $u_j^2 = e_j^2$ (Alwin and Jackson, 1979). Consider the case of the same three measures described above, in which there is a specific source of variation in each of the measures, as follows (returning to the notation used above):

$$Y_1 = T_1 + S_1 + E_1$$
$$Y_2 = T_2 + S_2 + E_2$$
$$Y_3 = T_3 + S_3 + E_3$$

Assume for present purposes that specific variance is orthogonal to both the true score and error scores, as is the case in standard treatments of the common-factor model.

The picture in Figure 4.1 (b) illustrates this case. Note that the *classical true-score assumptions do not hold*. Here the true scores for the several measures are not *univocal* and therefore cannot be considered *congeneric* measures. In case (b) there are disturbances in the equations linking the true scores. Consequently the reliable sources of variance in the model are not perfectly correlated. In the language of *common-factor analysis*, these measures contain more than one factor. Each measure involves a *common factor* as well as a *specific factor*, that is, a reliable portion of variance that is independent of the common factor. Unless one can assume measures are *univocal*, or build a more complex array of common factors into the model, measurement error variance will be overestimated, and item-level reliability underestimated (see Alwin and Jackson, 1979). This can be seen by noting that the random error components included in the disturbance of the measurement model in Figure 4.2 (b) contain both random measurement error (the Es) and specific sources of variance (the Ss). Consistent with the common-factor model representation, which assumes that the unique part of Y_g is equal to $U_g = S_g + E_g$, we can define the variance of Y_g as

$$VAR(Y_g) = VAR(T_g) + VAR(S_g) + VAR(E_g)$$
$$= VAR(T_g) + VAR(U_g),$$

and $VAR(U_g)$ is known to equal the sum of specific and measurement error variance.

A second issue that arises in the application of multiple indicators is that there may be correlated errors, or common sources of variance, masquerading as true scores. Imagine that respondents' use of a common agree-disagree scale is influenced by a "method" factor, e.g., the tendency to agree vs. disagree (called "yea-saying" or "acquiescence" in the survey methods literature). In this case there are two common factors at work in producing T_g. Recalling our earlier discussion of *non-random measurement error* (see Chapter 2), we can formulate the model as follows: $Y_g = T_g + E_g$, where $T_g = T_g^* + M_g$. This indicates that the covariance among measures, and therefore the estimates of reliability, are inflated by the operation of common method factors.

The *multiple-measures* model can, thus, be thought of as a special case of the *multiple-indicators* model in which the latent true scores are linear combinations of one another, i.e., perfectly correlated. Unfortunately, unless one knows that the multiple-measures model is the correct model, interpretations of the "error" variances as solely due to measurement error are inappropriate. In the more general case (the multiple-indicators case) one needs to posit a residual, referred to as "specific variance" in the factor analytic tradition, as a component of a given true score to account for its failure to correlate perfectly with other true scores aimed at measuring the same construct. Within a single cross-sectional survey, there is no way to distinguish between the two versions of the model, that is, there is no available test to detect whether the congeneric model fits a set of G variables, or whether the common-factor model is the more appropriate model.

To summarize our discussion up to this point, it is clear that two problems arise in the application of the CTST model to cross-sectional survey data. The first is that specific variance, while reliable variance, is allocated to the random error term in the model, and consequently, to the extent specific variance exists in the measures the reliability of the measure is underestimated. This problem could be avoided if there were a way to determine whether the congeneric (multiple-measures) model is the appropriate model, but as we have noted, within a single cross-sectional survey, there is no way to do this. The second problem involves the assumption of a single common factor, which is a problem with either version of the model shown in Figure 2.1. In this case the problem involves the presence of common method variance, which tends to inflate estimates of reliability. The latter problem is one that is also true of the multiple-measures model in that it is just as susceptible to the multiple-factor problem as the multiple-indicators model. The problem is actually more general than this, as it involves any source of multiple common factors, not simply common method factors.

In the remainder of this chapter and in the next I discuss two unique approaches to dealing with these problems. The first—the *multitrait-multimethod measurement design*—handles the second problem by attempting to identify sources of common method variance by explicitly including them as factors in the design of the measures. The second—the *quasi-simplex* approach using longitudinal data—handles the first problem by focusing on single questions and essentially doing away with the distinction between specific and common variance. In the latter case the specific portion of the observed score variance is allocated to the true component and not to measurement error (see Chapter 5).

4.2   MULTITRAIT-MULTIMETHOD APPROACHES

Interestingly, although the Campbell and Fiske (1959) MTMM approach has focused primarily on the concept of *validity*, the essential elements of their approach have been used to motivate the investigation of the reliability of measurement and its components. An argument can be made—one to which I return later in the chapter—that *there is a basic flaw in the application of these ideas to an understanding of reliability of survey measurement*, but this depends in part on the assumptions one is willing to make about what constitutes a true MTMM design. The central issue here, to which I return below, is whether one is willing to equate reliability with the concept of communality in the common factor model, therein assuming that all random error is measurement error and that no specific variance exists.

Before getting into these more controversial aspects of the application of this method, let me begin with a clarification of concepts and terminology. Consistent with the CTST definition of reliability, as used in previous chapters, Campbell and Fiske (1959, p. 83) define *reliability* as "the agreement between two attempts to measure the same trait through maximally similar methods." If there is nothing more that survey methodologists agree on with respect to assessing measurement error, virtually everyone agrees with this basic psychometric notion that reliability represents the consistency between replicate measures. Whether these measures are implemented as a part of the same interview, or in different interviews, or in some other way, nothing could be more basic to an understanding of the concept of reliability. Where people disagree is in what constitutes "maximally similar methods."

Distinct from this notion of *reliability* is the concept of *validity*, which, again consistent with the CTST definition (see Chapter 3), is "represented in the agreement between two attempts to measure the same trait through maximally different methods" (Campbell and Fiske, 1959, p. 83). At the time Campbell and Fiske wrote their classic paper this was a radical idea because this approach challenged the "split-half reliability" approach as more like an estimate of validity than reliability, compared to an immediate test-retest reliability using identical measures (see Chapter 5). The *convergence* among dissimilar measures of the same thing was clearly evidence of validity, but convergence alone was not enough to establish validity, they argued. It was also important that measures *not correlate* with other measures that *do not measure the same thing*, and to describe this they coined the term *discriminant validity*.

The major impediment to establishing validity of measures, they argued, was the existence of *nonrandom* error. The closer two measurements are in time, space, and structure, the more highly they should be expected to correlate as a result of such contextual similarity. They refer to this systematic or nonrandom contribution to the correlation among measures embodying such similarity generically as *method variance*. They proposed the use of a multitrait-multimethod (MTMM) matrix to address this problem, that is, the measurement of K variables (called *traits*) measured by each of Q methods, generating G = KQ observed variables. The central problem that their MTMM matrix approach addressed is the extent to which it is possible to make inferences about basic trait (or variable) relationships on the basis of intercorrelations among measures of those variables. The MTMM approach was presented as a way of assessing evidence for common variation among variables over and above that due to common method variation or covariation (Campbell and Fiske, 1959, p. 84). The approach did not focus on the estimation of reliability.

4.2.1   The MTMM Matrix[3]

As noted, the general MTMM matrix involves the measurement of K traits by each of Q methods of measurement. The correlation matrix in Table 4.1 presents

---

[3]This discussion relies heavily on the author's earlier treatment of the same material, reproduced here with permission of the publisher (see Alwin, 1974, pp. 81–82).

Table 4.1.   Multitrait-multimethod matrix for three traits and three methods

	Method 1			Method 2			Method 3		
Trait	X	Y	Z	X	Y	Z	X	Y	Z
1 · X	$r_{X_1 X_1}$								
1 · Y	$r_{X_1 Y_1}$	$r_{Y_1 Y_1}$							
1 · Z	$r_{X_1 Z_1}$	$r_{Y_1 Z_1}$	$r_{Z_1 Z_1}$						
2 · X	$\mathbf{r_{X_1 X_2}}$	$r_{X_2 Y_1}$	$r_{X_2 Z_1}$						
2 · Y	$r_{X_1 Y_2}$	$\mathbf{r_{Y_1 Y_2}}$	$r_{Y_2 Z_1}$	*Monomethod Blocks*					
2 · Z	$r_{X_1 Z_2}$	$r_{Y_1 Z_2}$	$\mathbf{r_{Z_1 Z_2}}$						
3 · X				$\mathbf{r_{X_3 X_2}}$	$r_{X_3 Y_2}$	$r_{X_3 Z_2}$	$r_{X_3 X_3}$		
3 · Y	*Heteromethod Blocks*			$r_{X_2 Y_3}$	$\mathbf{r_{Y_2 Y_3}}$	$r_{Y_3 Y_2}$	$r_{X_3 Y_3}$	$r_{Y_3 Y_3}$	
3 · Z				$r_{X_2 Z_3}$	$r_{Y_2 Z_2}$	$\mathbf{r_{Z_2 Z_3}}$	$r_{X_3 Z_3}$	$r_{Y_3 Z_3}$	$r_{Z_3 Z_3}$

Note: Values in validity diagonals (MTHM) are in boldface type. From (Alwin, 1974, p. 81).

the general form of the MTMM matrix for K = 3 (traits X, Y, and Z) and Q = 3 (methods I, II, and III). The correlations among traits all of which are measured by the same method are included in the *monomethod blocks*—in Table 4.1 there are three such monomethod blocks. There are two types of entries in the typical monomethod block: the monotrait-monomethod (MTMM) values and the hetero-trait-monomethod (HTMM) values. The monotrait values refer to the reliabilities of the measured variables—a topic to which we return below—and the heterotrait values are the correlations among the different traits within a given method of measurement. There are .5K(K-1) HTMM values in the lower triangle of each mono-method block.

Correlations among trait measures assessed by different methods constitute the *heteromethod blocks*, which also contain two types of entry: the monotrait-heteromethod (MTHM) values and the heterotrait-heteromethod (HTHM) values.

The MTHM values are also referred to as *validity values* because each is a correlation between two presumably different attempts to measure a given variable. The HTHM values are correlations between different methods of measuring different traits. In Table 4.1 there are .5Q(Q-1) heteromethod blocks, which are symmetric submatrices, each containing K MTHM values in the diagonal.

Given the MTMM matrix in Table 4.1 and the terminology outlined here, Campbell and Fiske (1959, pp. 82–83) advance the following criteria for *convergent* and *discriminant* validity. First, "the entries in the validity diagonal should be significantly different from zero and sufficiently large to encourage further examination of validity." That is, if two assessments of the same trait employing different methods of measurement do not converge, then there is probably little value in pursuing the issue of validity further. Second, "a validity diagonal value [MTHM values] should be higher than the values lying in its column and row in the heterotrait-heteromethod triangles." In other words, a "validity value" for a given trait should be higher than the correlations between that measure and any other measure having neither trait nor method in common. Third, a measure should "correlate higher with an independent effort to measure the same trait than with measures designed to get at different traits that happen to employ the same method." Thus, if different traits measured by the same method correlate more highly with one another than any one of them correlates with independent efforts to measure the same thing, then there is a clear domination of method variation in producing correlations among measures. Finally, "a fourth desideratum is that the same pattern of trait interrelationship be shown in all of the heterotrait triangles of both the monomethod and heteromethod blocks."

The Campbell and Fiske (1959) criteria for establishing the construct validity of measures have become standards in psychological and educational research (see APA, 2000). Indeed, it was the failure of conventional approaches to validation in the psychometric literature, which emphasized primarily convergence criteria (e.g., Cronbach and Meehl, 1955), that stimulated the efforts of Campbell and Fiske to propose a broader set of criteria. They viewed the traditional form of validity assessment—convergent validity—as only preliminary and insufficient evidence that two measures were indeed measuring the same trait. As operational techniques these criteria are, however, very difficult to implement and are not widely employed. While most investigators agree that there is a commonsense logic to these criteria, some fundamental problems exist with their application. An alternative approach that developed in the early 1970s (see Werts and Linn, 1970; Jöreskog, 1974, 1978; Althauser and Heberlein, 1970; Althauser, Heberlein, and Scott, 1971; Alwin, 1974)—formulated originally within a path-analytic framework—was to represent the MTMM measurement design as a common factor model that could be estimated using confirmatory factor analysis [or what we today refer to as *structural equation models* (SEM)]. Later on the technique was found to be useful by survey methodologists interested in partitioning the variance in survey measures collected within a MTMM design into components of variance (e.g., Andrews, 1984; Scherpenzeel, 1995; Scherpenzeel and Saris, 1997; Alwin, 1997). I cover the details of this approach, its advantages and limitations, in the remainder of the chapter, but first I discuss how the MTMM matrix approaches the question of reliability of measurement.

4.2.2 The MTMM Matrix and the Reliability of Measurement

In whatever way one wishes to conceptualize the relationships among variables in the MTMM matrix, it is important to recognize that this approach focuses essentially on nonrandom errors and as such the issues addressed are conceptually independent of the issue of random measurement error or unreliability (Alwin, 1974, p. 82). Campbell and Fiske (1959, pp. 82–83) allowed for reliability in the sense that they included entries in the *monomethod blocks*—see Table 4.1—that represent the same correlation of two measures involving the same trait and the same method, that is, what they called the "monotrait-monomethod (MTMM) values," or the "reliabilities of the measured variables." They pointed out that "the evaluation of the correlation matrix formed by intercorrelating several trait-method units must take into consideration the many factors which are known to affect the magnitude of correlations . . . a value in the validity diagonal must be assessed in light of the reliabilities of the two measures involved . . . " (Campbell and Fiske, 1959, p. 102).

Rarely, if ever, is this point acknowledged; that is, that the relationships among variables in the MTMM design be evaluated in light of the reliability of measurement. Campbell and Fiske (1959) proposed no systematic way of taking unreliability into account, although several writers considered ad hoc solutions to this problem. Indeed, in only a few of several examples in their paper did Campbell and Fiske (1959) actually include reliability estimates, and in most of these cases the estimates were based on ICR or split-half methods. Jackson (1969) and Althauser and Heberlein (1970) suggested that the MTMM matrix might be corrected for attenuation using conventional correction formulas (see Lord and Novick, 1968, pp. 69–73), but in most cases there is no independent estimate of reliability. Indeed, in most applications of the MTMM approach I have seen there is no possibility of including multiple measures of the same traits (in the sense of replicate measures, discussed earlier in this chapter), but rather reliability is conceptualized entirely within the framework of a common-factor model thought to underlie the MTMM design, and it is assumed (rightly or wrongly) that the disturbances for the observed trait-method combinations in these models reflect measurement error. It is to the discussion of these models, and a clarification of the nature of these assumptions and interpretations, that I now turn.

4.3 COMMON-FACTOR MODELS OF THE MTMM DESIGN

There is an increasing amount of support for the view that shared method variance inflates ICR estimates. One approach to dealing with this is to reformulate the

CTST along the lines of a multiple-common-factor approach and to include sources of systematic variation from *both* trait variables and method factors. With *multiple indicators* of the same concept, as well as different concepts measured by the same method, it is possible to formulate a *multitrait-multimethod* (MTMM) common-factor model. In general, the measurement of K traits measured by each of Q methods (generating $G = KQ$ observed variables) allows the specification of such a model. An example of the conventional factor analytic MTMM model is given in Figure 4.2, wherein I have specified a model for three concepts, each of which is measured by two methods (see Groves, 1989). We should perhaps highlight the fact that since the measurement of each concept is carried out, not by replicate measures, but by *multiple indicators*, the model shown in Figure 4.2 depicts sources of specific variance in each trait method combination, inasmuch as the disturbances on the observed trait-method combinations are conceptualized as $U_g = S_g + E_g$. These components of variance cannot be estimated using conventional methods, but it is important to include them here, so as to indicate that ultimately the random error terms identified in common factor MTMM models include both measurement error and specific variance.

Following from our discussion of *nonrandom measurement errors* in Chapter 3, we can formulate an extension of the common factor representation of the CTST given there as

$$Y = \Lambda_{T*} T^* + \Lambda_M M + U$$

where Y is a $(G \times 1)$ vector of observed random variables, T^* is a $(K \times 1)$ vector of "trait" true score random variables, M is a $(Q \times 1)$ vector of "method" true score random variables, and U is a $(G \times 1)$ vector of error scores. The matrices Λ_{T*} and Λ_M are $(G \times K)$ and $(G \times Q)$ coefficient matrices containing the regression relationships between the G observed variables and the K and Q latent trait and method latent variables. Note that with respect to the CTST model given above, $\Lambda T = \Lambda_{T*} T^* + \Lambda_M M$. The covariance structure for the model can be stated as

$$\Sigma_{YY} = \left[\Lambda_{T*} | \Lambda_M\right] \Phi_T \left[\Lambda_{T*} | \Lambda_M\right]' + \Theta^2$$

where Φ_T has the following structure:

$$\Phi_T = \left[\begin{array}{c|c} \Phi_{T*} & 0 \\ \hline 0 & \Phi_M \end{array} \right]$$

Note that the specification of the model places the constraint that the trait and method factors are uncorrelated (Browne, 1984). The estimation of this model permits the decomposition of reliable variance in each observed measures into "valid" and "invalid" parts (Andrews, 1984; Saris and Andrews, 1991).

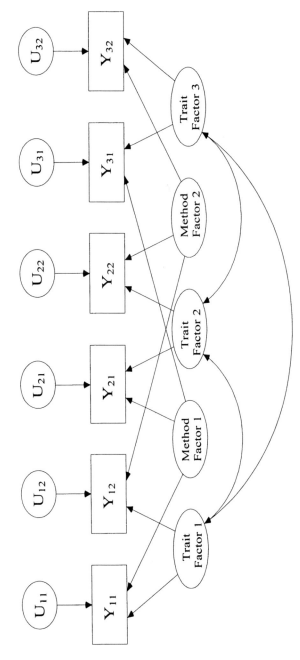

Figure 4.2. Path diagram for a common-factor representation of the multitrait-multimethod design for six observed variables.

4.3.1   Decomposition of Variance/Covariance

For purposes of illustration, let us assume that any given survey measure has three statistically independent components of variation, as follows: (1) a true source of variation representing the trait or latent variable being measured; (2) a second true source of systematic variation, due to the method characteristics associated with the question; and (3) a random error component. This setup follows earlier specifications of the multitrait-multimethod design for assessing errors of measurement in surveys (see Alwin, 1974, 1989; Alwin and Jackson, 1979; Andrews, 1984; Groves, 1991; Saris and Andrews, 1991).

Note that this model violates the basic assumption of CTST that the measures are univocal, i.e., that each measures one and only one source of true variation. Let Y_{ij} represent a measure assessing trait T_i with method characteristics M_j, and random error U_{ij}, such that

$$Y_{ij} = T^*_i + M_j + U_{ij}$$

Note that, consistent with our previous discussion, we have formulated the random error component as U_{ij} in order to highlight the fact that in multiple indicator models, the random error component contains both random measurement error and specific sources of variance [see the measurement model in Figure 4.1 (b)]. We assume for purposes of the present discussion that the unique part of Y_g is equal to $U_g = S_g + E_g$, and that the variance of the random error component, $VAR(U_g)$, equals the sum of specific and measurement error variance, $VAR(S_g) + VAR(E_g)$.

If it were possible to replicate Y_{ij} within the space of a single interview in such a way that the replicate measure was unaffected by memory or consistency motivation, then the covariance properties of the two measures would provide the basis for reliability estimation. In such a case the reliability of the measure Y_{ij} would assess the consistency of jointly measuring T_i and M_j. In this hypothetical case, let $i = 1$ and $j = 1$, so that we are talking about Y_{11} and its replicate. The product-moment correlation, $COR(Y_{11},Y_{11})$, expresses the following ratio:

$$= COV(Y_{11}, Y_{11}) / VAR(Y_{11})$$
$$= [VAR(T^*_1) + VAR(M_1)] / VAR(Y_{11})$$

which equals the reliability of measurement for Y_{11}. The components of these elements are shown in Table 4.2 under case (a).

This example shows that it is only when one assumes a measure is univocal does the reliability estimate provide an unambiguous interpretation of components of variance. In this case one is forced to ignore the distinction between the two "reliable" components, namely, T^*_i and M_j, thinking of reliability as expressing the consistency of measurement, regardless of the source of that consistency. While

Table 4.2. Decomposition of variances and covariances for measures involving all combinations of trait and methods

	Common Trait	Different Trait
Common Method	**Case (a)**	**Case (c)**
	$Y_{11} = T_1^* + M_1 + U_{11}$	$Y_{11} = T_1^* + M_1 + U_{11}$
	$Y_{11} = T_1^* + M_1 + U_{11}$	$Y_{21} = T_2^* + M_1 + U_{21}$
	$VAR(Y_{11}) = VAR(T_1^*) + VAR(M_1) + VAR(U_{11})$	$VAR(Y_{11}) = VAR(T_1^*) + VAR(M_1) + VAR(U_{11})$
	$VAR(Y_{11}) = VAR(T_1^*) + VAR(M_1) + VAR(U_{11})$	$VAR(Y_{21}) = VAR(T_2^*) + VAR(M_1) + VAR(U_{21})$
	$COV(Y_{11}Y_{11}) = VAR(T_1^*) + VAR(M_1)$	$COV(Y_{11}Y_{21}) = COV(T_1^*T_2^*) + VAR(M_1)$
Different Method	**Case (b)**	**Case (d)**
	$Y_{11} = T_1^* + M_1 + U_{11}$	$Y_{11} = T_1^* + M_1 + U_{11}$
	$Y_{12} = T_1^* + M_2 + U_{12}$	$Y_{22} = T_2^* + M_2 + U_{22}$
	$VAR(Y_{11}) = VAR(T_1^*) + VAR(M_1) + VAR(U_{11})$	$VAR(Y_{11}) = VAR(T_1^*) + VAR(M_1) + VAR(U_{11})$
	$VAR(Y_{12}) = VAR(T_1^*) + VAR(M_2) + VAR(U_{12})$	$VAR(Y_{22}) = VAR(T_2^*) + VAR(M_2) + VAR(U_{22})$
	$COV(Y_{11}Y_{12}) = VAR(T_1^*)$	$COV(Y_{11}Y_{22}) = COV(T_1^*T_2^*)$

this is entirely possible, and perhaps desirable, two things should be emphasized. First, as we have repeatedly pointed out, this is a purely hypothetical example, as it is extremely rare to exactly replicate a single measure within the same interview. Second, even if we could make such observations, we should note that the CTST model does not assist in understanding this problem, since it assumes measures are univocal. In other words, it is only if one recognizes that what is being assessed in such a case is the reliability of the sum, $T^*_i + M_j$, is the reliability interpretation appropriate.

As noted, the classical true-score model is impractical and unrealistic because it assumes something that is rarely feasible, namely the replication of a particular measure in a cross-sectional design. It is more commonly the case that either (1) the same trait is assessed using a different method; (2) a similar, but distinct, trait is assessed using a question with identical method characteristics, e.g., a rating-scale format with the same number of scale points that has the same effect on all traits that employ it; or (3) a similar, but distinct, trait is assessed using a question with a different method. These possibilities are depicted as cases (b), (c), and (d), respectively, in Table 4.2.

In any one of these additional cases, it might be assumed that the product-moment correlation between the relevant measures expresses the reliability of measurement. This, of course, is not necessarily the case. The facts of the matter are given in Table 4.2, where the components of variance/covariance for each of these logical combinations of two measures vary in commonality or "sameness" of trait and method. As the evidence in Table 4.2 indicates, there is considerable variability in what might be treated as the estimate of the "true" variance, if the product-moment correlation between the two measures in each case is used to estimate reliability. Technically, the only case in which the variance/covariance information involving the two measures can be unambiguously manipulated to form an estimate of reliability is, as suggested above, case (a), where the measures are true replicates. All other cases in the table are approximations, and in general I would argue that more often than not they will be wrong. Case (c) represents the situation, for example, that is perhaps the most typical in survey research—the use of a set of measures from a battery of questions that each measures a similar, but distinct, concept, with the same method. Here, as is apparent from the results presented in the table, the covariances among the measures contain method variance.

Finally, note that the MTMM approach capitalizes on the combination of cases (b), (c), and (d) and an analysis /decomposition of the resulting covariance structure in order to partition the variance in a given measure, Y_{ij} , as follows:

$$VAR(Y_{ij}) = VAR(T^*_i) + VAR(M_j) + VAR(U_{ij})$$

In general, the model can be shown to be identified in cases where one has at least two distinct traits measured by at least two distinct methods, but even here there are certain assumptions that must be made in order to identify the model (see Alwin, 1974; Alwin and Jackson, 1979; Saris and Andrews, 1991).

4.4 CLASSICAL TRUE-SCORE REPRESENTATION OF THE MTMM MODEL

In the previous chapter I discussed the relationship between random and systematic errors and how the CTST model could be clarified by considering the following extension of the classical true-score model: $y_{gp} = \tau^*_{gp} + \eta_{gp} + \varepsilon_{pg}$, where η_{gp} is a source of systematic error in the observed score, τ^*_{gp} is the true value, uncontaminated by systematic error, and ε_{gp} is the random error component discussed above. This model directly relates to the one given in the CTST model, in that $\tau_{gp} = \tau^*_{gp} + \eta_{gp}$. I, thus, presented the idea that the variable portion of measurement error contains two types of components, a random component, ε_{gp}, and a nonrandom or systematic component, η_{gp} (see Chapter 3, Section 3.3).

Taking this model and formulating for a population of persons, we wrote the model as: $Y_g = T_g + E_g$ for the gth measure of Y, where $T_g = T_g^* + M_g$. The path diagram in Figure 4.3 illustrates the structural relations between the components of Y_g for the MTMM common-factor model discussed in the foregoing. This model reformulates the common-factor model for the MTMM design given in the previous section. In a fundamental sense the models are identical and the fact that T_g^* and M_g are independent contributors to T_g allows us to rewrite the model this way. The population properties of Y_g, T_g, and E_g are the same as those given in the previous discussion, namely that the expectation of E_g is zero, from which it follows that $E[Y_g] = E[T_g]$. In addition, $E[T_g] = E[T_g^*] + E[M_g]$, $VAR[T_g] = VAR[T_g^*] + VAR[M_g]$, and $COV[T_g^*,M_g] = 0$. In other words, as formulated here, in the population model the true score is an additive combination of two uncorrelated components—a "trait" component and a "method" component. As illustrated in the following example, the decomposition of the true-score variance into the proportions of variance attributable to these two components for any trait-method combination can be implemented using a multitrait-multimethod measurement design along with confirmatory factor analysis.

Note that here we have included a specific component of variance in the true-score model for the MTMM model depicted in Figure 4.3, and consistent with CTST have formulated the disturbance of the observed scores as entirely due to random measurement error. However, consistent with our previous discussion I emphasize that the random error component may include specific variance and it would probably be better to formulate the disturbance as U_{ij} in order to highlight the fact that in multiple indicator models, the random error component contains both random measurement error and specific sources of variance. In other words, in this formulation of the model, the true score for a given trait-method combination is perfectly determined by its trait and method components and any specific variance that might be present in these true scores are allocated to the disturbance. In other words, as with the ICR estimates of reliability, due to its reliance on cross-sectional data, the MTMM approach is vulnerable to the criticism that the model may misspecify what is actually happening within the survey interview. We return to this issue in our critique of the MTMM approach to reliability estimation. We turn first to a brief review of the growing body of research that utilizes this approach, followed by an example

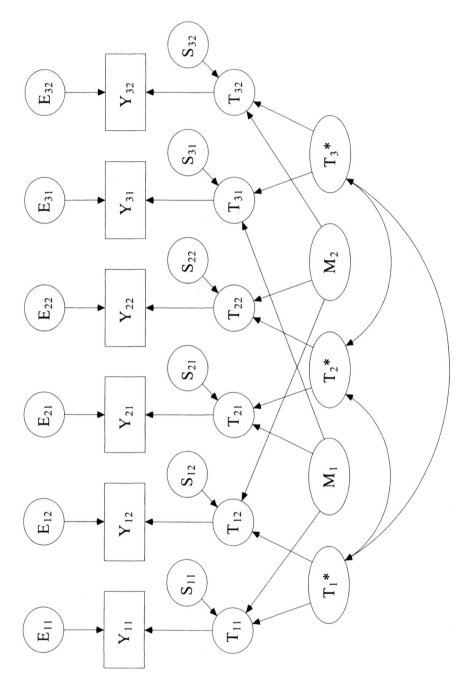

Figure 4.3. Path diagram for a classical true-score representation of the multitrait-multimethod design for six observed variables.

of how this approach may be used to decompose the true-score variance into "valid" and "invalid" parts.

4.5 THE GROWING BODY OF MTMM STUDIES

There are several studies that summarize the results of the application of the MTMM to the detection of method variance in large-scale surveys. Frank Andrews (1984) pioneered the application of the MTMM measurement design to large-scale survey data in order to tease out the effects of method factors on survey responses. His research suggested that substantial amounts of variation in measurement quality could be explained by seven survey design characteristics, here listed in order of their importance in his meta-analysis of effects on data quality:

- The number of response categories
- Explicit offering of a Don't Know option
- Battery length
- Absolute versus comparative perspective of the question
- Length of the introduction and of the question
- Position of the question in the questionnaire
- The full labeling of response categories

Andrews' (1984, p. 432) work suggested that several other design features were insignificant in their contributions to measurement error. He indicated that the mode of administering the questionnaire, e.g., telephone, face-to-face, or group-administered, was unimportant, stressing the encouraging nature of this result. He noted also that the use of explicit midpoints for rating scales, whether the respondent was asked about things they had already experienced or about predictive judgments with respect to the future, or the substantive topic being asked about in the survey question had only slight effects on data quality (Andrews, 1984, pp. 432–433).

Andrews' (1984) innovative application of the MTMM approach—based on Heise and Bohrnstedt's (1970) factor analytic assumptions—to the evaluation of data quality in surveys has been extended by a number of researchers. Saris and van Meurs (1990) presented a series of papers from a conference on the evaluation of measurement instruments by meta-analysis of multitrait-multimethod studies. These papers (including Andrews' original paper) addressed several topics of interest to researchers seeking to implement this research strategy. Saris and Andrews (1991) review many of these same issues, and Groves (1991) summarizes the applicability of the model to assessing nonrandom errors in surveys. In addition several other researchers have contributed to the legacy of Andrews' work and have made important contributions to understanding the impact of method factors to response variance (e.g., Andrews and Herzog, 1986; Rodgers, 1989; Rodgers et al., 1988; Rodgers et al., 1992). More recently, Scherpenzeel (1995) [see also Scherpenzeel and Saris (1997)] extended Andrews' (1984) work to several survey studies developed in The Netherlands. There is, unfortunately, no available archive of reliability results from MTMM studies (see Chapter 12).

4.5.1   Clarification of Terminology

Before proceeding further, there is some terminology that requires clarification—specifically the terms "validity" and "invalidity"—as there has been a proliferation of terminology associated with the interpretation of MTMM analyses. From my perspective, this is probably an unfortunate choice of terms, given that the concept of measurement *validity* normally refers either to evidence (1) that the measure is revealing the information the investigator intends, i.e., construct validity; or (2) that the measure predicts other variables expected on the basis of theory, i.e., criterion or predictive validity. Regardless of these reservations, using the factor analytic approach to the interpretation of relationships in the MTMM matrix (see Werts and Linn, 1970; Jöreskog, 1974, 1978; Alwin, 1974, 1997; Andrews, 1984), these terms have been used to refer to the contributions of the trait (validity) and method (invalidity) to the overall reliability (or communality) of a given variable. This usage is also consistent with the terms that Heise and Bohrnstedt (1970) used to describe components of reliability.

To review the above discussion, the modified true score model for the MTMM design is written as follows for a set of measures in which each measure is a unique combination of a given trait and a given method. Let Y_{ij} represent a measure assessing the ith trait T^*_i with the jth method M_j, and random error component U_{ij} associated with the measurement of a particular combination of a given trait and a given method, such that

$$Y_{ij} = \lambda_i T^*_i + \lambda_j M_j + U_{iij}$$

Note that, as above, we have written the random error component as U_{ij} in order to highlight the fact that in multiple indicator models, the random error component contains both random measurement error and specific sources of variance [see the measurement model in Figure 4.1 (b)]. Note also, as I clarify below, that this model may be viewed as the *reduced form* of a model that formulates the trait factor T^*_i and the method factor M_j as components of the true score T_i.

The λ-coefficients in the above model are factor pattern coefficients representing the regression relationships between the observed scores and the latent variables. In the standardized form of the model, they become "path coefficients" and interpretable as factor loadings. Following the terminology suggested by Heise and Bohrnstedt (1970), I refer to the standardized λ-coefficient linking a given measure to reliable trait variation as the *validity coefficient* and to the standardized λ-coefficient representing reliable method variation as the *invalidity coefficient*. In other words, both "valid" and "invalid" sources of variation contribute to reliability, and the purpose of this modeling strategy is to separate these two components (see Alwin, 1974).

Given this model, along with the assumption that the three components are independent, the decomposition of the population variance of a given measure (i.e., a particular trait-method combination in some population of interest) into components of reliability, validity, and invalidity of the indicator can be written as follows:

$$1.0 = \underbrace{\left[\frac{VAR[Y_{ij}]}{VAR[Y_{ij}]}\right]}_{} = \overbrace{\left[\frac{\lambda_i^2\, VAR[T_i^*]}{VAR[Y_{ij}]}\right]}^{Validity} + \overbrace{\left[\frac{\lambda_j^2\, VAR[M_j]}{VAR[Y_{ij}]}\right]}^{Invalidity} + \left[\frac{VAR[U_{ij}]}{VAR[Y_{ij}]}\right]$$

$$\underbrace{\hspace{7cm}}_{Reliability} \qquad \underbrace{\hspace{2cm}}_{Unreliability}$$

These identities show, based on the factor analytic decomposition of variance, that (1) the *reliability* of a given measure is equal to the sum of the true trait and method variance expressed as a proportion of the total observed response variance, (2) the *validity* of a measure is equal to the variance of the trait variance as a proportion of the total variance, and (3) the *invalidity* of a measure is equal to the variance of the method variance as a proportion of the total variance. The objective of the analysis of MTMM data following this tradition is to estimate the reliability of measures and its "valid" and "invalid" components using multiple methods of measurement in the assessment of multiple traits.

In contrast to these conventions, the "true score" approach—or what we have here referred to as the "modified CTST model"—conceives of trait and method factors as components of the true score. This model is formulated as follows: $Y_g = T_g + E_g$, where $T_g = T_g^* + M_g$ (see Chapter 3, as well as the prior discussion of this formulation in the present chapter). In this reformulation of the CTST model the trait and method factors are components, not of the observed score (as in the case of the factor analytic model), but of the true score. This approach, which is consistent with Saris and Andrews (1991), therefore analyzes the trait and method contributions to *true score variance* rather than *reliability*.

The path diagram in Figure 4.4 for a hypothetical trait-method combination clarifies the notation we use to describe these components. We note, first, that the results here are in standardized form, such that the coefficients in the diagram can be interpreted as path coefficients. Note also, that we use the notation that is often used in the factor analysis literature. In brief, the coefficients in this figure are defined according to the following constraints:

$$1.0 = b_{T^*}^2 + b_M^2$$
$$1.0 = h^2 + u^2$$
$$h^2 = [b_{T^*}\, h\,]^2 + [b_M\, h\,]^2$$

Note that the interpretation of these parameters and the decomposition of response variance involved is straightforward. The reliability in this case—denoted h^2—corresponds to the quantity representing the *communality* in the common-factor analysis tradition. It represents the proportion of the variance in the observed measure that is accounted for by the common factors in the model—in this case a trait factor and a method factor. The quantity u^2 denotes the variance of the random error—called *uniqueness* in the common-factor analysis tradition. As indicated in

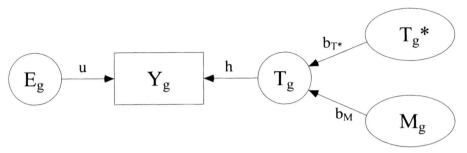

Figure 4.4. Path diagram showing the relationship between the common-factor and classical true-score representations of parameters in the multitrait-multimethod model for a single measure.

the discussion above, the reliability (or communality) can be partitioned into two orthogonal parts: $\lambda_{T*}^2 = [b_{T*} h]^2$ and $\lambda_M^2 = [b_M h]^2$, the first of which is due to the trait factor and interpreted as "validity" of measurement, and the second of which is due to the method factor, interpreted as "invalidity" of measurement.

Using the modified CTST formulation, Saris and Andrews (1991) introduce a new terminology—referring to "true-score validity" and "true-score invalidity"—in contrast to the manner in which we use the terms "validity" and "invalidity" above. It is perhaps useful therefore to reproduce the exact definitions of these terms in order to avoid any confusion that may be introduced by different habits. The following (see also Table 4.6) gives a listing of the ways in which key terms are used here:

(1) Reliability h^2
(2) Index of reliability h
(3) True-score validity b_{T*}
(4) Indicator validity $h\,b_{T*}$
(5) True-score invalidity b_M
(6) Indicator invalidity $h\,b_M$

The reader will see from these definitions that the key difference between indicator validity (invalidity) and true-score validity (invalidity) simply involves a rescaling by the index of reliability. In other words, by dividing the more traditional factor analytic conception of validity and invalidity by the square root of reliability, one arrives at components of validity that are disattenuated in the sense that they are free of the influence of the level of reliability. This has certain advantages, as we shall see from the example presented below. This redefinition of "validity" and "invalidity" as components of true variance (rather than observed variance) can improve the interpretation of the results, but users of this technique will have to judge for themselves. It is important to note that this rescaling of the factor loadings of the factor model does not affect the ability of the model to account for the data. In the words of Saris and Andrews (1991, p. 591), "it is rather arbitrary whether we use the one or the other parameterization," since "the fit of the model will always be the same for both."

4.6   AN EXAMPLE

It is a relatively rare occurrence where one finds data in which the design requirements of the MTMM approach are met (but see Andrews, 1984; Scherpenzeel and
Saris, 1997). Here I present an example of the use of the MTMM approach consisting of a comparison of measures using 7- and 11-category rating scales in the
measurement of life satisfaction collected by Angus Campbell and Philip Converse
(see Campbell and Converse, 1980). This comparison is based on a 1978 national
survey conducted by the Survey Research Center of the University of Michigan in
which 17 domains of life satisfaction were each measured using these two types of
rating scales. The 1978 Quality of Life (QoL) survey consists of a probability sample
(n = 3,692) of persons 18 years of age and older living in households (excluding
those on military reservations) within the coterminous United States. Interviews
were conducted during June through August 1978. The original sample of approximately 4,870 occupied housing units, constituting two independently chosen multistage area probability samples, was used to represent the noninstitutionalized adult
population of the United States. The overall completion rate was approximately 76
percent. Sampling and other procedural details are given in Campbell and Converse
(1980).[4]

The design of this survey included multiple measures of several domains of life
satisfaction. This design permits the analysis of both "trait" and "method" components of variation in each measure, given that each domain was assessed using multiple response formats. Seventeen domains of satisfaction were assessed, satisfaction
with the respondent's community, neighborhood, place of residence (dwelling unit),
life in the United States today, education received, present job (for those persons
who were employed), being a housewife (for unemployed women), ways to spend
spare time, personal health, family's present income, standard of living, savings and
investments, friendships, marriage (for those married), family life, self as a person,
and life as a whole.

All of these measures were assessed using both 7-point and 11-point response
scales. Three of these domains—place of residence, standard of living, and life
as a whole—were rated using three separate scales. The three methods used were
(1) a 7-point "satisfied-dissatisfied" scale, (2) a 7-point "delighted-terrible" scale,
and (3) an 11-point "feeling thermometer." The order of presentation of measurement
approaches was, unfortunately, the same across topics and across respondents. This
confounds question context and measurement format, and the possibility of respondent conditioning could affect the results (see Schaeffer and Presser, 2003), a subject
to which we return in our critique of the MTMM approach below. Also, the methods
of measurement differed in the extent of verbal labelling of response options. The

---

[4]The following discussion presents a reanalysis of data discussed by Alwin (1997). The reanalysis
employs improved methods of handling missing data, namely the use of full-information maximum-
likelihood methods. The reanalysis also presents two sets of results: (1) the contribution of trait and
method factors to reliability, and (2) the contribution of trait and method factors to true-score variance.

7-point "satisfied-dissatisfied" scale labeled only the endpoints and midpoint, the 7-point "delighted-terrible" scale presented a fully labeled set of response options, and the 11-point feeling thermometer labeled only the endpoints. Thus, the issues of number of scale points and extensiveness of verbal labeling are also confounded in the results presented here, but this is undoubtedly typical of such multitrait-multimethod designs for measurement.

The data described above were analyzed using the structural equation models, implemented using the AMOS software (Arbuckle and Wothke, 1999). Because of the problematic nature of the scale of survey measures we employed two types of correlational methods—those based on Pearson correlational methods that assume interval-level continuous variables and polychoric correlational methods for ordinal-polytomous variables, as discussed in Chapter 3. Both assume there is an underlying continuous variable Y*, corresponding to the observed variable Y, that is normally distributed. In the first case it is assumed that there is uniform correspondence between the continuous variable, Y*, and the observed variable, that is, Y = Y*. In the second case it is assumed that the nature of the latent variable is also continuous, as above, but the observed variable is ordinal. In this case the polychoric correlation estimates the association between two Y* variables, given their cross-tabulation. What distinguishes these two cases is the crudeness of measurement at the level of the observed responses.

Further, in the case of the Pearson-based correlations, we employed two different methods for handling incomplete data. First, we estimated the model where the incomplete data were dealt with using full-information maximum-likelihood (FIML) direct estimation techniques. This approach, described in detail by Wothke (2000) and implemented in AMOS (Arbuckle and Wothke, 1999), yields model parameter estimates that are both consistent and efficient if the incomplete data are *missing at random* (MAR) (Wothke, 2000). This approach also allowed us to include the three measures that were asked only of subsets of respondents, namely satisfaction with job, housework and marriage. Second, we estimated the model using *listwise-present* data, in order to compare the Pearson-based results with the model estimates using polychoric correlations. Thus, three sets of estimates of the model are presented—in Tables 4.3, 4.4, and 4.5—for the Pearson FIML, Pearson listwise, and polychoric listwise correlations, respectively. Notice that in the two sets of results involving listwise-present data, we by necessity excluded the measures of the three constructs assessed in subsets of respondents. We wanted to compare the FIML and listwise estimates, as well as compare the Pearson- and polychoric-based estimates. Given the appropriateness of the FIML approach, there would have been no need to compute results based on the listwise approach; however, present-day technology does not permit the application of the FIML technique in the computation of polychoric correlations.

The covariance matrix among the variables was analyzed in all three cases using the CTST representation of the MTMM model given above. The overall goal of this approach is to partition the sample response variances of the measures into three parts: (1) reliable trait variance, (2) reliable method variance, and (3) unreliable variance (Alwin, 1974; Groves, 1989; Saris and Andrews, 1991). For each set of results three coefficients are presented for each trait-method combination: (1) the estimated reliability for each item, (2) the standardized factor pattern coefficient linking the trait factor to the true score in question, and (3) the standardized factor pattern coefficient linking the method factor involved to the true score in question. The definitions we use follow those given by Saris and Andrews (1991, pp. 581–582), in which the coefficients of *validity* and *invalidity* are scaled such that the square of the two loadings sum to 1.0. In the tables these are denoted as h^2, b_{T*}, and b_M, respectively (see notation used in Figure 4.4 discussed above), a slight variant of the notation used by Saris and Andrews (1991).

In Tables 4.3 and 4.4 we present the model estimates of h^2, b_{T*}, and b_M for the model based on Pearson correlations. For each of these sets of results we also present the sample-based estimate of the likelihood-ratio statistic, denoted CMIN, for the model, along with several measures of relative fit. The CMIN test statistic is evaluated using the χ^2 probability distribution with degrees of freedom (df). Because the likelihood-ratio statistic is directly dependent on the sample size, it cannot by itself be used to evaluate the fit of the model. Even trivially different models would be rejected because of a highly significant difference on the χ^2 probability distribution, owing to nothing more than the large sample size. The Bentler and Bonett (1980) "normed fit index" (NFI) permits the comparison of models free of the influence of sample size (see Brown and Cudeck, 1993). In addition, we also employ the RMSEA index in the evaluation of whether the results provide a suitable fit of the model to the data (Browne and Cudeck, 1993).

The two sets of results employing traditional Pearson correlational data—the FIML results in Table 4.3 and the "listwise deleted" or "complete cases" results in Table 4.4—show very similar patterns. This is as expected, given the overlap in sample cases. The main difference between the results involves the fact that the FIML results provide estimates for the three questions that were asked of only a subsample of respondents—the measures of satisfaction with job, housework, and marriage. Although the CMIN values are different in the two cases, owing to the different case bases, the relative fit to the data is virtually identical in the two cases. The NFI estimates of .96 and .97 reflect an unusually good fit. As noted, Browne and Cudeck (1993) indicate that a value of the root mean square error of approximation (RMSEA) of about .05 or less represents a good fit of the model in relation to the degrees of freedom, and with this model we have achieved that level of fit in each case.

The results in Table 4.4 may be compared with those presented in Table 4.5, which employ polychoric correlations in the estimation of the model. These two sets of results are based on essentially the same number of cases. The fit of the model to the data is somewhat better in the case of the Pearson correlational data, although in terms of the relative fit statistics the two sets of results are comparable. What is most evident in this comparison is that the estimates of reliability are, as expected, slightly higher using the polychoric compared to the Pearson correlations—averages of .681 versus .611, .751 versus .689, and .786 versus .826 for the 7-point satisfaction scale, the 7-point "delighted-terrible" scale, and the 11-point feeling thermometers, respectively.

In order to compare the three sets of results for the model, we examined a set of "matched pairs" for each comparable coefficient across a given pair of models

Table 4.3.   Reliability, validity, and invalidity of life satisfaction measures: 1978 Quality of Life Survey—
full-information maximum-likelihood (N = 3,692)

Concept	7-Point Satisfaction Scale			7-Point Delighted-Terrible Scale			11-Point Thermometer		
	h^2	$b_{T\bullet}$	b_M	h^2	$b_{T\bullet}$	b_M	h^2	$b_{T\bullet}$	b_M
Community	.523	.904	.427				.878	.991	.132
Neighborhood	.603	.923	.384				.896	.990	.141
Dwelling	.661	.963	.270	.740	.953	.304	.796	.988	.156
United States	.479	.901	.434				.583	.987	.163
Education	.526	.955	.297				.806	.965	.261
Health	.604	.973	.232				.944	.963	.271
Time	.612	.923	.386				.714	.965	.261
Friends	.544	.952	.305				.713	.969	.246
Family	.531	.914	.406				.842	.989	.149
Income	.707	.985	.171				.835	.874	.487
Standard of living	.689	.983	.185	.746	.896	.445	.797	.876	.482
Savings	.828	.991	.137				.768	.952	.307
Life	.628	.911	.413	.600	.909	.417	.711	.975	.222
Self	.625	.939	.344				.736	.981	.195
Job	.663	.961	.277				.733	.930	.367
Housework	.734	.957	.291				.664	.965	.261
Marriage	.687	.972	.236				.912	.993	.116
Average	.626	.947	.306	.695	.919	.389	.784	.962	.248
CMIN	2,933.8								
df	456								
NFI	0.96								
RMSEA	0.040								

(see Blalock, 1972, p. 233). In the analysis of differences in strategies of reliability estimation presented in this section we employ a direct pair-by-pair comparison by examining the difference score for each pair and testing the hypothesis that the mean of the pair-by-pair differences is zero. The key results reported here are (1) the comparison between 7- and 11-point scales for both polychoric and Pearson-based correlations, and (2) a comparison between the polychoric and Pearson-based results within the 7-point and 11-point estimates. These four sets of results compare the coefficients of "reliability," "validity," and "invalidity" for each indicator and are given in Table 4.6. Note that the table also presents the "true score validity" and "true score invalidity" as suggested by Saris and Andrews (1991).

The comparison of the Pearson-based and polychoric correlations illustrates an expectation we advanced in Chapter 3 and a finding that we will encounter again

Table 4.4. Reliability, validity, and invalidity of life satisfaction measures: 1978 Quality of Life Survey—Analysis of Pearson correlations based on complete cases (N = 2,987)

Concept	7-Point Satisfaction Scale			7-Point Delighted-Terrible Scale			11-Point Thermometer		
	h^2	$b_{T•}$	b_M	h^2	$b_{T•}$	b_M	h^2	$b_{T•}$	b_M
Community	0.547	0.882	0.471				0.890	0.993	0.122
Neighborhood	0.617	0.907	0.420				0.906	0.991	0.136
Dwelling	0.671	0.958	0.286	0.731	0.949	0.315	0.800	0.989	0.151
United States	0.479	0.891	0.454				0.579	0.985	0.170
Education	0.542	0.964	0.264				0.796	0.963	0.270
Health	0.578	0.971	0.241				0.961	0.955	0.295
Time	0.607	0.934	0.359				0.722	0.963	0.271
Friends	0.536	0.969	0.249				0.722	0.957	0.291
Family	0.558	0.934	0.358				0.771	0.974	0.226
Income	0.689	0.989	0.150				0.838	0.877	0.481
Standard of living	0.689	0.989	0.145	0.752	0.883	0.470	0.820	0.866	0.500
Savings	0.816	0.993	0.120				0.780	0.953	0.302
Life	0.610	0.934	0.357	0.584	0.911	0.413	0.702	0.964	0.267
Self	0.623	0.958	0.287				0.719	0.969	0.245
Average	0.611	0.948	0.297	0.689	0.914	0.399	0.786	0.957	0.266
CMIN	1,741.8								
df	312								
NFI	0.97								
RMSEA	0.039								

in Chapter 6, namely, that the level of estimated reliability depends upon whether Pearson or polychoric correlations are used as input to the analysis. Estimates of reliability and indicator validity are higher when the polychoric correlations are used as the basis for the computations. This is not the case with the estimates of *true-score validity*, *true-score invalidity* and *indicator invalidity*, however, and there are some other inconsistencies as well. We expect that the differences between the Pearson and polychoric correlations will be less pronounced in the case of the 11-point scales, and although there are some noticeable differences, they are generally quite small. Invalidity estimates—either indicator invalidity or true score invalidity—are trivial, as are estimates of true score validity. The differences between the types of correlational estimates with respect to indicator validity (either for 7-point or 11-point scales) are probably due entirely to its dependence on reliability.

With respect to the response form used in the case of the estimates of reliability, 11-point scales have higher reliabilities in all cases, whether the polychoric or Pearson-based estimates are used. This strongly supports the hypothesis based on information theory, which posits that questions with more response categories permit the transmission of information more reliably (see Alwin, 1992, 1997). In most cases the 11-point scales have slightly higher validity coefficients and lower invalidity coefficients, indicating that traits measured using more response categories are

Table 4.5. Reliability, validity, and invalidity of life satisfaction measures: 1978 Quality of Life Survey—analysis of polychoric correlations based on complete cases (N = 2,984)

Concept	7-Point Satisfaction Scale			7-Point Delighted-Terrible Scale			11-Point Thermometer		
	h^2	b_{T*}	b_M	h^2	b_{T*}	b_M	h^2	b_{T*}	b_M
Community	0.648	0.850	0.528				0.939	0.991	0.132
Neighborhood	0.740	0.864	0.504				0.959	0.991	0.135
Dwelling	0.752	0.949	0.317	0.784	0.938	0.345	0.825	0.985	0.172
United States	0.549	0.902	0.433				0.627	0.980	0.198
Education	0.601	0.979	0.203				0.825	0.957	0.290
Health	0.645	0.976	0.216				0.971	0.954	0.300
Time	0.677	0.962	0.273				0.765	0.940	0.341
Friends	0.617	0.985	0.174				0.790	0.929	0.369
Family	0.615	0.960	0.279				0.838	0.951	0.310
Income	0.721	0.990	0.140				0.869	0.895	0.446
Standard of living	0.743	0.992	0.129	0.811	0.875	0.484	0.845	0.881	0.472
Savings	0.858	0.996	0.093	0.656	0.901	0.434	0.824	0.944	0.331
Life	0.670	0.959	0.285				0.731	0.934	0.356
Self	0.699	0.981	0.193				0.755	0.941	0.338
Average	0.681	0.953	0.269	0.751	0.905	0.421	0.826	0.948	0.299

CMIN 2,333.9
df 312
NFI 0.96
RMSEA 0.047

Table 4.6. Comparison of reliability, validity and invalidity estimates by response form and type of correlations analyzed

Concept	Notation	RESPONSE FORM		t-statistic[1]	p-value
		7-Point Satisfaction	11-Point Thermometer		
MTMM Model Using Polychoric Correlations					
Reliability	h^2	0.681	0.826	5.14	0.000
Index of Reliability	h	0.824	0.907	5.27	0.000
True Score Validity	$b_{T\cdot}$	0.953	0.948	0.24	0.817
Indicator Validity	$hb_{T\cdot}$	0.786	0.861	2.72	0.017
True Score Invalidity	b_M	0.269	0.299	0.47	0.645
Indicator Invalidity	hb_M	0.220	0.271	0.93	0.371
MTMM Model Using Pearson Correlations					
Reliability	h^2	0.611	0.786	5.61	0.000
Index of Reliability	h	0.780	0.885	5.80	0.000
True Score Validity	$b_{T\cdot}$	0.948	0.957	0.47	0.644
Indicator Validity	$hb_{T\cdot}$	0.741	0.847	3.88	0.002
True Score Invalidity	b_M	0.297	0.266	0.52	0.610
Indicator Invalidity	hb_M	0.228	0.236	0.17	0.870

Concept	Notation	TYPE OF CORRELATION ANALYZED		t-statistic	p-value
		Pearson	Polychoric		
MTMM Model for 7-Point Satisfaction Scale					
Reliability	h^2	0.611	0.681	10.83	0.000
Index of Reliability	h	0.780	0.824	10.59	0.000
True Score Validity	$b_{T\cdot}$	0.948	0.953	0.85	0.408
Indicator Validity	$hb_{T\cdot}$	0.741	0.786	10.74	0.000
True Score Invalidity	b_M	0.297	0.269	1.87	0.085
Indicator Invalidity	hb_M	0.228	0.220	0.60	0.556
MTMM Model for 11-Point Satisfaction Scale					
Reliability	h^2	0.786	0.826	8.68	0.000
Index of Reliability	h	0.885	0.907	8.48	0.000
True Score Validity	$b_{T\cdot}$	0.957	0.948	2.06	0.060
Indicator Validity	$hb_{T\cdot}$	0.847	0.861	6.99	0.000
True Score Invalidity	b_M	0.266	0.299	2.79	0.015
Indicator Invalidity	hb_M	0.236	0.271	3.805	0.002

[1]All tests are two-tailed t tests for matched pairs with 13 df, testing the hypothesis that the mean difference is zero.

more highly correlated with the underlying trait than is the case with those measured by 7-point scales. However, at the true-score level, there are no statistically significant differences in validity or invalidity, suggesting that differences between the two forms of measurement are entirely at the level of measurement unreliability. Although the coefficients of invalidity are generally lower in the case of the 11-point scales, the differences are not statistically significant. This runs counter to the hypothesis that questions with more response categories are more vulnerable to the effects of systematic response error, such as response sets. The results decidedly support the use of 11-point scales over 7-point scales in the measurement of life satisfaction, if reliability is used as the criterion of evaluation. With respect to the criterion of validity, results here support the conclusion that there is no difference between the two forms of measurement. These results are consistent with those presented by Alwin (1997), although we have approached the data somewhat differently.

4.6.1   Interpretation

This example illustrates the advantages of an MTMM investigation focusing on the reliability of single survey questions and avoiding the complexities involved in formulating this set of issues within the framework of internal consistency estimates of reliability, e.g., coefficient α, which depend not only on the extent of item-level reliability, but on the number of items and the unidimensionality of scale components as well (Cronbach, 1951). Second, this study has made the comparison of response-scale length with respect to one domain of content, self-assessed satisfaction with various aspects of life, and has not confounded the assessment of the relation between the number of response categories and reliability with the types of concepts being measured (cf. Andrews, 1984). Third, although this study has compared only two scale lengths, 7 and 11 categories, the comparison focuses on a critical issue in the debate about whether it is efficacious to extend more traditional approaches. The 7-category scale is by far the longest type of response scale used in most survey measurement of subjective variables. Scales with more than seven categories may be impractical in many situations, and there is an issue of whether such decisions can be justified in terms of reliability of measurement. Of course, the choice of response scales should also be based on construct validity and the theoretical appropriateness of a particular type of measure may be overriding concerns. Increasingly, matters of practical significance are important, given the relationship of survey costs to the amount of information obtained in surveys. As noted earlier, there are two methodological issues that detract from the clarity of these results. The first, a criticism advanced by Schaeffer and Presser (2003), is that the order of presentation of measurement forms was the same across topics across respondents. The 7-point satisfaction scales were given first, the thermometers were given second, somewhat later in the questionnaire, and the "delighted-terrible" scales last, near the end of the questionnaire. The nonbalanced nature of the design does not permit distinguishing between form of measurement and measurement context, in that there may be some form of conditioning effect. For example, respondents may have the answers to the earlier questions still relatively available in memory, and this may affect their responses on the later tasks, namely the thermometer. This would work to increase the true correlations among measures in the monotrait-hetereomethod block, and the net result would be stronger trait correlations for the later set of measures and less influence of method factors. However, in this case the very last set of measures employed are the 7-point "delighted-terrible" scales, and the hypothesis is disconfirmed in that case, i.e., they do not have lower method components, nor do they have higher validity coefficients. In light of this evidence, this explanation for our main results hardly seems credible.

A second problem of interpretation in this study involves the fact that the three forms of measurement included different degrees of labeling. The 7-point "satisfied vs. dissatisfied" scale labeled the endpoints and midpoint, the 11-point thermometer labeled only the endpoints, whereas the 7-point "delighted-terrible" scale presented fully labeled set of response options. There is clearly confounding between number of response categories and the nature of the verbal labels attached. However, within the present context it is extremely difficult to sort out. Past research may be the only basis for guidance. Past research, however, has shown that the extensiveness of labeling does not appear to be related to the reliability of measurement (Andrews, 1984; Alwin and Krosnick, 1991).

If these alternative explanations can be ruled out in the present case, the results suggest that in the measurement of satisfaction with various domains of life, 11-point scales are clearly more reliable than comparable 7-point scales. Moreover, measures involving 11-point scales have higher correlations with underlying trait components and lower correlations with method components underlying the measures. In addition to being more reliable, then, 11-point scales are no more vulnerable to response sets, as conceptualized here in terms of shared method variance, when compared with 7-point scales. These findings support the conclusion that questions with more response categories may be preferable to fewer, when feasible, in that they produce more reliable and valid measures. Reductions in measurement errors in this sense can result in more powerful statistical decision-making, less biased correlation and regression estimates, and greater confidence in the usefulness of the data. We return to the issue of the reliability of questions with different numbers of response categories in Chapter 9.

4.7 CRITIQUE OF THE MTMM APPROACH

The MTMM model has clear advantages over the multiple indicators approach inasmuch as it attempts to model one source of specific variance. However, a skeptic could easily argue that it is virtually impossible to estimate reliability from cross-sectional surveys because of the failure to meet the assumptions of the model and the potentially biasing effects of other forms of specific variance. While I am a strong supporter of the MTMM tradition of analysis and have written extensively about its advantages (see Alwin, 1974, 1989, 1997; Alwin and Jackson, 1979), these designs are very difficult to implement given the rarity of having multiple traits each measured simultaneously by multiple methods in most large-scale surveys. Moreover, due to the assumptions involved, the interpretation of the results in terms of reliability estimation is problematic. To be specific, the reliability models for cross-sectional data place several constraints on the data that may not be realistic.

There are two critical assumptions that we focus on here—first the assumption that the errors of the measures are independent of one another, and second that the measures are *bivocal*, that is, there are only two sources of reliable variation among the measures, namely, the trait variable and the method effect being measured. These assumptions rule out, for example, the operation of other systematic factors that may produce correlations among the measures. For example, in cross-sectional survey applications of the MTMM model it may be impossible to assume that measurement errors are independent, since even though similar questions are not given in sequence or included in the same battery, they do appear in relative proximity. The assumptions for reliability estimation in cross-sectional studies rule out the operation of other types of correlated error. And while the MTMM model allows for correlated errors that stem from commonality of method, this does not completely exhaust the possibilities. Obviously, it does not rule out memory in the organization of responses to multiple efforts to measure the same thing. Respondents are fully cognizant of the answers they have given to previous questions, so there is the possibility that memory operates to distort the degree of consistency in responses, consistency that is reflected in the *validity* component of the MTMM model.

To make the argument somewhat more formally, the reader should recall that the MTMM approach relies on the assumptions of the common factor model in equating the communality with reliability (see also Heise and Bohrnstedt, 1970). In the common-factor model the variance of any measure is the sum of two variance components—communality and uniqueness. A variable's *uniqueness* is the complement to the common parts of the data (the *communality*) and is thought of as being composed of two independent parts, one representing *specific variation* and one representing *random measurement error* variation. Using the traditional common factor notation for this, this can be restated as follows. The variable's *communality*, denoted h_j^2, is the proportion of its total variation that is due to common sources of variation. It's *uniqueness*, denoted u_j^2, is the complement of the communality, that is, $u_j^2 = 1.0 - h_j^2$. The uniqueness is composed of specific variance, s_j^2, and random error variance, e_j^2. Specific variance is *reliable* variance, and thus the reliability of the variable is not only due to the common variance, but to specific variance as well. In common factor analytic notation, the reliability of variable j can be expressed as $r_j^2 = h_j^2 + s_j^2$. Unfortunately, because specific variance is thought to be independent (uncorrelated with) sources of common variance, it becomes confounded with measurement error variance. Because of the presence of specific variance in most measures, it is virtually impossible to use the traditional form of the common factor model as a basis for reliability estimation (see Alwin, 1989; Alwin and Jackson, 1979). The problem is that the common factor model typically does not permit the partitioning of u_j^2 into its components, s_j^2 and e_j^2. In the absence of specific variance (what we here refer to as the "multiple measures" model), classical reliability models may be viewed as a special case of the common factor model, but in general it is risky to assume that $u_j^2 = e_j^2$ (Alwin and Jackson, 1979).

There have been other criticisms of the MTMM approach. One of these, advanced by Schaeffer and Presser (2003), is that the order of presentation of measurement forms is typically the same across respondents and that the method of analysis does not take this into account. This may appear on the face of it to be a trivial objection, given that "the order of presentation" issue affects all studies of method effects in surveys [e.g., the question form studies by Schuman and Presser (1981)] and virtually all survey results generally. It may be viewed as particularly problematic in "within-subjects" designs, for the reasons given above, although hardly any implementation of the survey method of which I am aware randomizes the order of presentation of survey questions to the respondent. However, as noted above, the nonbalanced nature of the design does not permit distinguishing between form of measurement and measurement context, in that there may be some form of conditioning effect. For example, respondents may have the answers to the earlier questions still relatively available in memory, and this may affect their responses on the later tasks. This argument is compatible with the objections given above regarding the potential for memory to heighten the overall assessed consistency of measurement, i.e., reliability, in the MTMM design.

4.8 WHERE ARE WE?

We argued from the beginning of this book that in order to assess the extent of measurement error *for single survey questions* it was necessary not only to develop a theoretical model for conceptualizing the behavior of measurement errors in survey questionnaires, but also to develop designs that permit the estimation of model parameters specifying the linkages between the concepts being measured and errors of measurement. This chapter began by reviewing some of the limitations of the ways in which the classical theoretical model for reliability (CTST) is applied to survey data (developed more fully in Chapter 3). I argued specifically that for purposes of evaluating the reliability of standard measurement approaches used in survey research, it is important that we move beyond estimates of internal consistency reliability (ICR) and adopt a strategy that focuses on the reliability of *individual survey measures.*

The following four elements of the classical ICR model are problematic for purposes of examining the *reliability of single-survey questions*: (1) the assumption of the existence of multiple, or replicate, measures of the same true score; (2) its failure to distinguish multiple sources of systematic variance, that is, it does not separate "valid" true-score variation from "invalid" true-score variation; (3) its failure to distinguish multiple sources of random variance, that is, it does not separate "unreliable" sources of variation from random sources of "reliable" variation; and (4) the assumption of statistical independence of measurement errors across measures.

The discussion in this chapter led to the observation that the analysis of measurement errors in cross-sectional survey data could be improved by an extension of the classical model to include the conceptualization of systematic as well as random errors in modeling the reliability of individual survey questions. Within this type of design the chapter has considered ways in which *multiple-indicators* designs can be estimated in cross-sectional data. Specifically, the multitrait-multimethod measurement design was presented as a possible solution to one aspect of these problems, namely the incorporation of systematic sources of error connected to methods of measurement. However, this approach is quite unique and it is extraordinarily rare that one can find studies with these characteristics. Moreover, by specifying one source of specific variance—that associated with the method of measurement—it does not take care of other types of specific variance. There is room for other approaches that may have several advantages over the MTMM approach (see Saris and Andrews, 1991). We explore one of these—the quasi-simplex approach using longitudinal data—in the following chapters.

CHAPTER FIVE

Longitudinal Methods for Reliability Estimation

The investigation of response unreliability is an almost totally undeveloped field, because of the lack of mathematical models to encompass both unreliability and change.
James Coleman, "The mathematical study of change" (1968, p. 475)

The topic of measurement error estimation is a complex one, and there are, in general, no simple solutions, except where the assumption of independence of errors on repeated measuring can be made.
Harvey Goldstein, *Multilevel statistical models* (1995, p. 142)

In the previous chapter we concluded that while there are a number of advantages to using cross-sectional survey designs to generate estimates of measurement reliability, there are several key disadvantages as well. A principal disadvantage stems from the inability to design replicate measures in cross-sectional studies that meet the requirements of the CTST model, that is, the dual assumptions of *univocity* and the assumption of independence of errors. Further, I made an effort to point out the limitations of the use of *multiple indicators* in reliability estimation—that is, the use of approximations to replicate measures that include multiple sources of reliable variance—stemming from the confounding of random measurement error with reliable specific variance. The problems of such designs for reliability estimation have been recognized for some time and the limitations to the reliability interpretations placed on "multiple indicators" models are increasingly becoming better known (e.g., see Alwin and Jackson, 1979; Goldstein, 1995). Even when one is able to employ the multitrait-multimethod (MTMM) approach in cross-sectional designs (i.e., measures of the same trait or concept that vary key features of the measures), and even if one is willing to extend the logic of the CTST model to include multiple sources of "true variance," the problems do not go away (see Chapter 4). The assumption of the independence of errors continues to be problematic. Specifically, in order to interpret the parameters of such models in terms of reliability of measurement, it is essential to be able to rule out the operation of memory and other forms of consistency motivation, in the organization of responses to multiple measures of the same construct.

In this chapter I discuss an alternative to cross-sectional designs for reliability estimation. This approach uses the multiple-wave reinterview design, variously referred to as the test-retest, repeated measures, or panel design, involving *multiple measures* obtained over time (see Moser and Kalton, 1972, pp. 353–354).[1] Such longitudinal designs also have problems for the purpose of estimating reliability, and here we present some of the ways of coping with these problems. I begin with a discussion of the simple test-retest (i.e., two-wave panel) approach, which in the present context uses a single reinterview in order to capture replicate measures (see Siegel and Hodge, 1968). As we shall see, in order for this approach to be useful in the estimation of the reliability of measurement, we must assume that there is no change in the underlying quantity being measured. With two waves of a panel study, the assumption of no change, or even perfect correlational stability, is unrealistic. This chapter discusses how this problem can be overcome by incorporating three or more waves of data and modeling the change, thus separating unreliability and true change (Coleman, 1968; Heise, 1969; Jöreskog, 1970; Wiggins, 1973). A second major problem with the test-retest method is that if the interval between measurements is too short the consistency in measurement may be due to memory and related processes. This problem can be remedied through the use of design strategies that select reinterview measurements that reduce the likelihood of memory effects.

5.1   THE TEST-RETEST METHOD

In order to introduce the idea of using longitudinal data in the estimation of the reliability of survey data we begin with a discussion of the classic test-retest method. In the test-retest case one has two measures, Y_t and Y_{t-1}, separated in time, specifically measures of T_t and T_{t-1}. In this case the correlation between the two measures is often taken as an estimate of reliability of both measures (e.g., Siegel and Hodge, 1968). Although this approach has been in wide use in some fields of the social and behavioral sciences, it has been suspect for many years among survey methodologists, due to some of the rather obvious problems. For example, Moser and Kalton (1972, pp. 353–354) had this to say about test-retest estimates of reliability:

> ... even if persons were to submit themselves to repeat questioning, a comparison of the two sets of results would hardly serve as an exact test of reliability, since they could not be regarded as independent. At the retest, respondents may remember their first answers and give consistent retest answers, an action which would make the test appear more reliable than is truly the case. Alternatively, the first questioning may make them think more about the survey subject ... or events occurring between the two tests may cause them to change their views on the subject. In any of these circumstances, the test and

---

[1]It is important to note, consistent with my usage in the previous chapter, that by *multiple measures* I refer specifically to identical measures replicated over occasions of measurement.

retest scores are not exactly comparable, so the difference between the two is a mixture of unreliability and change in the characteristic itself; the effect of this is that an underestimate of reliability is obtained.

Interestingly, the focus of this critique is based on one of the same sets of concerns we have raised about cross-sectional approaches to reliability estimation, viz., the failure to meet the assumption of independence of errors [see the quote from Goldstein (1995) above].

In the present discussion I clarify somewhat more formally the problems that Moser and Kalton (1972) and others have pointed to (1) the operation of memory or other forces that produce artificially high degrees of consistency and (2) the existence of change in the true score over the period between measurements, the effect of which is to attenuate estimates of reliability based on correlations between measures. The path diagrams in Figure 5.1 show a "replicate measures" (test-retest) model for two measures of a true score $(P = 2)$. Each path diagram in this figure [models (a) through (d)] is a different version of the two-wave panel model, constructed with the purpose of illustrating several possible scenarios. For purposes of discussion, we assume that in all cases Y_2 is a replicate of Y_1 (i.e., the repetition of an exactly worded question) asked in a reinterview of the same population studied at wave 1. Under some conditions, the correlation between Y_1 and Y_2 is an estimate of the reliability of measurement. The purpose of this discussion is to clarify under which conditions this is the case and under which conditions it is not the case.

Consider the model represented by the path diagram in Figure 5.1, model (a). Here the measurement model is $Y_1 = T_1 + E_1$ and $Y_2 = T_2 + E_2$, where the assumptions of CTST hold, that is, $COV(E_1,T_1) = COV(E_2,T_2) = COV(E_1,E_2) = 0$ in the population. Assume further that T_1 and T_2 are related by the principle of "essential tau-equivalence," that is, $T_2 = c + T_1$, where c is a fixed constant (see Lord and Novick, 1968, p. 50). In other words, T_1 and T_2 are linearly related and are correlated at 1.0. The proof of this is straightforward:

$$COR[T_1,T_2] = COV[T_1,T_2] / VAR[T_1]^{1/2} VAR[T_2]^{1/2}$$
$$= COV[T_1, c + T_1] / VAR[T_1]^{1/2} VAR[c + T_1]^{1/2}$$
$$= VAR [T_1] / VAR[T_1]^{1/2} VAR[T_1]^{1/2}$$
$$= 1.0$$

Note that if the constant $c = 0$, that is, if $T_1 = T_2 = T$, we may skip a few steps in working through the above algebra. Under this model, the correlation between Y_1 and Y_2 can be written as follows:

$$COR[Y_1,Y_2] = COV[Y_1,Y_2] / VAR[Y_1]^{1/2} VAR[Y_2]^{1/2}$$
$$= COV[T_1,T_2] / VAR[Y_1]^{1/2} VAR[Y_2]^{1/2}$$
$$= VAR [T_1] / VAR[Y_1]^{1/2} VAR[Y_2]^{1/2}$$

In this case, since $COV[Y_1,Y_2] = COV[T_1,T_2] = VAR[T]$, then $\rho^2[Y_1] = COV[Y_1,Y_2] / VAR[Y_1]$ and $\rho^2[Y_2] = COV[Y_1,Y_2] / VAR[Y_2]$.

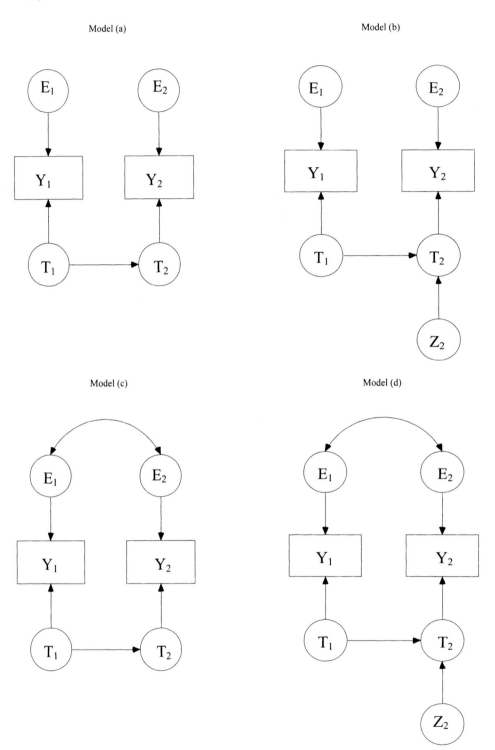

Figure 5.1.   Path diagrams for models of the two-wave panel design.

If it can be further assumed that $VAR[E_1] = VAR[E_2]$, that is, if the measures meet the definition of parallel measures (see Chapter 3), then the following equivalence holds:

$$COR[Y_1,Y_2] = VAR\ [T_1]\ /\ VAR\ [Y_1] = \rho^2$$

This can be seen as follows:

$$
\begin{aligned}
COR[Y_1,Y_2] &= VAR\ [T_1]\ /\ VAR[T_1 + E_1]^{1/2}\ VAR[T_2 + E_2]^{1/2} \\
&= VAR\ [T_1]\ /\ VAR[T_1 + E_1]^{1/2}\ VAR[c + T_1 + E_2]^{1/2} \\
&= VAR\ [T_1]\ /\ VAR[T_1 + E_1]^{1/2}\ VAR[T_1 + E_2]^{1/2} \\
&= VAR\ [T_1]\ /\ [VAR[T_1]^{1/2} + VAR[E_1]^{1/2}] \\
&\quad \times [VAR[T_1]^{1/2} + VAR[E_2]^{1/2}]
\end{aligned}
$$

Thus, if one is able to realistically assume that the true score of the variable at time 2 is the same except for no more than a constant shift in the distribution, this removes much of the problem, and estimates of the reliability of measurement at each wave can be obtained, as shown above. Further, if the measurement error variances are equal, then the simple correlation between Y_1 and Y_2 estimates the reliability of measurement of Y.

The classical literature on reliability estimation refers to "the correlation between truly parallel measurements taken in such a way that the person's true score does not change between them is often called *the coefficient of precision*" (Lord and Novick, 1968, p. 134). In this case the only source contributing to measurement error is the unreliability or imprecision of measurement. The assumption here, as is true in the case of cross-sectional designs, is that "if a measurement were taken twice and if no practice, fatigue, memory, or other factor affected repeated measurements," the correlation between the measures reflects the precision, or reliability, of measurement (Lord and Novick, 1968, p. 134). In practical situations in which there are in fact practice effects, fatigue, memory, or other factors contributing to the correlation between repeated measures, the test-retest correlation is not the appropriate estimate of reliability.

Indeed, I would argue that the case of model (a) of Figure 5.1 represents a relatively rare phenomenon—but it does clarify the assumptions one is making when using the simple test-retest correlation to estimate the reliability of measurement. All other versions of the two-wave reinterview design in Figure 5.1 include one or both of the following: (1) correlated measurement errors due to the operation of memory and related phenomena and (2) true nonconstant change in the underlying true score. It can be shown that either or both of these conditions prevent the use of the simple correlation between Y_1 and Y_2 as an estimate of reliability.

Consider the model represented by the path diagram in model (b) of Figure 5.1. Here $Y_1 = T_1 + E_1$ and $Y_2 = T_2 + E_2$, and the assumptions $COV(E_1,T_1) = COV(E_2,T_2) = COV(E_1,E_2) = 0$ hold in the population. In this case, however, the underlying true-score distribution is not completely stable, or at least not stable up to a constant of change, as above. In this case we assume for purposes of exposition that T_2 can be written as a linear regression function of T_1, that is

$$T_2 = \beta_0 + \beta_{21} T_1 + Z_2$$

where $E[Z_2] = E[T_1,Z_2] = 0$. This feature of the model results in an attenuation of the correlation $COR[Y_1,Y_2]$, preventing it from being used as an estimate of reliability. This can be seen as follows:

$$\begin{aligned} COR[Y_1,Y_2] &= COV[Y_1,Y_2] /VAR[Y_1]^{1/2} VAR[Y_2]^{1/2} \\ &= COV[T_1,T_2] /VAR[Y_1]^{1/2} VAR[Y_2]^{1/2} \\ &= \beta_{21} VAR[T_1] /VAR[Y_1]^{1/2} VAR[Y_2]^{1/2} \end{aligned}$$

In the typical case β_{21} is less than 1.0, reflecting an "autoregressive" feature of change in the true score distribution from time 1 to time 2 (Alwin, 1988; McArdle and Nesselroade, 1994). In model (b) of Figure 5.1 this involves a type of "regression to the mean," indicating that the correlation between T_1 and T_2 is < 1.0. If $COR[T_1,T_2]$ is < 1.0, then it can be seen that $COR[Y_1,Y_2]$ is not an appropriate estimate of reliability.

It can be shown that if $VAR[Z_2] = 0$, $VAR[T_2] = \beta_{21}^2 VAR[T_1]$ and $COR[T_1,T_2] = 1.0$. This follows from the definition of β_{21} in the bivariate case, that is

$$\beta_{21} = COV[T_1,T_2] / VAR[T_1]$$

and rewriting $VAR[T_2]$ as

$$\begin{aligned} VAR[T_2] &= COV[T_1,T_2]^2 VAR[T_1] /VAR[T_1]^2 \\ &= COV[T_1,T_2]^2 /VAR[T_1] \end{aligned}$$

leaves us with the identity

$$\begin{aligned} 1.0 &= COV[T_1,T_2] /VAR[T_1]^{1/2} VAR[T_2]^{1/2} \\ &= COR[T_1,T_2] \end{aligned}$$

If $VAR[Z_2] > 0$, then it follows that $COR[T_1,T_2] < 1.0$ and $COV[T_1,T_2] < VAR[T_1]^{1/2}$ $VAR[T_2]^{1/2}$. Finally, since $VAR[T_1]$ will not normally exceed $COV[T_1,T_2]$ in this case, we can safely conclude that when $VAR[Z_2] > 0$, β_{21} will be less than 1.0. We can see, then, that in this case $COR[Y_1,Y_2]$ will underestimate both $\rho^2[Y_1]$ and $\rho^2[Y_2]$, due to the attenuation caused by true change in the underlying true-score distribution.

As shown above, in model (b) the test-retest correlation under-estimates reliability because of attenuation due to "regression to the mean," that is, changes among units of observation in their relative position in the true score distribution. This "true change" aspect to this model was commented on by Moser and Kalton (1972) as one of the serious flaws of the method. In other cases the test-retest correlation overestimates reliability. Consider the model represented by the path diagram in Figure 5.1, model (c). Here $Y_1 = T_1 + E_1$ and $Y_2 = T_2 + E_2$, where the assumptions $COV(E_1,T_1) = COV(E_2,T_2) = 0$ hold. However, as indicated by the model, the correlation between the errors is nonzero. In this case the presence of the correlated errors inflates the estimate of reliability. For simplicity let us assume that there is no change in the underlying true score, i.e., $T_1 = T_2 = T$ [as is depicted in Figure 5.1, model (c)]. In this case the correlation of Y_1 and Y_2 can be written as follows:

$$
\begin{aligned}
COR[Y_1,Y_2] &= COV[Y_1,Y_2] \,/\, VAR[Y_1]^{1/2} \, VAR[Y_2]^{1/2} \\
&= [VAR[T] + COV[E_1,E_2]] \,/\, VAR[Y_1]^{1/2} \, VAR[Y_2]^{1/2} \\
&= VAR[T] \,/\, VAR[Y_1]^{1/2} \, VAR[Y_2]^{1/2} \\
&\quad + COV[E_1,E_2]] \,/\, VAR[Y_1]^{1/2} \, VAR[Y_2]^{1/2}
\end{aligned}
$$

It is perhaps easier to see how the correlation of the errors inflates $COR[Y_1,Y_2]$ as an estimate of reliability if we add the assumption of equal error variances, $VAR[E_1] = VAR[E_2] = VAR[E]$, which also implies that $VAR[Y_1] = VAR[Y_2] = VAR[Y]$. Here the expression for $COR[Y_1,Y_2]$ reduces to

$$
\rho^2 + [COV[E_1,E_2]] \,/\, VAR[Y]
$$

and it is clear that the test-retest correlation overestimates the reliability of measurement as a function of the "correlated error" term.

The situation is even more complex in the fourth and final example given in Table 5.1, namely, the case of model (d). In this case both processes are operating, that is, correlated errors contributing to the test-retest correlation being an overestimate of reliability and attenuation due to true change in the latent true-score distribution contributing to it being an underestimate. With both phenonmena involved, $COR[Y_1,Y_2]$ is an unknown mixture of processes that contribute to upward and downward biases as an estimate of reliability, and there is no way to identify the true nature of the processes involved. I assume that these facts are self-evident, based on the previous exposition with respect to models (a), (b), and (c), and that there is no need for further proof of this assertion.

5.2 SOLUTIONS TO THE PROBLEM

The above discussion is intended to provide a formal statement of the practical problems with using the test-retest method. Moser and Kalton (1972, p. 354) indicate that the longer the interval between the test and retest, the less is the risk of the memory effect but the greater is the risk of intervening events causing respondents to change their views. Their solution to the problem, however, is intractable. They suggest that to solve the problem one needs "to choose an interval long enough to deal adequately with the first risk and yet short enough to deal adequately with the second." In the following we propose an alternative approach, based on the work of Coleman (1968), Heise (1969), and others, that makes it possible to lengthen the retest interval long enough to rule out the effects of memory, but that at the same time permits the separation of unreliability from true change in the underlying true-score distribution via the construction of a model that accounts for that change.

5.2.1   The Quasi-Markov Simplex Model

The analysis of panel data must be able to cope with the fact that people change over time, so that models for estimating reliability must take the potential for individual-level change into account. Given these requirements, techniques have been developed for estimating measurement reliability in panel designs where $P \geq 3$, in which change in the latent variable is incorporated into the model. With this approach there is no need to rely on multiple indicators within a particular wave or cross section in order to estimate the measurement reliability, and there is no need to be concerned about the separation of reliable common variance from reliable specific variance. That is, there is no decrement to reliability estimates due to the presence of specific variance in the error term; here specific variance is contained in the true score. This approach is possible using modern SEM methods for longitudinal data, and is discussed further in this and subsequent chapters (see also Alwin, 1989, 1992; Alwin and Krosnick, 1991; Saris and Andrews, 1991).

The idea of the simplex model began with Guttman (1954), who introduced the idea of the simplex structure for a series of ordered tests, which demonstrated a unique pattern of correlations. Following Anderson's (1960) discussion of stochastic processes in multivariate statistical analysis, Jöreskog (1970) summarized a set of simplex models, the parameters of which could be fit using confirmatory factor analytic methods and for which tests of goodness of fit could be derived. He made a distinction between "perfect simplex" and "quasi-simplex" models—the former being those in which measurement errors were negligible and the latter being those allowing for substantial errors of measurement. The Markov simplexes he discussed were scale-free in the sense that they could be applied in cases where the units of measurement were arbitrary (see Jöreskog, 1970, pp. 121–122). In the survey methods literature the model has come to be known as the "quasi-simplex model" (Saris and Andrews, 1991), or simply the "simplex model" (Alwin, 1989).

As we noted above, the limitation of the test-retest approach using a single reinterview is that it must assume there is no change in the underlying quantity being measured. To address the issue of taking individual-level change into account, this class of autoregressive or quasi-simplex models specifies two structural equations for a set of P over-time measures of a given variable Y_t (where $t = 1, 2 \ldots P$) as follows:

$$Y_t = T_t + E_t$$
$$T_t = \beta_{t,t-1} T_{t-1} + Z_t$$

The first equation represents a set of measurement assumptions indicating that (1) over-time measures are assumed to be τ-equivalent, except for true score change, and (2) measurement error is random. The second equation specifies the causal processes involved in change of the latent variable over time. Here it is assumed that Z_t is a random disturbance representing true score change over time. This model is depicted in Figure 5.2.

This model assumes a lag-1 or Markovian process in which the distribution of the true variables at time t is dependent only on the distribution at time $t-1$ and not directly dependent on distributions of the variable at earlier times. If these assumptions do not hold, then this type of simplex model may not be appropriate. In order to estimate

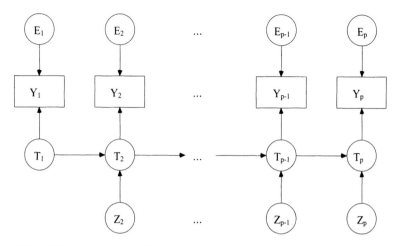

Figure 5.2. Path diagram of the quasi-Markov simplex model—general case (P > 4).

such models, it is necessary to make some assumptions regarding the measurement error structures and the nature of the true change processes underlying the measures. All estimation strategies available for such three-wave data require a lag-1 assumption regarding the nature of the true change. This assumption in general seems a reasonable one, but erroneous results can result if it is violated. The various approaches differ in their assumptions about measurement error. One approach assumes equal reliabilities over occasions of measurement (Heise, 1969). This is often a realistic and useful assumption, especially when the process is *not* in dynamic equilibrium, i.e., when the observed variances vary with time. Another approach to estimating the parameters of the above model is to assume constant measurement error variances rather than constant reliabilities (Wiley and Wiley, 1970). Where P = 3, either model is just-identified, and where P > 3 the model is overidentified with degrees of freedom equal to .5 [P (P + 1)] − 2P. The four-wave model has two degrees of freedom, which can be used to perform likelihood-ratio tests of the fit of the model.

There are two main advantages of the reinterview design for reliability estimation: (1) the estimate of reliability obtained includes all reliable sources of variation in the measure, both common and specific variance; and (2) under appropriate circumstances it is possible to eliminate the confounding of the systematic error component discussed earlier, if systematic components of error are not stable over time. In order to address the question of stable components of error, the panel survey must deal with the problem of memory, because in the panel design, by definition, measurement is repeated. So, while this overcomes one limitation of cross-sectional surveys, it presents problems if respondents can remember what they say and are motivated to provide consistent responses. If reinterviews are spread over months or years, this can help rule out sources of bias that occur in cross-sectional studies. Given the difficulty of estimating memory functions, estimation of reliability from reinterview designs makes sense only if one can rule out memory as a factor in the covariance of measures over time, and thus, the occasions of measurement must be separated by sufficient periods of time to rule out the operation of memory.

5.2.2 The Matrix Form of the Quasi-Markov Simplex Model

We can write the above model more compactly for a single variable assessed in a multiwave panel study as

$$Y = \Lambda_Y T + E$$
$$T = B T + Z$$
$$= [I - B]^{-1} Z$$

Here Y is a $(P \times 1)$ vector of observed scores; T is a $(P \times 1)$ vector of true scores; E is a $(P \times 1)$ vector of measurement errors; Z is a $(P \times 1)$ vector of disturbances on the true scores; Λ_Y is a $(P \times P)$ identity matrix; and B is a $(P \times P)$ matrix of regression coefficients linking true scores at adjacent timepoints. For the case where $P = 4$ we can write the matrix form of the model as

$$\begin{pmatrix} Y_1 \\ Y_2 \\ Y_3 \\ Y_4 \end{pmatrix} = \begin{bmatrix} 1 & 0 & 0 & 0 \\ 0 & 1 & 0 & 0 \\ 0 & 0 & 1 & 0 \\ 0 & 0 & 0 & 1 \end{bmatrix} \times \begin{pmatrix} T_1 \\ T_2 \\ T_3 \\ T_4 \end{pmatrix} + \begin{pmatrix} E_1 \\ E_2 \\ E_3 \\ E_4 \end{pmatrix}$$

$$\begin{bmatrix} 1 & 0 & 0 & 0 \\ -\beta_{21} & 1 & 0 & 0 \\ 0 & -\beta_{32} & 1 & 0 \\ 0 & 0 & -\beta_{43} & 1 \end{bmatrix} \times \begin{pmatrix} T_1 \\ T_2 \\ T_3 \\ T_4 \end{pmatrix} = \begin{pmatrix} Z_1 \\ Z_2 \\ Z_3 \\ Z_4 \end{pmatrix}.$$

The reduced form of the model is written as:

$$Y = \Lambda_Y [I - B]^{-1} Z + E$$

and the covariance matrix for Y as

$$\Sigma_{YY} = \Lambda_Y [I - B]^{-1} \Psi [I - B']^{-1} \Lambda_Y + \Theta^2$$

where B and Λ_Y are of the form described above, Ψ is a $(P \times P)$ diagonal matrix of variances of the disturbances on the true scores, and Θ^2 is a $(P \times P)$ diagonal matrix of measurement error variances. This model can be estimated using any number of structural equation modeling approaches.

5.3 ESTIMATING RELIABILITY USING THE QUASI-MARKOV SIMPLEX MODEL

Heise (1969) developed a technique based on three-wave *quasi-simplex* models within the framework of a model that permits change in the underlying variable being measured. This same approach can be generalized to multiwave panels, as described

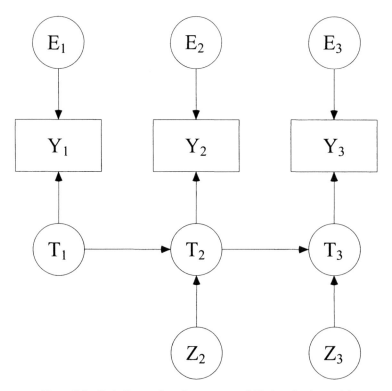

Figure 5.3. Path diagram for a three-wave quasi-Markov simplex model.

above. The three-wave case is a special case of the above model that is just-identified, and it is therefore instructive to illustrate that the approach produces an estimate of reliability of measurement. Consider the quasi-simplex model for three occasions of measurement of Y_g, which is depicted graphically in Figure 5.3. The measurement model linking the true and observed scores is given in scalar form as follows:

$$Y_1 = T_1 + E_1$$
$$Y_2 = T_2 + E_2$$
$$Y_3 = T_3 + E_3.$$

The structural equation model linking the true scores at the three occasions is given in scalar form as

$$T_1 = Z_1$$
$$T_2 = \beta_{21} T_1 + Z_2$$
$$T_3 = \beta_{32} T_2 + Z_3$$

Recall that these models invoke all the assumptions of the CTST models, i.e., $E[E_t]$ = $E[E_t T_t]$ = 0, and the Markovian assumption of the simplex models, i.e., $E[Z_t]$ = $E[Z_t T_{t-1}]$ = 0. We can write the reduced form of the measurement model as follows:

$$Y_1 = Z_1 + E_1$$
$$Y_2 = [\beta_{21} T_1 + Z_2] + E_2$$
$$Y_3 = [\beta_{32} T_2 + Z_3] + E_3$$
$$= [\beta_{32} [\beta_{21} T_1 + Z_2] + Z_3] + E_3$$

With this set of equations in hand, it is possible to write the variances and covariances of the model in terms of the parameters that compose them. These quantities are as follows:

$$VAR(Y_1) = VAR[Z_1] + VAR[E_1]$$
$$COV(Y_1,Y_2) = \beta_{21} VAR[Z_1]$$
$$COV(Y_1,Y_3) = \beta_{32} \beta_{21} VAR[Z_1]$$
$$VAR(Y_2) = VAR[T_2] + VAR[E_2]$$
$$= \beta_{21}{}^2 VAR[Z_1] + VAR[Z_2] + VAR[E_2]$$
$$COV(Y_2,Y_3) = \beta_{32} [\beta_{21}{}^2 VAR[Z_1] + VAR[Z_2]]$$
$$VAR(Y_3) = VAR[T_3] + VAR[E_3]$$
$$= \beta_{32}{}^2 [\beta_{21}{}^2 VAR[Z_1] + VAR[Z_2]] + VAR[Z_3] + VAR[E_3]$$

Because the true scores and errors of measurement are independent, the variances of the observed scores may be expressed as the sum of the variances of the true and error scores. In the above expressions for these quantities, we have expanded the equations for $VAR(T_2)$ and $VAR(T_3)$ in terms of the parameters of the quasi-simplex model.

Wiley and Wiley (1970, p. 112) show that by invoking the assumption that the measurement error variances are equal over occasions of measurement, the $P = 3$ model is just-identified, and parameter estimates can be defined. They suggest that measurement error variance is "best conceived as a property of the measuring instrument itself and not of the population to which it is administered." Following this reasoning one might expect that the properties of one's measuring instrument would be invariant over occasions of measurement and that such an assumption would be appropriate. The solution to the model's parameters under this assumption is as follows:

$$\beta_{32} = COV(Y_1,Y_3) / COV(Y_1,Y_2)$$
$$VAR(E_t) = VAR(Y_2) - [COV(Y_2,Y_3) / \beta_{32}]$$
$$VAR(Z_1) = VAR(Y_1) - VAR(E_t)$$
$$\beta_{21} = COV(Y_1,Y_2) / VAR(Z_1)$$
$$VAR(Z_2) = VAR(Y_2) - [\beta_{21} COV(Y_1,Y_2) + VAR(E_t)]$$
$$VAR(Z_3) = VAR(Y_3) - [\beta_{32} COV(Y_2,Y_3) + VAR(E_t)].$$

Following the CTST model for reliability, the reliability for the observed score Y_t is the ratio of the observed variance, that is, $VAR(T_t) / VAR(Y_t)$, and this model permits the calculation of distinct reliabilities at each occasion of measurement using the above estimates of $VAR(E_t)$ and $VAR(T_t)$. It is also possible to calculate "standardized" values for the stability parameters β_{21} and β_{32} by applying the appropriate ratios of standard deviations (see Wiley and Wiley, 1970, p. 115).

It should be pointed out, which I elaborate on more extensively later on, that the assumption of the equality of error variances is not essential for obtaining an estimate of reliability—namely, the reliability of wave-2 measurement—and this assumption is invoked primarily to obtain estimates of reliability for waves 1 and 3, in addition to the wave-2 estimate. Later on I will show that in addition to the Wiley-Wiley approach there are two other approaches to estimating reliability using this model, that is, two other sets of assumptions; but all three are equivalent with respect to the identification of wave-2 reliability.

While the Wiley-Wiley approach to identifying the parameters of this model—the assumption of *homogeneity of error variances* over time—appears on the face of it to be a reasonable assumption, it is somewhat questionable in practice. As we have repeatedly stressed in this and previous chapters, *measurement error variance is a property both of the measuring device and the population to which it is applied.* It may therefore be unrealistic to believe that it is invariant over occasions of measurement. If the true variance of a variable changes systematically over time because the population of interest is undergoing change, then the assumption of constant error variance necessitates a systematic change in the reliability of measurement over time, an eventuality that may not be plausible. For example, if the true variance of a variable increases with time, as is the case with many developmental processes, then, by definition, measurement reliability will increase over time. Under the assumption of constant measurement error variance, as the true variance grows the error variance becomes a smaller and smaller part of the observed score variance.

Thus, it can be seen that the Wiley-Wiley assumption requires a situation of dynamic equilibrium—constant true-score variance over time—one that may not be plausible in the analysis of developmental processes. Such a state of affairs is one in which the true variances are essentially homogeneous with respect to time. In order to deal with the possibility that true variances may not be constant over time, Heise (1969) proposed a solution to identifying the parameters of this type of quasi-simplex model that avoids this problem. He assumed that the *reliability of measurement* of variable Y is constant over time. Because of the way reliability is defined, as the ratio of true to observed score variance, Heise's model amounts to the assumption of a constant ratio of variances over time. This model is frequently considered to be unnecessarily restrictive because it involves a strong set of assumptions compared to the Wiley-Wiley model. However, it is often the case that it provides a more realistic fit to the data (see Alwin and Thornton, 1984; Alwin and Krosnick, 1991b; Alwin, 1989).

Heise's model is stated for the observed variables in standard form, which is an inherent property of the model. The model linking the true and observed scores is given as

$$Y_1 = \rho_1 T_1 + E_1$$
$$Y_2 = \rho_2 T_2 + E_2$$
$$Y_3 = \rho_3 T_3 + E_3.$$

where each of the ρ_j coefficients is the *index of reliability*, i.e., $VAR(T_t)^{1/2} / VAR(Y_t)^{1/2}$, the square root of reliability. In other words, the model "standardizes" both observed and unobserved variables. This has the effect of discarding any meaningful differences in the observed variances of the measures over time, and in some cases this may be desirable in order to obtain an estimate of reliability.

The structural equation model linking the true scores at the three occasions is identical to the set of equations given above, except that all true and observed variables are in standard form. Following the same algebra used above to reproduce the variance/covariance structure for the observed variables (in this case a matrix of correlations), we may write the correlations among the observed variables as follows:

$$COR(Y_1,Y_2) = \rho_1 \, \pi_{21} \, \rho_2$$
$$COR(Y_1,Y_3) = \rho_1 \, \pi_{21} \, \pi_{32} \, \rho_3$$
$$COR(Y_2,Y_3) = \rho_2 \, \pi_{32} \, \rho_3$$

Heise (1969, p. 97) shows that by assuming an equivalence of the ρ-coefficients the solution to the model's parameters is straightforward:

$$\rho = [COR(Y_1,Y_2) \, COR(Y_2,Y_3) / COR(Y_1,Y_3)]^{1/2}$$
$$\pi_{21} = COR(Y_1,Y_3) / COR(Y_2,Y_3)$$
$$\pi_{32} = COR(Y_1,Y_3) / COR(Y_1,Y_2)$$

where π_{21} and π_{32} are the standardized versions of the β-coefficients in the metric form of the model. These coefficients are interpreted by Heise as *stability coefficients*, that is, the extent to which the true scores of the variable at a later state is dependent on the distribution of true scores at an earlier time.

It is important to reiterate that this model involves a lag-1 or Markovian assumption, that is, there is no direct effect of the true state of the unobserved variable at time 1 on its true state at time 3. Thus, the effect of T_1 on T_3 is simply given as the product $\pi_{21} \, \pi_{32}$. Wiley and Wiley (1970, p. 115) show that Heise's computation of the stability coefficients as standardized quantities is expressible in terms of their model. The model makes no assumptions about the equivalence of remeasurement lags. Obviously the magnitude of stability will depend on the amount of time that has elapsed between intervals, but the estimates of reliability are likely to be immune to the amount of time between waves of the study (see Alwin, 1989).

It is important to reiterate that the Heise (1969) estimate of reliability for this model, namely, $\rho^2 = COR(Y_1,Y_2) \, COR(Y_2,Y_3) / COR(Y_1,Y_3)$ can be shown to equal the wave-2 reliability using the Wiley and Wiley (1970) approach. Note, first, that this estimate can be rewritten as

$$\begin{aligned}
\rho^2 &= COV[Y_1 Y_2] \, COV[Y_2 Y_3] \, VAR[Y_1]^{1/2} \, VAR[Y_3]^{1/2} \\
&\quad / COV[Y_1 Y_3] \, VAR[Y_1]^{1/2} \, VAR[Y_3]^{1/2} \, VAR[Y_2] \\
&= COV[Y_1 Y_2] \, COV[Y_2 Y_3] / COV[Y_1 Y_3] \, VAR[Y_2] \\
&= \beta_{21} \, VAR[Z_1] \, \beta_{32} \, VAR[T_2] / \beta_{21} \, \beta_{32} \, VAR[Z_1] \, VAR[Y_2] \\
&= VAR[T_2] / VAR[Y_2]
\end{aligned}$$

Recall next that the estimate of $VAR[T_2]$ in the Wiley-Wiley model (see the development above) is written as follows:

$$VAR[T_2] = COV[Y_1Y_2] \, COV[Y_2Y_3] \, / \, COV[Y_1Y_3]$$

From this it can be seen that wave-2 reliability, ρ^2, using the Wiley-Wiley model is written as

$$COV[Y_1Y_2] \, COV[Y_2Y_3] \, / \, COV[Y_1Y_3] \, VAR[Y_2]$$

which, as above, reduces to $VAR[T_2] \, / \, VAR[Y_2]$, and the conclusion follows from this.

There is a third approach to the question of identifying the parameters of these simplex measurement models for multiwave data that takes a somewhat more conservative approach. This approach assumes neither that measurement error variance is constant over time, nor that reliability is equal at all timepoints. Werts, Jöreskog, and Linn (1971; see also Werts, Linn and Jöreskog, 1977, 1978; and Jöreskog, 1974) show that just-identified and overidentified models may be estimated for such panel data without such restrictions, if one is content to identify the wave-2 reliability only. If, for example, we were to consider the three-wave problem discussed above somewhat differently, it would be possible to obtain the information we need about the reliability of measurement but nothing about intertemporal stability. If one is content to leave the reliability and stability parameters confounded for Y_1 and Y_3, it is straightforward to estimate the reliability of Y_2.

Suppose we respecify the above model, shown in Figure 3.3, as a single-factor model in which the latent variable is the true score at time 2, as follows:

$$Y_1 = \alpha_{12} \, T_2 + E_1$$
$$Y_2 = \alpha_{22} \, T_2 + E_2$$
$$Y_3 = \alpha_{32} \, T_2 + E_3.$$

As indicated in the previous chapter, this model is not identifiable unless some form of constraint is introduced to give the latent variable a metric. Thus, if we let $\alpha_{22} = 1.0$, then it is possible to solve for the remaining parameters as follows:

$$COV(Y_1,Y_2) \, COV(Y_1,Y_3) \, / \, COV(Y_2,Y_3) = \alpha_{12}{}^2 \, VAR(T_2)$$
$$COV(Y_1,Y_2) \, COV(Y_2,Y_3) \, / \, COV(Y_1,Y_3) = VAR(T_2)$$
$$COV(Y_1,Y_3) \, COV(Y_2,Y_3) \, / \, COV(Y_1,Y_2) = \alpha_{32}{}^2 \, VAR(T_2)$$

Given these equations it is then possible to solve for the α_{12}, α_{32}, and $VAR(T_2)$ quantities. Note that the α_{12} and α_{32} coefficients in this model confound true-score stability with measurement reliability. The error variance, $VAR(E_2)$, can be obtained residually, however, and the unbiased estimate of the reliability of measurement at wave 2 can be obtained as follows:

$$\rho_2^2 = [COV(Y_1,Y_2) \, COV(Y_2,Y_3)] \, / \, [COV(Y_1,Y_3) \, VAR(Y_2)]$$

In fact, the three approaches discussed here all agree on the estimate of the reliability of the time-2 measure (see Werts et al., 1971).

We conclude from this that it is just the outer measures of the three-wave model for which one must impose some constraint in order to estimate measurement reliability at other waves. The utility of this set of observations is quite readily seen for multiwave panel models in which more than three waves of data are available. Generalizing to multiple-wave panel studies, Werts et al. (1971) state that, "when the assumptions of the Wiley-Wiley structural model are given, error variances, true score variances, and unstandardized regression weights between corresponding true scores are identified for all but the first and last measures. For this reason it appears unnecessary to make either the equal reliability or the equal error variance assumption for inner measures" (Werts, et al., 1971, p. 111). While this is true, it should be stressed that all three approaches outlined above produce identical estimates for the wave-2 reliability parameter, and as our results show (see Chapter 6), there are only trivial differences between this estimate and the wave-1 and wave-3 reliability estimates obtained using the Wiley and Wiley (1970) approach.

5.4   CONTRIBUTIONS OF THE LONGITUDINAL APPROACH

I consider the longitudinal approach to assessing reliability of survey measurement to represent a strong alternative to the cross-sectional designs for reliability estimation. Indeed, it may in fact be preferable under the right circumstances. The multiple-wave reinterview design discussed in this chapter goes well beyond the traditional test-retest design (see Moser and Kalton, 1972, pp. 353–354), and the limitations of the test-retest design can be overcome by incorporating three or more waves of data separated by lengthy periods of time. Specifically, employing models that permit change in the underlying true score using the quasi-Markov simplex approach allows us to overcome one of the key limitations of the test-retest design (Alwin, 1989, 1992; Heise, 1969; Humphreys, 1960; Jöreskog, 1970; Saris and Andrews, 1991; Werts, Breland, Grandy, and Rock, 1980).[2] Moreover, through the use of design strategies with relatively distant reinterview intervals, the problem of consistency due to retest effects or memory could be remedied, or at least minimized.

The key elements of my argument with respect to the contributions of the longitudinal approach hinge on how such designs remedy the limitations of the cross-sectional survey designs in the estimation of measurement reliability. As I have argued here, one major disadvantage of the cross-sectional approach derives from the inability to design replicate measures that can be used in cross-sectional studies that meet the requirements of the statistical models for estimating reliability. Specifically, given the substantial potential for sources of specific variance when "nonreplicate" measures are used, it is difficult to make the assumption of *univocity*, that is, that the question measures one and only one thing. This assumption is not required in

---

[2]For an alternative perspective, see Coenders, Saris, Batista-Foguet, and Andreenkova (1999).

the longitudinal case—questions can be multidimensional. Finally, given the fact that reliability estimation is based on the idea of replicate measures, it is difficult to assume that in cross-sectional designs the errors are independent, an assumption that is critical to the statistical models for estimating reliability.

In the previous chapter (Chapter 4) I introduced the distinction between *multiple measures* and *multiple indicators*. There I identified the features of the "multiple-measures model" with the assumptions of the CTST model, namely, that measures are *univocal* and measure the same true score. In other words, the measures contain no *specific* variance (see Section 4.1.1). That discussion clarified the fact that the true scores and error scores of the model can have a variety of different properties—under this model measures could be *parallel, tau-equivalent,* or *congeneric.* The key is that the measures are replicate measures—all measure the same true score and no other— regardless of scale and regardless of restrictions on the error variances. This model also makes the assumption that the errors of measurement are independent, which places high demands for the use of replicate measures. By contrast, we identified the features of the "multiple-indicators model" as the embodiment of the assumptions of the *common-factor model*—a somewhat more general model allowing for more that one source of reliable variance in the measures, even in the single "common factor" case. In the common-factor tradition, even if all of the indicators measure the same common factor, the model allows for the possibility that each indicator contains something unique to (i.e., not in common with) the others. The common-factor model, in other words, explicitly includes specific variance as part of the model.

In any given case both models will fit the data identically, and interpretation of the results depends on other considerations. The problem is that normally *specific variance* cannot be uniquely identified, and since it shares some of the same properties as *unreliability* of measurement (specifically, that it is uncorrelated with the common factor), it is aggregated with measurement error. In the common-factor model representation it is essentially indistinguishable from measurement error. In short, if one can satisfy the assumptions of the "multiple-measures model" (i.e., the CTST assumptions) it is possible to use the common-factor model to estimate reliability. If not—and the best that is possible is the "multiple indicators" model—the interpretation of the "uniqueness" as made up solely of measurement error is questionable, and the model cannot be used to estimate reliability. As noted, either model is consistent with the data, and one's ability to interpret the variance of the unique part of the data (specific variance and random measurement error) depends on other considerations.

My argument is that (with the possible exception of the MTMM approach), because of the inability to replicate measures and the high likelihood of specific variance in survey measures, the covariation of multiple questions in surveys is better described by the multiple-indicators model rather than the multiple-measures model. The limitations of cross-sectional designs for reliability estimation can be seen in Figure 5.4 (a), where I have presented a path diagram for multiple indicators of the same concept gathered in a cross-sectional survey. The assumption of the independence of errors continues to be problematic—specifically, in order to interpret the parameters of such models in terms of reliability of measurement, it is essential to be

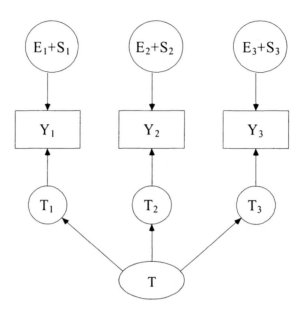

(a) Multiple Indicators

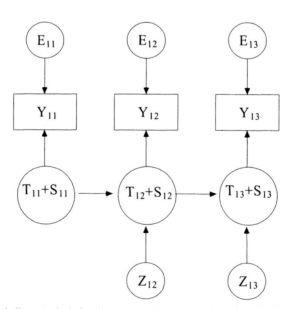

(b) Simplex Model for a Single Indicator

Figure 5.4. Path diagrams depicting the presence of specific variance in multiple-indicators and simplex models.

able to rule out the operation of memory, or other forms of consistency motivation, in the organization of responses to multiple measures of the same construct.

One point I wish to emphasize with respect to Figure 5.4 (a) is that to the extent the "errors" on the indicators truly contain specific variance (the Ss), then standard reliability formulations based on classical true score models will *underestimate* reliability [see Alwin and Jackson (1979) for further discussion of this issue]. This is one of the reasons that standard treatments of Cronbach's α (e.g., Lord and Novick, 1968) indicate that the coefficient is a lower bound estimate on reliability. Reviewing this issue briefly, as I pointed out in Chapter 4, consistent with the common-factor model representation, which assumes that the unique part of Y_g is equal to $U_g = S_g + E_g$, we can define the variance of Y_g as

$$VAR(Y_g) = VAR(T_g) + VAR(S_g) + VAR(E_g)$$
$$= VAR(T_g) + VAR(U_g),$$

and $VAR(U_g)$ is known to equal the sum of specific and measurement error variance. Standard reliability formulations based on CTST would assume that $VAR(S_g)$ is nonexistent and the expression such as

$$VAR(T_g) / [VAR(T_g) + VAR(U_g)]$$

would be an appropriate estimate of the reliability of the measure Y_g. In fact, the expression

$$[VAR(T_g) + VAR(S_g)] / [VAR(T_g) + VAR(U_g)]$$

is more appropriate. However, since it is not possible within this framework to partition $VAR(U_g)$ into its components, $VAR(S_g)$ and $VAR(E_g)$, to the extent $VAR(S_g)$ exists, $VAR(U_g)$ will overestimate the amount of measurement error variance.

Assuming the goal is to estimate the reliability of Y_g, regardless of its content, the model in Figure 5.4 (b) may be a more appropriate one. This is the quasi-Markov simplex model for multiple over-time measures discussed in this chapter. Note that the notation here differs somewhat from that used in Figure 5.4 (a) for the multiple-indicators model. There the subscripts refer to an index that runs over the G indicators, whereas in Figure 5.4 (b) the subscripts index both indicator and time, specifically Y_{gt}, the gth indicator observed at the tth occasion of measurement. In the example given, the simplex model employs three measures of Y_1 gathered over successive waves of the survey, whereas the multiple-indicators model employs a single measure of Y_1 in a one-time survey, coupled with measures of Y_2 and Y_3, which may or may not be replicate measures. My argument is quite straightforward—if one can design surveys in a manner consistent with Figure 5.4 (b), one has a better opportunity for modeling a quantity that is interpretable as *reliability* than is the case with the model depicted in Figure 5.4 (a). It is quite simply a matter of where one allocates the (reliable) specific variance, to the portion considered true variance, or to the portion considered error variance. The choice ought to be clear!

5.5   COMPONENTS OF THE SURVEY RESPONSE

I have argued throughout this book that in order to assess the extent of measurement error *for single survey questions,* it was necessary not only to develop a theoretical model for conceptualizing the behavior of measurement errors in survey question-naires but also to develop designs that permit the estimation of model parameters specifying the linkages between the concepts being measured and errors of measure-ment. In the previous chapters I reviewed some of the limitations of the ways in which the classical theoretical model has been applied to survey data. Our discussion of the factor analytic decomposition of response variance into reliable and unreliable parts led to the observation that the analysis of measurement errors in survey data could be improved by an extension of the classical model to include the conceptualization of systematic as well as random errors. We argued along the way that responses to survey questions could be conceptualized quite broadly to include several compo-nents of error, and that any approach to estimating the reliability of survey measures had to consider the desirability of incorporating the reliability of specific variance as well as common variance into the overall picture of reliability estimation. If one is interested in the reliability of survey measurement—specifically, the reliability of survey questions—then one must include in that reliability estimate all sources of reliable variance, *not just reliable common variance.*

Two basic designs have been considered: (1) cross-sectional measurement designs and (2) longitudinal or panel measurement designs. Within the first type of design, the chapter discussed ways in which both *multiple-measures* and *multiple-indicators* designs can be estimated in cross-sectional data. The multitrait-multimethod mea-surement design was presented and discussed in the previous chapter (Chapter 4) as a possible solution to one aspect of these problems, namely, the incorporation of systematic sources of error connected to methods of measurement. Although I favor the application of such designs in cross-sectional data, I raised several issues about the potential limitations to interpreting the parameters of those models in terms of reliability of measurement; and although the assumptions of these models are clear, the seriousness of the consequences of violating them is unclear. As an alternative to the cross-sectional design for reliability estimation, we have considered the rein-terview design for reliability estimation and have argued that such designs may have the potential to overcome some of the limitations of the cross-sectional approaches. We are only beginning to realize the potential of longitudinal measurement research designs to inform these matters, and in this final section of the chapter I present a brief review of the nature of the components of survey measurement error that can be isolated in empirical applications of these designs (see also Alwin, 1989).

Let us return to the notation for a *fixed person* that we used in our initial discussion of CTST (see Chapter 3) and reformulate the original model including more than one source of "true" variance, more than one type of measurement error, and observations of true scores that change over time. Let us assume, however, in contrast to the discus-sion of the CTST model in Chapter 3, that we have only one measure Y (i.e., G = 1), but instead that observations of Y are obtained over repeated intervals. The model assumes that Y_t is a measure of the continuous latent random variable T_t, at time t.

An *observed score* y_{tp} for a fixed person p at time t is defined as a random variable for which a range of values for person p can be observed from a propensity distribution for person p relating a probability density function to possible values of Y at time t. The *true score* is defined as in Chapter 2, and from this we define *measurement error* for a given observation as the difference between the true score and the particular score observed for p on Y_t, i.e., $\varepsilon_{tp} = y_{tp} - \tau_{tp}$. Note that a different error score would result had we sampled a different y_{tp} from the propensity distribution of person p, and an infinite set of replications will produce a distribution for ε_{tp}. One further assumption is made about the errors of measurement in the measurement of y_{tp} and this is that ε_{tp} is uncorrelated with errors of measurement of any other Y_t. As noted above, this is called the assumption of *measurement independence* (Lord and Novick, 1968, p. 44) and in the present context this assumption is crucially related to the length of the reinterview period between observations of Y_{tp}.

The purpose of the present discussion is to rewrite the individual components of the particular score in terms of components of variation implied by this model. In Table 5.1 we present a list of the components of variance in an individual response observed for p on Y_t that is, the components of y_{tp}. As this scheme shows, the literature on survey measurement error suggests that the simple variation in responses might be summarized with a somewhat more complex measurement model than is used in most discussions of CTST. In the most general case we can formulate the observed score as follows, using the notation presented in this table:

Response $(y_t) =$ True common-trait variable (τ_t^*)
+ Constant measurement errors (μ_t)
+ Specific-trait variable (υ_t)
+ Method invalidity variable (η_t)
+ Random measurement errors (ε_t)

Thus, an individual's response to a given survey question measured at time t, y_t, with response categories $y(1), y(2) \ldots y(r)$, assessed in a given population, can be conceived as a function of the following components:

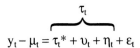

$$y_t - \mu_t = \overbrace{\tau_t^* + \upsilon_t + \eta_t}^{\tau_t} + \varepsilon_t$$

Table 5.1. Components of the survey response in longitudinal designs

Components for Fixed Person	Population Random Variables
True variable of interest (τ_t)	T_t
True common trait (τ_t^*)	T_t^*
True specific trait (υ_t)	S_t
True method invalidity (η_t)	M_t
Constant measurement errors (μ_t)	C_t
Random measurement errors (ε_t)	E_t

where the μ_t term represents constant errors, and the remaining terms are sources of variation. At a given time t, we may conceive of η_t, υ_t, and ε_t as random with respect to τ_t^*; however, the only necessary assumption is that ε_t is uncorrelated with all other components of y_t.

If we extrapolate to a population of such persons and define random variables for the population, reflecting these theoretical sources of response variation (see Table 5.1), the variance of Y_t may, thus, be written as

$$VAR(Y_t) = VAR(T_t) + VAR(E_t)$$
$$= VAR(T_t^*) + VAR(S_t) + VAR(M_t) + VAR(E_t)$$

Note that by definition S_t is independent of T_t^*, but it is possible that M_t is correlated with T_t^* and S_t. The convention is to assume that they are not. The challenge for measurement error analysis is to separate $VAR(E_t)$ from $VAR(S_t)$ and $VAR(M_t)$, so that the estimates of reliability include only measurement error in the E_t term.

5.6   WHERE TO FROM HERE?

In the remainder of the book I present the results of an empirical study of survey measurement error using the longitudinal methods described in this chapter. I present the methods used in the present study in Chapter 6, including a description of the samples, available measures, statistical designs for reliability estimation, and the problem of attrition in longitudinal designs. The remaining chapters present our results on the potential effects of *topic and source of survey reports* on the reliability of measurement (Chapter 7), the issue of the *organizational context of questions* and errors of measurement (Chapter 8), our findings regarding *the role of question characteristics* in the reliability of measurement (Chapter 9), and the relationship of *respondent characteristics*—education and age—to the reliability of measurement (Chapter 10). This study is based on the hope that empirical research can elaborate and exploit to a greater extent some of the designs covered in the literature for studying measurement error. Such empirical developments can further inform the processes that produce measurement errors in ways that will ultimately improve the collection of survey data and the inferences drawn from them. Following the discussion of results in these five chapters, I conclude the book with two additional chapters—one dealing with approaches to reliability estimation when the data are categorical (Chapter 11) and a final chapter dealing with conclusions and future directions (Chapter 12).

CHAPTER SIX

Using Longitudinal Data to Estimate Reliability Parameters

> I often say that when you can measure what you are speaking about and express it in numbers you know something about it; but when you cannot measure it, when you cannot express it in numbers, your knowledge is of a meager and unsatisfactory kind: it may be the beginning of knowledge, but you have scarcely, in your thoughts, advanced to the stage of science, whatever the matter may be.
>
> Lord Kelvin, *Electrical units of measurement* (Thomson, 1883)

There is growing recognition that measurement errors pose a serious limitation to the validity and usefulness of the information collected via survey methods. In addition to the threats posed by coverage or sampling frame problems, sampling errors, and survey nonresponse, measurement errors can render the survey enterprise of little value when errors are substantial. The main objectives of this project are to raise several questions about the nature and extent of measurement error, to quantify one aspect of those errors—random errors of measurement or unreliability—and to investigate the conditions under which such errors are more (or less) likely to occur. In this chapter I present an approach to these matters and lay the groundwork for the presentation of several sets of results that examine these questions in subsequent chapters.

As I argued in the previous chapters, most attention to reliability in survey research is devoted to item analysis and scale construction, or econometric or psychometric adjustments for measurement error using multiple indicators within the framework of LISREL or related approaches. While these procedures are invaluable and highly effective in reducing the impact of measurement errors on substantive inferences, they are inherently limited with respect to assessing reliability of survey measurement and have not been informative regarding the nature and sources of measurement errors. Specifically, these approaches generally cannot address the question of the reliability of single survey items because they are focused on composite scales or on latent constructs specified for multiple indicators. We view this as a serious limitation. First, survey methodologists are increasingly interested in the reliability of specific survey questions (e.g., a question about household income, or a question about an attitude toward United States military involvement in country X) rather than in the behavior of batteries of questions or multiple indicators. Second, these approaches, in general, do not help us identify the potential sources of measurement error that may be linked to the nature of the information sought or the formal properties of the survey question.

6.1   RATIONALE FOR THE PRESENT STUDY

While the most common approach to estimating reliability of survey data involves the calculation of ICR estimates like Cronbach's α (Cronbach, 1951), the limitations of this approach are in all likelihood not very well known. To quote one prominent statistician (Goldstein, 1995, p. 142):

> The common procedure, especially in education, of using "internal" measures based upon correlational patterns of test or scale items, is unsatisfactory for a number of reasons and may often result in reliability estimates that are too high.

Goldstein goes on to recommend the use of the type of longitudinal structural equation methods of the type specified in the previous chapter for purposes of estimating reliability of measurement. Indeed, the longitudinal methods discussed in the previous chapter open up a number of possibilities for assessing the reliability of single survey questions via the use of data for questions appearing repeatedly in extant longitudinal studies. The increasing prevalence of longitudinal data creates an opportunity to explore the reliability of survey measurement in ways that heretofore have been unrealized.

Moreover, virtually all discussions of measurement error in the survey methods literature focus on cross-sectional data and ignore the models developed for longitudinal data (e.g., Biemer and Stokes, 1991; Groves, 1991; Saris and Andrews, 1991). Many of the available reports on measurement models contain valuable discussions of random and nonrandom components of error in cross-sectional data, but none of them focus on the limitations we have identified here, nor is there any recognition of the great value of longitudinal data for reliability estimation. There are exceptions to this—Marquis and Marquis (1977), Alwin (1989), Saris and Andrews (1991)—which provide discussions of reliability components derived from structural equation models of autoregressive processes.

This and subsequent chapters present the results of a project whose aim was to assemble a database for survey questions, consisting of question-level information on reliability and question characteristics for nearly 500 variables from large-scale longitudinal surveys of known populations in the United States The objective was to assemble information on measurement reliability from several representative longitudinal surveys, not only as an innovative effort to improve knowledge about the strengths and weaknesses of particular forms of survey measurement but also to lay the groundwork for developing a large-scale database on survey measurement reliability that can address basic issues of data quality in the social, economic, and behavioral sciences. This chapter presents the essential features of the study design and methods used in this research. I discuss these methods in four parts. First, I describe the longitudinal data sets selected for inclusion in the present analysis; second, I summarize the measures available in these studies and the conceptual domains represented; third, I discuss the variety of statistical estimation strategies available for application here; and fourth, I discuss the problem of panel attrition and its consequences for reliability estimation are addressed.

6.2 SAMPLES AND DATA

The design requirements set for this study included the following: (1) the use of longitudinal studies with at least three waves of measurement that (2) are representative of known populations and (3) have reinterview intervals of at least 2 years. The last requirement is implemented in order to rule out the effects of memory in longitudinal studies. In many panel studies the reinterview intervals are rarely distant enough to rule out memory as a source of consistency in reliability estimates. Our experience indicates (Alwin, 1989, 1991, 1992; Alwin and Krosnick, 1991a, 1991b) that panel studies with short reinterview intervals (1–4 months) have substantially higher levels of reliability than do those with longer intervals, such as two years. Even so, some have argued that repeated measurements at the beginning and end of a 20-minute questionnaire can be considered free of memory effects and therefore a basis for estimating reliability (e.g., Coenders, Saris, Batista-Foguet, and Andreenkova, 1999; Saris and van den Putte, 1988; van Meurs and Saris, 1990).

I selected data from six panel studies that met these requirements; four were based on probability samples of the United States, two were large-scale surveys of a sample originating in the Detroit metropolitan area (all data collection in these surveys was conducted by the Survey Research of the University of Michigan): (1) the 1956–1958–1960 National Election Study (NES) panel, (2) the 1972–1974–1976 NES panel, (3) the 1992–1994–1996 NES panel, (4) the 1986–1989–1994 American's Changing Lives (ACL) panel study, (5) the 1980–1985–1993 Study of American Families (SAF) sample of mothers, and (6) the 1980–1985–1993 Study of American Families (SAF) sample of children. We describe the sample and data from each of these panel studies in the following paragraphs.

6.2.1 National Election Panel Studies

The National Election Study (NES), conducted by the University of Michigan's Center for Political Studies, provides three panel data sets, which were linked to the biennial national NES cross sections. Every 2 years since 1952 (except 1954), the Survey Research Center of the University of Michigan's Institute for Social Research (ISR) has interviewed a representative cross section of Americans to track national political participation. In presidential election years, a sample is interviewed before the election and immediately afterward. In non-presidential election years, only postelection surveys are conducted. Data were obtained from face-to-face interviews with national full-probability samples of all citizens of voting age in the continental United States, exclusive of military bases, using the Survey Research Center's multistage area sample (see Miller and NES, 1993). The sample sizes typically range between 1,500 and 2,000. On several occasions the NES conducted long-term panel studies in connection with the biennial cross sections. We employ three of these here (1) the 1956–1958–1960 panel, (2) the 1972–1974–1976 panel, and (3) a 1992–1994–1996 panel.

The main content of the NES questions include a range of factual and subjective content, including political party affiliations, beliefs about political issues, approval ratings of political leaders, ideological self-assessments, self-assessments of trust in government, and attitudes toward political parties, social groups, and government policies. A wide range of measures were used, and various types of response scales, including the Election Study feeling thermometers and seven-point scales, which are known to vary in their reliability (see Alwin, 1992, 1997; Alwin and Krosnick, 1991b).

Of the respondents interviewed in 1956, 1,132 were reinterviewed in 1958, and again in 1960 (see Campbell, Converse, Miller, and Stokes, 1971). The questionnaires used in the 1958 and 1960 panel studies were the same as those used in the 1958 and 1960 cross-sectional studies, and there was considerable overlap between the 1956, 1958, and 1960 studies. Of the respondents interviewed in 1972, 1,296 were successfully reinterviewed in 1974 and in 1976. Again, the questionnaires used in these reinterview studies were the same as those used for the cross-sectional samples interviewed at those times, and there was considerable overlap between the 1972, 1974, and 1976 surveys. In 1992 the NES launched another panel study, focusing primarily on contemporary politics, namely the success of the Clinton coalition. The main content of the panel focused on standard NES questions, similar to those in earlier surveys; however, several additional questions were included on evaluations of the economy, attitudes toward the media, special interests, government, and a range of beliefs about presidential legislative proposals. The 1992–1994–1996 NES panel began with a multistage area probability sampling design used to select 1,522 cases for the 1992 cross section from all eligible citizens residing in housing units in the 48 coterminous states who would be of voting age on or before the 1992 election day. Of these 1,522 potential respondents, 1,126 gave a preelection interview, for a response rate of 75.0%, and of these 1,126, 1,005 also gave a postelection interview, for a reinterview response rate of 89.3%. Then in 1994, 759 of these 1,005 1992 pre- and postelection survey respondents were reinterviewed, giving a 1994 follow-up response rate of 75.5%. Of the 759 panel respondents reinterviewed in 1994, 597 were also reinterviewed in 1996, for a follow-up response rate of 78.7%. In 1994 the multistage area probability sampling design sampled 1,437, of which there were 1,036 new respondents in 1994, for a 72.1% response rate. Only respondents with completed interviews in 1994 were eligible for reinterview in 1996, and in addition to the 597 panel cases in 1992 who were reinterviewed in 1994 and 1996, there were 719 of the 1994 new cross-sectional cases who also provided a 1996 interview (Center for Political Studies, 1994). Later in the chapter we discuss the implications of patterns of attrition in the NES and other panel studies for the statistical estimation of the measurement reliability.

6.2.2 The Americans' Changing Lives Panel

The Americans' Changing Lives (ACL) project is a three-wave panel study of the American population, begun in 1986, with a focus on psychological determinants of health and effective functioning [House, (no date)]. Wave 1 of the panel consists of face-to-face interviews with a national probability sample of 3,617 adults aged 25 years and over, with oversampling of blacks and persons over 60. The household-level response rate for the first wave was 70%. The second wave of ACL, conducted in early 1989 about 2.5 years after the first wave, includes 2,867 of the first wave (some 83% of the survivors), also interviewed in person. The third wave, conducted in early 1994, consisted of a brief 25-minute telephone reinterview with approximately 2,700 respondents from the first wave. The interval between the second and third waves is also about 2.5 years. All data collection on this project have been conducted by ISR's Survey Research Center. Full interviews were obtained with about 2,700 respondents in the third wave, which represents about 86% of all surviving first-wave respondents. Approximately 2,400 third-wave respondents will have complete data from all three waves. Except as noted, all the ACL data have been appropriately weighted to adjust for differential initial selection probabilities, and differential nonresponse, so that the data can be used to generalize to the national population 25 years and older. The ACL study provides a range of variables measured in all three waves, focusing primarily on health and well-being.

6.2.3 The Study of American Families

The Study of American Families (SAF) is a longitudinal panel study of mothers and children. It is based on a probability sample of the children drawn from the July 1961 birth records of white, first-, second-, and fourth-born children (in approximately equal numbers) in the Detroit metropolitan area (Wayne, Oakland, and Macomb counties). It was originally known as the Family Growth in Detroit Study and sometimes referred to as a Detroit Area Study, but since 1980 it has gone by its present name (Thornton and Binstock, 2001). The study spans a period of 31 years, from 1961 through 1993. Throughout this study, the original panel members and their 1961-born children were reinterviewed regardless of their geographic location, and while it is not a probability sample of the United States, by 1993 respondents resided in more than 20 states in the United States spread from coast to coast. The families participating in the SAF are represented by members of a birth cohort (those born in 1961) and their mothers. They are a unique resource for present purposes. Because the children come from the same birth cohort, they were exposed to the same historical period and societal influences at the same time in the life cycle. The timing of the interviews was also propitious, since the children were first interviewed at age 18, just when they were completing the transition to adulthood, again at age 23, and for a third time at age 31. For purposes of estimating the reliability of measurement, the data set is even richer for the sample of mothers, since these women were interviewed 8 times: twice in 1962, and in 1963, 1966, 1977, 1980, 1985, and 1993.

Since 1977, the SAF data have been collected by the Survey Research Center of the University of Michigan's ISR. The retention rates for this panel have been unusually high throughout the course of the project. Interviews were obtained from 92% of the sampled families in the initial 1962 wave. By 1993, 929 mothers and 932 children, representing some 85% of the families originally interviewed in 1962. There is a full set of interviews across all waves of the study from both mothers and children in 867 families, representing 82% of the original families interviewed. Recent efforts to obtain the cooperation of these families has yielded extremely high rates of retention—approaching 100%. Except for some of the early interviews with the mothers, which were conducted by telephone, all data were collected in face-to-face interviews. The SAF provides a range of measures available for at least three waves on respondents interviewed in 1993. These include attitudes about family issues and substantial information on employment, living arrangements, and socioeconomic achievement. Attitudes were measured on a variety of topics, including measures of fertility aspirations and expectations, use of birth control, attitudes toward abortion, sex-role attitudes, and attitudes toward cohabitation, marriage and divorce.

6.3   DOMAINS OF MEASUREMENT

The focus of this study is on the single survey question, our primary unit of analysis. In order to be included in the analysis of reliability the questions from the six panel studies described above had to be repeated in two subsequent waves. Here we specify the major conceptual domains in which there are replicated measures that will permit the serious approach to the investigation of reliability that we undertake here. Because of space limitations, we will not give the exact question wording; however, these can be obtained from the original source materials, that is, questionnaires and codebooks available from the several studies involved. *We should stress that in order to be included in the present analysis, the wording of survey questions had to be exact across all waves of the study;* otherwise the question was excluded from further consideration. We do provide variable descriptions in the Appendix, along with source information, detailed content codes, sample sizes, and reliability estimates. We proceed here with an outline of the major conceptual domains, indicating the source of our reliability estimates for that domain. Because of the uniqueness of the panel datasets on which we draw, most of these domains reflect areas where there is very little known about reliability issues. Taken together, this set of studies represents what is perhaps the richest potential for a comprehensive assessment of reliability of survey measurement.

At the most abstract level the variables from these surveys fall into two general categories: *facts* vs. *nonfacts*—the latter containing subjective variables, such as attitudes, beliefs, and values. The major distinction here is that facts can presumably be verified against objective records, whereas beliefs, attitudes, and values, for example, are by definition a matter of personal judgment. Even in the case of facts such *objective* information must often be estimated, in which case there is some ambiguity in the distinction between facts and beliefs. For most purposes we consider five basic content categories, as shown in Table 6.1. In the case of facts, we further divide the category into those questions that involve self-reports and those that involve proxy reports. We also further subdivide self-descriptions into (1) those that are self-assessments or self-evaluations and (2) those that are self-perceptions, without any implied evaluative dimension, i.e., a statement of "what is" about the self.

References to subjective variables, such as beliefs, attitudes, and values, are often confused in the language surrounding survey measures, but the meaning of these terms can be clarified. Beliefs, for example, are cognitive representations of "what is"—basic information that produce states of "expectancy" about the physical and social environment. The "subjects" of beliefs—what beliefs are about—include a wide variety of phenomena, including the physical world, the self, other persons, and society. Examples of belief statements are: "A child should work hard and learn as much as he/she can." "A working mother can establish just as warm and secure a relationship with her child as a woman who does not work." "Children should be seen and not heard." All of these statements embody a common characteristic—all

Table 6.1. Types of content involved in survey measures

1. *Facts:* Objective information regarding the respondent or members of the household, e.g., information on the respondent's characteristics, such as the date of birth, amount of schooling, amount of family income, and the timing, duration, and frequencies of certain behaviors.

2. *Beliefs:* Perceptions or subjective assessments of states and/or outcomes regarding the respondent or others, e.g., information on respondent's perceptions of how political parties stand on various issues.

3. *Attitudes:* Evaluative responses to particular objects or actors, assumed to exist along a positive/negative continuum of acceptance, favorability, or agreement. Attitudes, for example, on policy issues or political leaders, are frequently used and measured along a dimension of approval or disapproval.

4. *Values:* Subjective evaluations of the importance of certain end states of existence, or preferred modes or means of attaining them.

5. *Self-descriptions:* Subjective assessments or evaluations of the state of the respondent within certain domains, e.g., a common form of self-description in political studies is party affiliation, or assessments of interest in politics, or the extent to which the respondent cares about who wins an election.

express information about the subject, which is taken by the actor—the "believer"—to be true. Of course, they may or may not be true, or for that matter, there may be no way of verifying the truth value of all that is believed. Whatever their source, beliefs become a relatively stable part of the individual's cognitive organization.

Values are defined as essential "standards" that govern behavioral choices. They may be viewed as a subclass of beliefs, in that values are assertions about what is good or desirable. They are stable expressions of individual and collective beliefs about what ends to seek, the standards used to choose among desired end states, and means to achieve them. For example, the statement "human life should be preserved" expresses a value affirming the sanctity of human life. Another statement, "personal freedom is good," also expresses a value, asserting the importance of the abstract notion that liberty and choice for individuals is a desired end-state. Values are not very frequently measured in surveys, but there are a few instances of values among the collection of measures assembled here.

Often confused with beliefs and values is the concept of *attitude*. The attitude concept is often described as "the most distinctive and indispensable concept in contemporary social psychology" (Allport, 1968). Attitudes are predispositions to respond or behave in particular ways toward social objects, along a positive or negative dimension (e.g., approval vs. disapproval, approach vs. avoidance, satisfaction vs. dissatisfaction). Attitudes are often thought to have emotional, cognitive, and behavioral dimensions, all of which are evaluative in nature. Such evaluations are often easily manipulated and are subject to situational factors. Some researchers have concluded that there is little evidence that stable, underlying attitudes can be said to exist (e.g., Abelson, 1972; Wicker, 1969), and even among those who accept the theoretical legitimacy of the attitude concept, there is considerable skepticism that the concept applies to all members of the population (Converse, 1964, 1970, 1974). While the attitude concept is an important one in its own right, in part because it is often easier to measure than other aspects of cognitive functioning, it is often viewed as epiphenomenal because it is derived from things more basic, namely values and beliefs. This should trouble the analyst seeking to measure attitudes using the survey interview.

There are some clear differences between *attitude* on one hand and *values* on the other. First, virtually all attitude theorists admit that the concept of attitude is derived from other things, and therefore is "not a basic, irreducible element of personalty" (Rokeach, 1970, p. 112). Attitudes are dispositional in the same sense as we have defined values, but values are more basic than attitudes, and are often thought to underlie them. What Rokeach (1970, p. 112) defines as an *attitude*—"a relatively enduring organization of beliefs around an object or situation predisposing one to respond in some preferential manner"—some would refer to as a *value*. Also, although attitudes are conceived of with respect to specific objects and situations, values can be thought of as "abstract ideals, positive or negative, not tied to any specific attitude object or situation, representing a person's beliefs about ideal modes of conduct and ideal terminal goals" (Rokeach, 1970, p. 124). Attitudes and self-evaluations are similarly often confused, since the latter can be thought of as attitudes toward one's self. Therein is the critical difference imposed here—attitudes are those evaluative variables in which the object of the response is not the self.

We included *all* questions in these six panel studies that met our basic design requirements. Questions were excluded that were primarily measures of latent classes rather than continuous latent variables, e.g., sex, marital status, race, and religious preference. We return to the question of estimating the reliability of measures of latent classes in a subsequent discussion. Open-ended questions yielding nonordered categories were excluded, unless there was an available metric in which to scale the responses. We therefore excluded open-ended questions inquiring, for example, about "What R likes about the Republican Party?" We included measures of occupational prestige, where the occupation and industry codes could be converted to a standard prestige or socioeconomic index (SEI) score (Hauser and Warren, 1997).

Questions were excluded if the actual question wording was different across waves. A substantial number of questions could not be used because the study staff changed the question wording from wave to wave. Or, in several cases questions were excluded because the questions sought updated information from the previous interview, e.g., " . . . since the last interview has your health changed?" Similarly, questions were excluded if the number of response options used was different across waves. Also, questions were excluded if the "object" of the questions was not the same across waves (e.g., a question in the election studies asked about the approval or disapproval of presidential performance, but the presidents were different across the waves of the study). This latter exclusionary principle removed any variables recording interviewer characteristics, such as the age, education, or years of experience of the interviewer. Since the interviewers are typically not the same person, this is a clear case where the object of the question is different. Finally, questions were also excluded if it was clear from the skip patterns that the subsamples responding to them would be small. Ultimately, our requirement in the present study was that our reliability estimates needed at least 50 cases to be included, and in some cases these questions were excluded before they got into our question pool. In other cases we were not aware of the available sample sizes until we actually analyzed the data. Employing the exclusionary rules listed here resulted in 543 "analyzable" questions, that is, questions that survived these exclusionary principles. Further, 63 of these 543 questions were excluded for reasons associated with the analysis—either because of small sample sizes (n = 43) detected at the time of the analysis or because of difficulties with the fit of the model (n = 20). The Appendix lists the 483 questions that were ultimately analyzeable. This appendix does not list all of the survey questions excluded for all the major reasons given above.

Table 6.2 presents the counts of the number of questions used in the present study. These numbers are given by panel study and by categories of source/content. This table includes the 483 replicated survey questions available from the six panel studies. The information obtained in the questions come from four different sources: (1) self-reports, (2) proxy reports, (3) interviewer reports, and (4) organization reports. Although most of our attention in subsequent analyses focuses on the self- and proxy report measures, we include the other reports here in order to illustrate some interesting differences in reliability of measurement by the source of the information obtained.

Ultimately, we will analyze variation in reliability across these measures due to (1) the source of information, (2) the context of the question, (3) the content of the

Table 6.2. Source and content of survey measures included in analysis by panel study

Source/Content	NES 50s	NES 70s	NES 90s	ACL	SAF CH	SAF MO	Total
Self reports							
Facts	13	10	15	29	2	10	79
Beliefs	7	44	25	13	17	11	117
Values	7	6	13	1	7	9	43
Attitudes	1	25	35	1	6	8	76
Self-assessments	1	---	3	12	9	2	27
Self-perceptions	15	14	13	31	5	15	93
Proxy reports							
Facts	1	9	2	---	---	1	13
Interviewer reports							
Facts	1	3	---	---	---	---	4
Beliefs	---	5	6	11	---	---	22
Organization reports							
Facts	1	6	2	---	---	---	9
Total n	47	122	114	98	46	56	483

question, and (4) the formal attributes of the question (see Chapters 7, 8, and 9). For reasons that will become clear as we proceed, for most purposes we limit our analysis of methods of reliability estimation to self- and proxy-report measures. Table 6.3 gives the final breakdown of the 426 self- and proxy report measures by panel study used in the analyses reported here and in subsequent chapters.

Across these several categories of types of variables, the measures can be classified according to the specific type of content measured. The domains covered by these studies, along with an indication of their major sources among the panel studies used here are: (1) Labor market behavior and aspirations (ACL, SAF, and NES)—employment status, educational experiences, and occupational aspirations; (2) Housing and residential mobility—ownership of housing, housing costs, difficulties with housing costs, migration, and residential preferences; (3) Socioeconomic position (SAF, ACL, and NES)—educational attainment, occupational status attainment, income, and wealth; (4) Socioeconomic values (SAF)—consumption aspirations, expectations about future standard of living; (5) Health and health behavior (ACL)—the number of chronic conditions, an index of functional health, self-ratings of health, and a measure of activity limitations due to health problems. ACL health questions were developed in consultation with ACL's physician consultants; (6) Chronic stress and psychological well-being (ACL)—standard depression scales and measures of job stress, financial stress, and family stress; (7) Productive activities (ACL)—nine broad activity categories: regular paid work, irregular paid work, child care, housework, home maintenance, formal volunteer work, informal help to family and friends, care to a person with chronic problems, and helping a person with acute problems, measured in terms of the hours spent annually in each type of productive activity; (8) Self-assessed well-being—overall life satisfaction and

Table 6.3. Source and content of self- and proxy reports of survey measures by panel study

Content	Source						Total
	NES 50s	NES 70s	NES 90s	ACL	SAF CH	SAF MO	
Facts[1]	13	13	13	29	2	9	79
Beliefs	7	44	24	13	17	11	116
Values	7	6	13	1	7	9	43
Attitudes	1	25	35	1	6	8	76
Self-assessments	1	0	2	11	9	2	25
Self-perceptions	13	12	11	31	5	15	87
Total n	42	100	98	86	46	54	426

[1]Includes both self and proxy reports.

self-efficacy (ACL), and in the NES surveys trust in government or political alienation; (9) Social relationships and socioenvironmental supports (ACL)—frequency of contact and positive and negative aspects of relationships with five key others: spouse, mother, father, grown children, and friends and relatives; measures of confidants and network density are also included; (10) Attitudes toward family issues (SAF)—fertility aspirations/expectations, use of birth control; attitudes toward abortion; sex-role attitudes; attitudes toward cohabitation, marriage, and divorce; and behavioral reports of the division of household labor and finances; and (11) Electoral behavior and political attitudes (NES)—five major categories: political efficacy and beliefs about government responsiveness, attitudes, and beliefs about specific government policies, political orientations toward social and political groups, e.g., racial minorities, unions, political parties, self-assessed ideological ratings, and political party identification. This is by no means an exhaustive list of the variables in these conceptual domains, but it does give evidence of the wide range of content, typical of that studied in many social surveys.

6.4 STATISTICAL ESTIMATION STRATEGIES

We employ two different estimation strategies to the models discussed in the previous chapter: the *equal reliability* (Heise, 1969) and *equal error variances* models (Wiley and Wiley, 1970; Werts, Jöreskog, and Linn, 1971; Werts, Linn and Jöreskog, 1977, 1978; Jöreskog, 1970; 1974, 1978). As indicated in the previous chapter, both of these estimation strategies assume that measurement error is random and that there is linear regression to the mean in the true scores over time. These models have been shown to be useful in reliability estimation (Alwin, 1989, 1992), and there is evidence of their robustness (Palmquist and Green, 1992; Coenders et al., 1999). Recall from the discussion in the previous chapter that the main difference between the two approaches has to do with the nature of the constraints imposed on the model. The formal properties of the model are the same—the Heise model uses a correlation matrix as input whereas the Wiley-Wiley approach uses a covariance matrix. It is entirely possible to think of the Heise model as simply the Wiley-Wiley model applied to correlational (as opposed to covariance) data. The consequences of the two strategies are different, however, in that the Heise approach results in the assumption of equal reliability whereas the Wiley-Wiley strategy is less restrictive, allowing unequal reliabilities over waves, but requiring equal error variances.

We obtained information from the panel data described above that will permit either set of estimates. In the case of the Heise approach there is a single estimate of reliability, whereas in the Wiley-Wiley approach, there are P reliability estimates in a P-wave model, so given the data sets employed we will have three estimates for each measure from those models. We employ a simple average of the three in the following analyses. Also, as we pointed out in the previous chapter, the time-2 reliability is identical in the two models and can be shown to be an unbiased estimate of measurement reliability, given the assumptions of the model. Thus, we expect the Heise and Wiley-Wiley estimates to be relatively similar and will depart from one another only when there is heterogeneity in variance across waves. Nonetheless, we investigate the veracity of this assumption later in this chapter.

We apply these strategies to the data under different sets of assumptions about the scale of the observed variables. For survey questions with 15 or fewer categories, we will also make use of tetrachoric and polychoric (tetrachoric in the case of dichotomous variables and polychoric for all others) correlations as input into the Heise model. We employ two approaches to these correlations, one in which the thresholds are unconstrained across waves—we call these estimates "Heise(1) estimates"— and one in which we constrain equal thresholds across waves—which we call "Heise(2) estimates." In the latter case, where thresholds are constrained equal, Jöreskog (1994) has defined an asymptotic covariance matrix that can be used as input into the Wiley-Wiley model. Thus, we have one estimate that can be referred to as a "Wiley-Wiley model applied to polychoric-based covariance estimates." Because this will generate two Heise estimates for polychoric correlations, this will generate five sets of reliability estimates that can then be subjected to meta-analysis, wherein we will predict variation in reliability on the basis of theoretical hypotheses and prediction.

We believe these strategies employ an exhaustive set of approaches involving the application of *state-of-the-art* structural equation method (SEM) estimation strategies, including both the use of standard Pearson correlation and covariance approaches, as well as newer tetrachoric and polychoric approaches for binary and ordinal data (see Jöreskog, 1990, 1994; Muthén, 1984–2004). These approaches can be implemented by Jöreskog and Sörbom's *PRELIS* and *LISREL8* software (1996a, 1996b) and M*plus* (Muthén and Muthén, 2001–2004). Both software packages include estimation using Jöreskog's (1994) polychoric-based asymptotic covariance matrix approach with thresholds constrained equal.

6.4.1   Analysis Strategy

As noted above, we employ five different approaches to the estimation that differ in the assumptions made about the scale properties of the observed measures, that is, whether they are measured on interval versus ordinal scales, and assumptions about the properties of patterns of measurement error variance. We developed the critical differences in these approaches in the discussion above and in the previous chapter. These estimation strategies were all based on the SEM approach, and they all assume a continuous underlying latent variable, consistent with classical true-score models. In Table 6.4 we display the array of approaches to reliability estimation for several classes of models using longitudinal data. The table cells are defined according to cross-classification of the nature of the observed and latent variables. These combinations determine in most cases the appropriateness of a particular estimation strategy, that is, whether a particular estimation strategy can be justified in a given case. We have already discussed the *continuous latent variable* approaches, involving both ordinal and continuous/interval observed variables, and although we focus primarily on the reliability of measuring continuous latent variables, we include in Table 6.4 approaches to estimating reliability for measures of *latent classes*, a topic to which we return below.

The five estimates we use for the quasi-simplex model involving continuous latent variables are as follows: (1) the Heise model applied to Pearson correlations; (2) the Wiley-Wiley model applied to the covariance matrix based on Pearson correlations; (3) the Heise model applied to polychoric correlations, using unequal thresholds across waves [Heise(1)]; (4) the Heise model applied to polychoric correlations, using equal thresholds across waves [Heise(2)]; and (5) the Wiley-Wiley model applied to Jöreskog's (1994) asymptotic covariance matrix based on polychoric correlations.

The purpose of the following section is to analyze the differences in these estimates and to draw conclusions regarding the most useful methods of estimation

Table 6.4. Reliability estimation strategies for longitudinal data

	Latent Variable			
	Categoric		Continuous	
	No Change Present	Change Present	Equal Reliability Model	Equal Error Variance Model
Observed Variable				
Categoric/discrete	Latent Class Model	Latent Markov Model		
Categoric/ordinal	Latent Class Model	Latent Markov Model	SEM--Polychoric Correlation Matrix	SEM--Polychoric Asymptotic Covariance Matrix
Continuous/interval			SEM--Pearson Correlation Matrix	SEM--Pearson Covariance Matrix

for present purposes. There are a few additional things to note when interpreting the analyses that follow. First, our analyses of differences in these five approaches to reliability estimation use weighted data where possible. The ACL data, and the 1950s and 1990s NES panels, employ sample weights to adjust for nonresponse and or disproportionate sampling strategies. The NES 1970s panel data release published no sample weights, so we use unweighted data exclusively in that case. The two SAF panels are based on simple random samples of specific populations and are therefore self-weighting samples. Second, we have tried to remove as much redundancy in the measures as possible. Our goal has been to estimate the reliability of single survey questions, but in some cases the data released use measures based on more than one question. Where we employ such *synthesized* variables, we exclude the constituent parts from the analysis. In other cases, the study published a measure *derived* from another, e.g., "bracketed age" was derived from "age." In such cases we employ the source variable and exclude the derived variable from our analyses, although redundant measures are included in the Appendix. This latter document provides a detailed list of the questions that were excluded because of redundancy and the reasons for their exclusion. Third, over and above the 483 questions analyzed by "continuous variable" methods, several variables were excluded because they were available only on small samples, or because of poor fit of the SEM models to the data. Finally, all of the estimates presented in this part of the chapter are based on "listwise present" cases, that is, in each of the studies described above the sample of people on which we based our reliability estimates was required to have "nonmissing" data at all three waves of the study. Later in the chapter we return to the issue of whether this approach provides biased estimates of reliability.

6.5 COMPARISON OF METHODS OF RELIABILITY ESTIMATION

In Table 6.5 we display the five different types of estimates for several groups of variables arranged by topic. This table illustrates some of the similarities and differences in results across the different estimates as well as across content areas. Generally speaking, the estimates based on polychoric correlational methods tend to be substantially higher. The table also suggests that facts can be more reliably measured than nonfacts. We return to the latter issue—our purpose here is to systematically examine the differences in these results by methods of reliability estimation, that is, the similarities and differences across the columns of Table 6.5.

The comparisons we wish to make in this section involve "matched pairs" in the sense that the "samples" of survey questions on which we have estimated reliabilty involve the same cases. One might be tempted to use a difference-of-means test to examine these differences, but this would be inappropriate (Blalock, 1972, p. 233). If n is the number of measures examined in this study, it should be apparent that we do not have two independent samples of 2n cases (n in each sample) that have been independently selected. In the analysis of differences in strategies of reliability estimation presented in this chapter we employ a direct pair-by-pair comparison by examining the difference score for each pair and testing the hypothesis that the mean of the pair-by-pair differences is zero.

Table 6.5. Five estimates of reliability for survey measures by type of content

| Content | Reliability Estimates | | | | |
	Pearson/ Heise[1]	Pearson/ Wiley-Wiley[2]	Polychoric/ Heise(1)[3]	Polychoric/ Heise(2)[4]	Polychoric/ Wiley-Wiley[5]
Facts	0.766 (72)	0.768 (72)	0.766 (40)	0.756 (37)	0.758 (37)
Beliefs	0.479 (116)	0.492 (116)	0.582 (114)	0.585 (114)	0.600 (114)
Values	0.555 (43)	0.552 (43)	0.653 (43)	0.661 (42)	0.664 (41)
Attitudes	0.603 (76)	0.594 (76)	0.676 (76)	0.665 (74)	0.666 (72)
Self-assessments	0.497 (25)	0.502 (25)	0.600 (25)	0.636 (25)	0.637 (25)
Self-perceptions	0.505 (87)	0.506 (87)	0.631 (87)	0.626 (86)	0.630 (83)
Proxy facts	0.808 (7)	0.820 (7)	1.000 (1)	1.000 (1)	1.000 (1)
Total	0.569 (426)	0.572 (426)	0.641 (386)	0.640 (379)	0.646 (373)

[1]Estimates based on Pearson correlations and the assumption of equal reliabilities (see Heise, 1969).
[2]Estimates based on the Pearson-based covariance matrix and the assumption of equal error variances (see Wiley and Wiley, 1970). The average of the three estimates is presented.
[3]Estimates based on polychoric correlations with unconstrained thresholds and the assumption of equal reliabilities. Excludes items with 16+ response categories.
[4]Estimates based on polychoric correlations with thresholds constrained equal and the assumption of equal reliabilities. Excludes items with 16+ response categories.
[5]Estimates based on the polychoric-based asymptotic covariance matrix with thresholds constrained equal (see Jöreskog, 1994) and the assumption of equal error variances. Excludes items with 16+ response categories and feeling thermometers.
Note: The number of questions upon which reliability estimates are based is given in parentheses.

Note that the standard error of the pair-by-pair difference is a positive function of the variances (across measures) of the two reliability estimates and a negative function of 2 times their covariance. Given the high degree of correlation between these estimates, we expect these standard errors to be quite small, and therefore it will take only small differences in reliabilities across methods of estimation to produce significant results. The upshot of this that when we reject the null hypothesis we will need to rely on other "side information" to interpret the results, but that the failure to reject the null hypothesis will normally mean that there are no real substantive differences between reliability estimates.

Table 6.6.   Comparison of approaches to measurement reliability: Pearson product-moment correlation-based estimates

| | | Reliability Estimates | | | |
Content	n	Heise	Wiley-Wiley	t	p-value
Facts	72	0.766	0.768	0.486	0.629
Beliefs	116	0.479	0.492	3.925	0.000
Values	43	0.555	0.552	0.643	0.524
Attitudes	76	0.603	0.594	2.616	0.011
Self-assessments	25	0.497	0.502	1.206	0.240
Self-perceptions	87	0.505	0.506	0.307	0.760
Proxy facts	7	0.808	0.820	1.680	0.144
Total	426	0.569	0.572	1.571	0.117
	F-ratio	38.277	36.434		
	p-value	0.000	0.000		
Facts	79	0.770	0.772	0.831	0.408
Non-facts	347	0.523	0.526	1.376	0.170
Total	426	0.569	0.572	1.571	0.117
	F-ratio	175.748	180.382		
	p-value	0.000	0.000		

The first comparison we make (shown in Table 6.6) is between the Heise and Wiley-Wiley estimates—estimates based on equal reliability vs. equal error variances—applied to standard Pearson correlation/covariance methods. These results show that there is little difference in reliability estimates by whether or not one assumes reliability is equal over waves. This provides further support for results reported previously (Alwin, 1989), which argued that the two approaches produced reliability estimates that were trivially different. These results seem to confirm this conclusion for Pearson-based correlation/covariance methods.

We can further examine this issue using the polychoric-based correlation/covariance estimates. In Table 6.7 we compare the Heise vs. Wiley estimates where the polychoric-based estimates are used. Note that whenever polychoric-based reliability estimates are compared the sample size drops because we did not calculate polychoric-based reliability estimates for questions that had sixteen or more categories.[1] This reduces factual measures by one-half. Again there is very little support for the argument that these two methods produce different results. Only in the case

---

[1]This was done for two reasons: (1) Jöreskog (1990) argues that variables with 16 or more categories can be considered nonordinal or interval variables, and (2) our efforts to estimate polychoric correlations for variables with more than 15 categories more often than not resulted in convergence problems.

Table 6.7. Comparison of approaches to measurement reliability: polychoric correlation-based estimates

| Content | n | Reliability Estimates | | t | p-value |
		Heise	Wiley-Wiley		
Facts	37	0.756	0.758	0.491	0.626
Beliefs	114	0.585	0.600	3.613	0.000
Values	41	0.667	0.664	0.557	0.581
Attitudes	72	0.680	0.666	2.863	0.006
Self-assessments	25	0.636	0.637	0.190	0.851
Self-perceptions	83	0.636	0.630	0.800	0.426
Proxy facts	1	1.000	1.000	---	---
Total	373	0.645	0.646	0.080	0.936
	F-ratio	8.886	6.316		
	p-value	0.000	0.000		
Facts	38	0.762	0.764	0.491	0.626
Non-facts	335	0.632	0.632	0.017	0.987
Total	373	0.645	0.646	0.080	0.936
	F-ratio	26.444	24.638		
	p-value	0.000	0.000		

of *beliefs* is there a statistically significant difference, but even in this case the difference is quite small. Our provisional conclusion, then, is that whether one uses Pearson- or polychoric-based approaches, the Heise and Wiley-Wiley strategies produce very similar results with no systematic differences.

As we noted earlier, we have employed two types of polychoric correlation estimates, those in which the thresholds are allowed to freely vary across waves [Heise(1)] and those that constrain the thresholds to be equal across waves [Heise(2)]. The comparison of polychoric-based results using equal vs. unequal thresholds is shown in Table 6.8. There is very little difference in these results in the measurement of facts and most categories of nonfacts, although the Heise(2) estimates are always systematically higher. The only case in which it appears to make a difference is the case of *self-assessments* where the equal thresholds estimate is significantly higher than the unequal thresholds estimate. This is the largest difference across columns that we have observed thus far, .60 versus .64. This is a difference worth explaining.

Tables 6.9 and 6.10 compare, for both the Heise and Wiley-Wiley approaches, how the Pearson- and polychoric-based estimates are different. Given that polychoric correlations are generally higher than Pearson correlations, ceteris paribus, we expected that the reliability estimates based on polychoric correlations would be systematically higher. In Table 6.9 we compare the Heise estimates where the

Table 6.8. Comparison of approaches to measurement reliability: polychoric correlation-based estimates—Heise estimates

Content	n	Heise (1)	Heise (2)	t	p-value
		Reliability Estimates			
Facts	37	0.751	0.756	0.658	0.515
Beliefs	114	0.582	0.585	0.848	0.399
Values	42	0.649	0.661	1.315	0.196
Attitudes	74	0.673	0.665	0.793	0.431
Self-assessments	25	0.600	0.636	3.101	0.005
Self-perceptions	86	0.627	0.626	0.097	0.923
Proxy facts	1	1.000	1.000	---	---
Total	379	0.636	0.640	1.125	0.261
	F-ratio	9.864	7.580		
	p-value	0.000	0.000		
Facts	38	0.757	0.762	0.658	0.515
Non-facts	341	0.622	0.626	1.003	0.317
Total	379	0.636	0.640	1.125	0.261
	F-ratio	31.487	26.573		
	p-value	0.000	0.000		

polychoric estimates allow unequal thresholds, i.e., Heise(1) estimates. In Table 6.10 we compare the Wiley estimates using Pearson based covariances with Wiley estimates based on Jöreskog's (1994) asymptotic covariance matrix.

In both cases it is clear that there are highly significant differences between reliability estimates for the quasi-simplex model depending on whether Pearson-based correlations/covariances or polychoric-based correlations/covariances are used. In each set of comparisons the difference averaged .10 on the 0–1 metric on which we assess reliability of measurement. These differences not only are statistically significant; they also imply major differences in results. Estimates of reliability based on polychoric correlations or covariances are uniformly higher than Pearson-based estimates. As noted, we expect there to be a convergence in these two estimates as the number of categories approaches 15, that is, as we approach "interval scale" measurement. We return to this issue when we examine the link between reliability of measurement and the number of response categories employed.

Recall that in the polychoric-based models, the model estimates the reliability of Y*, an "ideal" continuous variable thought to underlie the categories of the measured variable. In the case of an interval-scale variable—arbitrarily defined here as an ordered set of 16 or more categories (see Jöreskog, 1990)—Y and Y* are thought to be in very close correspondence. In the case of an ordinal variable with fewer than 16 categories, the latent variable Y* is continuous, but the observed variable is thought to be a coarsely categorized version of the latent variable. The purpose of

Table 6.9. Comparison of approaches to measurement reliability: Pearson versus polychoric correlation-based Heise estimates

Content	n	Heise[1]	Heise (1)[2]	t	p-value
		Reliability Estimates			
Facts	40	0.700	0.766	6.716	0.000
Beliefs	114	0.471	0.582	20.696	0.000
Values	43	0.555	0.653	15.929	0.000
Attitudes	76	0.603	0.676	10.555	0.000
Self-assessments	25	0.497	0.600	9.265	0.000
Self-perceptions	87	0.505	0.631	17.993	0.000
Proxy facts	1	0.976	1.000	---	---
Total	386	0.541	0.641	31.668	0.000
	F-ratio	19.839	11.384		
	p-value	0.000	0.000		
Facts	41	0.707	0.772	6.742	0.000
Non-facts	345	0.521	0.625	31.797	0.000
Total	386	0.541	0.641	31.668	0.000
	F-ratio	60.499	39.015		
	p-value	0.000	0.000		

[1]Estimates from Pearson-based correlations.
[2]Estimates from polychoric-based correlations.

the tetrachoric and polychoric correlation/covariance approaches is to estimate the correlations among such variables under the assumption that Y^* is continuous and normally distributed. On the assumption that coarse measurement at the level of the survey response will attenuate the associations observed among such variables and by correcting for the fact that we are observing crudely measured variables, we will disattenuate the observed associations among variables. However, in using the quasi-simplex models applied to polychoric-based correlations/covariances it must be made clear that we are estimating the reliability of Y^* rather than Y, that is, the "ideal" measure of Y rather than Y itself.

6.6 THE PROBLEM OF ATTRITION

The analysis of panel data, such as that proposed here, must deal with the inevitable problem of attrition. This can be problematic if the attrition is great, and/or if nonrespondents are systematically different in their reporting errors. In most panel studies there is substantial attrition, resulting in the cumulation of nonresponses over waves. In the best of cases the analysis will be based on less than one-half of the original sample (see Coenders et al, 1999). There is no way to effectively control for this

Table 6.10.   Comparison of approaches to measurement reliability: Pearson versus polychoric correlation-based Wiley-Wiley estimates

Content	n	Reliability Estimates Wiley-Wiley[1]	Wiley-Wiley[2]	t	p-value
Facts	37	0.692	0.758	6.163	0.000
Beliefs	114	0.485	0.600	16.096	0.000
Values	41	0.554	0.664	8.849	0.000
Attitudes	72	0.598	0.666	5.497	0.000
Self-assessments	25	0.502	0.637	8.806	0.000
Self-perceptions	83	0.507	0.630	10.356	0.000
Proxy facts	1	0.975	1.000	---	---
Total	373	0.542	0.646	21.624	0.000
	F-ratio	16.041	6.316		
	p-value	0.000	0.000		
Facts	38	0.700	0.764	6.195	0.000
Non-facts	335	0.524	0.632	20.967	0.000
Total	373	0.542	0.646	21.624	0.000
	F-ratio	52.077	24.638		
	p-value	0.000	0.000		

[1]Estimates from Pearson-based covariances.
[2]Estimates from polychoric-based covariances.

problem, except to perhaps realize the limits of one's generalizations. The only way we have to gauge the effect this has on our reliability estimates is to examine the consistency of the results under different approaches to addressing the problem of incomplete data, particularly the comparison of the "listwise" approach used above with other approaches to incomplete data.

Even though the data described above were collected by a social science survey organization well known for its high quality of data collection, there is a substantial amount of attrition. Tables 6.11–6.15 present the disposition of cases in the six panels used here. In our analysis we deal not only with "unit" nonresponse and "unit" attrition, but "item" nonresponse as well, so the numbers of cases given in these tables will not correspond exactly to the cases actually analyzed. The first three tables present patterns of "missingness" across waves of the NES panels. The footnotes to the tables (Tables 6.11, 6.12 and 6.13) describe in detail how we assembled these data sets from the NES cross sections. The NES is complicated given the existence of the repeated cross sections as well as the panel design component embedded within them. In the tables for the NES we include all cases with at least one wave of data, even if it is not part of the panel design. In the ACL and SAF studies, by contrast, new cross sections are not added at each wave and the majority of the cases are among those who provided a response at wave 1. The level of attrition in the ACL

Table 6.11.   Incomplete data patterns for NES 50s panel and
cross-section cases[1]

1956	1958	1960	Unwt N	Percent[2]
1	1	1	1,132	66.8
1	1	0	181	10.7
1	0	1	107	6.3
1	0	0	569	---
X	1	1	275	16.2
X	1	0	99	---
X	X	1	166	---
			2,529	100.0

1 = Data Present; 0 = Missing data;
X = No data because respondents had yet to enter study.

[1]Data for panel respondents participating in interviews conducted during 1956, 1958 and 1960 are taken from the American Panel Study:
1956, 1958, 1960 (ICPSR 7252), the American National Election
Study, 1958 (ICPSR 7215), and the American National Election Study,
1956 (ICPSR 7214). For panelists interviewed in 1960 and at least one
other time in 1956 or 1958, their data can be retrieved from the panel
file (ICPSR 7252). Data for panelists who did not receive interviews in
1960 but were a part of the cross-sectional study conducted in 1958 can
be retrieved from the 1958 cross-sectional study (ICPSR 7215) and the
1956 cross-sectional study (ICPSR 7214). The data for panelists not
interviewed in 1960 and not a part of the 1958 cross-sectional sample,
but still receiving at least one interview in 1956 and an interview in
1958 have been lost from current NES records.
[2]Percent of panel cases: cross-section only cases excluded.

study is probably typical for most panel studies, that is, the group with complete
cases across all three waves represents about two-thirds of the panel cases. The
levels of attrition in the SAF study represent the best that can probably be realized,
in that the investigators in this study have taken every step necessary to maintain
contact and foster the cooperativeness of the respondents (see Freedman, Thornton
and Camburn, 1980; Thornton, Freedman, and Camburn, 1982). Still, even where
special efforts have been taken to ensure high response rates over time, if incomplete
cases due to attrition are deleted from the analysis, then the sample is censored, with
possible consequences for variance estimates.

6.6.1   Estimation of Reliability in the Presence of Incomplete Data

The literature on missing data (e.g., Little and Rubin, 1987, 1989; Little and Schenker, 1995) distinguishes between data missing at random (MAR) and data missing
completely at random (MCAR), both of which are assumed to be missing independent of the true value of the missing variable. Both types of incomplete data are

Table 6.12. Incomplete data patterns for NES 70s panel and cross-section cases[1]

1972	1974	1976	Unwt N	Percent[2]
1	1	1	1,296	65.8
1	1	0	328	16.7
1	0	1	24	1.2
1	0	0	1,057	---
X	1	1	321	16.3
X	1	0	154	---
X	X	1	1,275	---
			4,455	100.0

1 = Data Present; 0 = Missing data;
X = No data because respondents had yet to enter study.

[1]Data for panel respondents participating in interviews conducted during 1972, 1974, and 1976 can be retrieved from the American National Election Series: 1972, 1974, 1976 (ICPSR 2407). All respondents interviewed in 1972 were required to have completed both the pre- and post-election interviews in order to be included in the panel study.
[2]Percent of panel cases; cross-section only cases excluded.

Table 6.13. Incomplete data patterns for NES 90s panel and cross-section cases[1]

1992	1994	1996	Unwt N	Percent[2]
1	1	1	597	40.4
1	1	0	162	11.0
1	0	1	0	0.0
1	0	0	246	---
X	1	1	719	48.6
X	1	0	317	---
X	X	1	398	---
			2,439	100.0

1 = Data Present; 0 = Missing data;
X = No data because respondents had yet to enter study.

[1]Panel respondents interviewed during 1992, 1994, and 1996 are drawn from the American National Election Studies, 1992–1997: Combined File (ICPSR 2407). Respondents interviewed in 1992 were required to have completed both pre- and post-election interviews in 1992 in order to be included in the panel study. Further, only respondents who received interviews in 1994 were eligible for reinterview in 1996.
[2]Percent of panel cases; cross-section only cases excluded.

Table 6.14. Incomplete data patterns for ACL panel cases[1]

1986	1989	1994	Unwt N	Percent[2]
1	1	1	2,223	73.1
1	1	0	644	21.2
1	0	1	175	5.8
1	0	0	575	---
			3,617	100.0

1 = Data Present; 0 = Missing data.

[1]Panel respondents were interviewed in 1986, 1989, and 1994. In Waves 2 and 3 no attempt was made to interview Wave 1 nonrespondents. The data for 1986 and 1989 is available from ICPSR study number 6438.
[2]Percent of panel cases; cross-section only cases excluded.

Table 6.15. Incomplete data patterns for SAF Mothers and SAF Children panel cases

SAF Mothers Panel Cases[1]					SAF Children Panel Cases[1]				
1980	1985	1993	Unwt N	Pct %	1980	1985	1993	Unwt N	Pct %
1	1	1	879	79.0	1	1	1	875	78.6
1	1	0	48	4.3	1	1	0	29	2.6
1	0	1	4	0.4	1	0	1	12	1.1
0	1	1	1	0.1	0	1	1	18	1.6
1	0	0	38	3.4	1	0	0	19	1.7
0	1	0	1	0.1	0	1	0	1	0.1
0	0	1	0	0.0	0	0	1	1	0.1
0	0	0	142	12.8	0	0	0	158	14.2
			1,113	100.0				1,113	100.0

1 = Data Present; 0 = Missing data.

[1]Panel respondents were first interviewed in 1962 as part of the Detroit Area Study, 1962: Family Growth in Detroit (ICPSR 7401). In order to be eligible for the study, Detroit-area Caucasian mothers had to have given birth to their first, second, or fourth child in 1961. Mothers were interviewed from 1962 through 1993 at seven possible interview points. In 1980, 1985, and 1993, both mothers and their children were interviewed. Data for these interviews can be retrieved from the Intergenerational Study of Parents and Children, 1962–1993: [Detroit] (ICPSR 9902).

possible when the major source of their missingness is attrition. With MAR data the missingness depends on other variables, whereas with MCAR data the missingness is statistically independent of all other variables. When the missingness is MCAR, then it is possible to consider the differences between retained and attrited cases as due to random factors and to therefore ignore such differences. In longitudinal research such as this, it is important to be able to assess the extent to which attrition biases the results. This is especially the case when results rely on "listwise present" cases, as in the analyses undertaken above.

We conducted sensitivity analyses using two alternative approaches. First, using the conceptual tools proposed in the literature on pattern-mixture models, we employed a multiple-group modeling strategy that incorporates information on incomplete data patterns into the measurement model (see Diggle, Liang and Zeger, 1994; Hedecker and Gibbons, 1997). Allison (1987) introduced a pattern-mixture approach into the structural equation method (SEM) literature wherein he proposed that the problem of incomplete data be dealt with by specifying a model in which multiple-group submodels are formulated for different patterns of missing data. The Allison (1987) approach has also been employed in the SEM literature to deal with attrition (see McArdle and Bell, 2000; McCardle and Hamagami, 1992), but its use is not widespread. Rather than attempting to either impute "estimates" of scores that are missing at particular waves, this approach constructs a model for each missing data pattern that uses all data available within that group. The basic idea is to treat each unique pattern of incomplete data as a separate group in a multiple-group SEM model, where variables from absent waves are treated as "unobserved" latent variables. A model is then estimated simultaneously in each group, where the model parameters are assumed equal across groups. For example, in applying the Allison approach to variables in the NES90s data, one defines seven groups for which the model holds equally, and where the pattern involves a "0," the observed variable at that location is specified as unobserved.

The several submodels are then estimated simultaneously, under the assumption that the model parameters are equal across groups. As an example, Figure 6.1 specifies the nature of the model we estimate for each of the patterns of nonresponse in the case of the typical measure from the 1990s NES series. This model employs the innovative idea that Allison (1987) had to formulate observed variables missing at a particular wave as latent variables. Note that the groups in Figure 6.1 correspond to the rows of Table 6.13, and following the diagrammatic conventions of SEM methods, we have employed circles for latent variables and boxes for observed variables. Note that where a wave is missing, the model specifies the missing observed variable as a latent variable. The approach requires that all the parameters of the submodels be constrained equal across groups.

Second, we also handle the incomplete data using fixed-information maximum-likelihood (FIML) direct estimation techniques, which analyzes the same cases included in the Allison model, but handles the missing data differently. The FIML approach is to simply treat attrition as incomplete data and employ "direct estimation" techniques that rely on *full-information maximum-likelihood* (FIML). This approach, described in detail by Wothke (2000) and implemented in AMOS (Arbuckle and Wothke, 1999), involves a single-group model in which absent waves are treated as incomplete or missing data. Note that this is *not* an "imputation" approach. If data are MAR, the FIML approach yields model parameter estimates that are both consistent and efficient (Wothke, 2000).

Another way to deal with MAR processes is to find those variables that are predictive of attrition. We have not done that in the models we have estimated here, although we would point out that if attrition is linked to initial wave-1 measures, then to this extent some of the MAR processes are controlled in this manner.

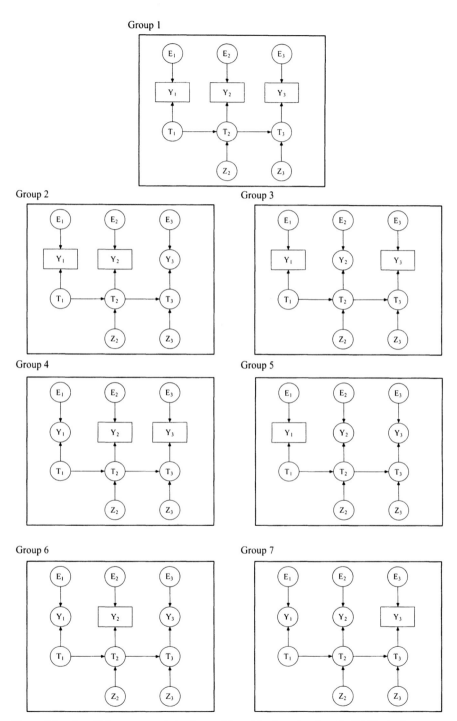

Figure 6.1. A multiple-group quasi-simplex model used for the estimation of reliability in the presence of incomplete data.

All of these approaches rely on the principles of maximum-likelihood estimation (Little and Rubin, 1987, 1989). There is a range of structural equation packages that can handle the Allison solution, e.g., *LISREL8* (Jöreskog and Sörbom, 1996b), M*plus* (Muthén and Muthén, 2001–2004), and *AMOS* (Arbuckle and Wothke, 1999). All of these packages provide maximum-likelihood approaches to estimation and permit the implementation of multiple-group specifications, which is important for the present application. These three software packages all permit direct input of raw data, but none currently allows one to apply sample weights to deal with sample design effects or poststratification weighting for nonresponse. Where we have wanted to employ sample weights, we have had to do our analyses in two steps—first we have calculated our correlation and covariance matrices under various weighting regimes outside the SEM software in SAS, and then read these matrices into the SEM programs. While it is possible to follow this strategy with the Allison approach, it is impractical in the present situation. Moreover, the FIML direct estimation approach cannot be implemented in this manner, so a major consequence of this is that we must use unweighted data to investigate the effects of attrition by comparing the results of different models. It turns out that, except in the ACL data set, it does not make any difference whether we employ weighted or unweighted data, but we return to this in the following discussion.

The main advantage of the listwise and Allison approaches, then, is that they can take sample weights and design effects into account, although this creates even more complex data management problems, as the covariance structures have to be imported into the model estimation procedure. In the present research we did not take sample weights into account in our analyses using the Allison approach.

One singular advantage of the Allison model is that theoretically it allows a test of the MCAR assumption. As Wothke (2000, p. 19) observes, if the Allison model fits the data, the data are MCAR. In other words, since the same model is imposed across all groups, i.e., both the group with complete data and the groups that involve varying patterns of attrition, the deviation from perfect fit reflects the failure of the parameters to be equal across groups. The lack of equality in parameters across groups defined by differing patterns of complete data suggests a violation of the MCAR assumption.

The analyses presented here (Tables 6.16–6.18) employ the average (over three waves) of the wave-specific reliability estimates based on the Wiley and Wiley (1970) approach, which we discussed earlier. These sensitivity analyses all rely on Pearson-based covariances among the variables and employ the self-report and proxy-report measures (N=426) only using the AMOS 5.0 software. In the results presented here we include only those variables for which there were no problematic solutions (at most 417 of 426).

As noted above, we were not able to employ sample weights where necessary in these sensitivity analyses because the available software for performing these analyses does not include procedures for employing sample weights. Three studies— ACL, NES50s and NES90s—employed sample weights, so we first compare our unweighted listwise results with the weighted listwise results (reported above) in

Table 6.16.   Comparison of listwise present Wiley-Wiley reliability estimates for weighted and unweighted data

		Reliability Estimates			
Panel	n	Weighted	Unweighted	t	p-value
ACL	86	0.575	0.562	2.286	0.025
NES 50s	42	0.603	0.600	0.923	0.362
NES 90s	98	0.604	0.600	1.060	0.292
Total	226	0.593	0.585	2.574	0.011

these three studies. A comparison of the two sets of results for each of these studies is given in Table 6.16. These two sets of results show there is very little difference in the mean reliability estimates—.593 versus .585. In the ACL data the differences are marginally significant.

The results of our analyses using the listwise, FIML, and Allison approaches are summarized in Table 6.17 via a comparison of the average reliability estimates for common measures using the three strategies. As these results indicate, the list-wise approach provides systematically higher estimates, a difference in reliability of approximately .01. This is true whether the comparison is with the FIML or Allison approaches. Given the fact that the standard errors of the differences examined here are very small (due to the high covariances among the estimates; see Table 6.18), even such small differences are statistically significant. There is, however, little or no difference between the estimates produced by the FIML and Allison approaches, which reinforces Wothke's (2000, p. 239) argument that the Allison multigroup approach can be regarded as an alternative to the FIML estimation with incomplete data. We see no reason to consider the results of the two methods to be any different, and either appears to provide an improvement to the listwise approach.

While the FIML and Allison approaches provide a powerful alternative to the list-wise approach employed in the previous analyses, the fact that their implementation in our analysis was limited to unweighted data may result in some hesitation in using them. Where sample weights are irrelevant, the FIML is probably the best solution

Table 6.17.   Comparison of Wiley-Wiley reliability estimates using different approaches to incomplete data[1]

		Reliability Estimates				
Comparison	n	Listwise	FIML	Allison	t	p-value
Listwise vs. FIML	425	0.568	0.554	---	5.691	0.000
Listwise vs. Allison	414	0.559	---	0.543	7.064	0.000
FIML vs. Allison	413	---	0.545	0.543	1.059	0.290

[1]All panels unweighted.

Table 6.18. Correlations and descriptive statistics among Wiley-Wiley reliability estimates using different approaches to incompete data[1]

	Weighted Listwise	Unweighted Listwise	Unweighted Allison	Unweighted FIML
Weighted Listwise	---	226	218	226
Unweighted Listwise	0.974	---	414	425
Unweighted Allison	0.935	0.962	---	413
Unweighted FIML	0.931	0.961	0.977	---
Mean	0.593	0.568	0.543	0.554
St. deviation	0.183	0.174	0.164	0.173
n of cases	226	426	414	425

[1]Correlations below diagonal; Ns above diagonal.

to the attrition problem from the perspective of efficiency, in part because it is a "single group" approach and because it does not require covariance structure input. The analyses therefore can be done in a single step with minimal model complexity; however, at the present time there is no AMOS option to include sample weights. The FIML approach as implemented by AMOS therefore has certain limitations. The Allison method, on the other hand, can employ weighted covariance matrices, but this can prove to be quite cumbersome in some cases because of the necessity of carrying out the analysis in two steps. In this case one would first calculate the covariances employing weighted data, and then given the multiple covariance matrices, the model can be estimated for multiple groups under proper constraints.

As noted, if the Allison model fits the data, the incomplete data can be considered to be MCAR (see Allison, 1987, p. 88). In other words, if the model imposed on all groups fits the data well, it means there is little variation in the fit of model parameters to groups involving varying patterns of attrition. On the other hand, the deviation from perfect fit reflects the failure of the model parameters to reproduce the covariance structure across groups, suggesting that measurement error parameters are different in groups defined by differing patterns of complete data. The latter suggests a violation of the MCAR assumption.

An inspection of these detailed estimates and our summary in Table 6.19 indicates that it is routinely the case that the "equality of reliability" Allison model is not rejected, that is, more often than not the conclusion at the level of single measures would be that the listwise solution is quite close to the optimal solution. As we noted earlier, the Allison model assumes that all the model parameters are equal across the groups defined by differing missing data patterns. When the parameters of this model are indistinguishable from the listwise results, the Allison model will fit the data very well—in our statistical test we will fail to reject it as significantly different statistically from a model that fits the data perfectly. These results are summarized in Table 6.19 for each of the several different panel studies suggest that for the most part the attrition is MCAR and that for the overwhelming majority of cases the listwise solution is relatively unbiased. Given the samples of cases here are relatively

Table 6.19. Goodness of fit information for Allison model for reliability estimation by panel study and content category

Panel	Content	n	N	Reliability Estimates		Goodness of Fit				Relative Fit	
				Listwise	Allison	χ^2	df	p-value	RMSEA	n with Reasonable Fit[1]	n with Close Fit[2]
NES 50s	Facts	11	1813.455	0.827	0.824	29.812	10.182	0.123	0.041	10	10
	Non-facts	29	1850.103	0.488	0.462	13.543	11.138	0.462	0.008	29	29
NES 70s	Facts	10	4040.700	0.834	0.829	156.634	12.000	0.003	0.040	9	8
	Non-facts	87	3502.690	0.497	0.490	28.836	11.966	0.165	0.017	87	87
NES 90s	Facts	10	2150.600	0.784	0.787	23.169	8.700	0.310	0.026	9	8
	Non-facts	85	2298.459	0.572	0.539	19.027	9.494	0.232	0.017	85	84
ACL	Facts	24	2911.542	0.624	0.596	124.504	9.708	0.084	0.053	18	15
	Non-facts	56	2950.661	0.500	0.484	137.021	9.232	0.031	0.049	50	40
SAF (combined)	Facts	7	1004.857	0.794	0.790	4.845	4.143	0.273	0.014	7	7
	Non-facts	88	948.205	0.521	0.514	11.501	7.091	0.328	0.019	87	84
Total	Facts	62	2560.839	0.739	0.727	83.032	9.371	0.136	0.040	53	48
	Non-facts	345	2325.899	0.521	0.505	38.273	9.600	0.226	0.022	338	324

[1]RMSEA values less than or equal to .08.
[2]RMSEA values less than or equal to .05.

large and that likelihood-ratio tests are heavily influenced by sample size, this is an even more important and impressive finding. The RMSEA measures of fit confirm these conclusions. Browne and Cudeck (1993) indicate that a value of the RMSEA of about .05 or less represents a good fit of the model in relation to the degrees of freedom, and on the average we have achieved that level of fit. Obviously, by averaging these results we have deemphasized the areas of our data where the Allison and listwise results are quite different, and we should point out that there are some instances where it would be a serious mistake to discount the biasing effects of attrition.

6.7 WHICH RELIABILITY ESTIMATES?

In this chapter we have gone to great lengths to explore several different dimensions of our reliability estimates. Some of the variations in strategy (e.g., equal reliability versus equal error variances) have yielded very few differences. Similarly, whether we use weighted or unweighted data proved to be inconsequential in the majority of cases. However, other variations in our reliability estimates have produced what appear to be substantively different and/or statistically significant results that need to be taken into account in subsequent analyses of these data. Specifically, we found there to be two important results from the extensive analysis undertaken here: (1) for variables with less than 16 response categories, there is an important difference—of about .10 points on the 0.0–1.0 reliability metric—between polychoric- and Pearson-based estimates; and (2) the approach to incomplete data yields an important, but generally nonsignificant, difference indicating that reliability estimates based on listwise-present data are biased slightly upward, a bias of about .01 reliability points on the average.

To summarize the results of the foregoing analyses, we note, first, that on the basis of the above analyses we are led to the conclusion that in these data it makes very little difference whether we constrain equal reliabilities or equal error variances over waves, and so, it should not matter which we use. This is a result that reinforces the wisdom of Heise's (1969) early approach to the separation of reliability and stability in panel data. In our subsequent analyses, thus, we will use the average of the P-wave Wiley-Wiley estimates, simply because it incorporates the most information; moreover, the Heise estimate is one of those three estimates (wave-2 reliability), so in this sense they are redundant estimates.[2] Second, we concluded there are significant differences between the Pearson- and polychoric-based correlational/covariance approaches, although we admit to some ambiguity in the interpretation of reliability. We noted, however, that there are only trivial differences between the two approaches to estimating polychoric correlations—equal versus unequal thresholds—and we conclude that this is an unimportant matter for present purposes.

---

[2]We do make exceptions to this general practice and in some instances we employ the Heise estimates (see Chapter 10).

In the analyses presented in subsequent chapters, we rely to some extent on all of these different approaches; and where it is relevant and makes sense to do so, we perform our analyses using each of them. However, in some cases the presentation of all of these different sets of results represents a distraction, and in most cases we limit our attention to a "hybrid" estimate discussed in the following paragraph. At the same time, there are other instances in which we are concerned that the conclusions we reach might be sensitive to the particular approach to reliability estimation chosen, and in these cases we do everything possible to explore a variety of estimation strategies.

Third, we believe our analyses of attrition yielded important results, and we employ these approaches in some of our subsequent analyses; however, for the most part we do not employ the overall results of these approaches in our examination of question and respondent characteristics. This is in part because the differences in results among the various approaches to incomplete data yielded such small differences, and also because we do not have polychoric-based estimates for the Allison and FIML approaches. For most of our purposes in the following analysis we will employ two basic approaches, both using listwise-complete data: (1) the Wiley-Wiley estimates obtained using the Pearson-based covariance approach in cases where the number of response categories exceeds 15, and (2) the polychoric-based Wiley-Wiley estimates where the number of response categories is 15 or fewer. In these two cases we use the Heise estimates only when for some reason the Wiley-Wiley estimate is absent. In the general case we, thus, constructed a "hybrid" reliability estimate—one that uses the polychoric-based reliabilities for those measures that have less than 16 response categories and Pearson-based reliabilities for measures that have 16 or more categories. The assumption is that for variables with 16 categories or more the polychoric and Pearson correlations will be more or less equivalent, so it is a natural extension of this logic to supplement the polychoric-based estimates with the Pearson correlations when the number of categories exceeds 15 (see Jöreskog, 1990). In Chapter 8 we present some comparisons of these two approaches by *number of response categories* that provides evidence to support this assumption.

6.8 CONCLUSIONS

In this chapter we have described the study design and methods for investigating the reliability of survey measurement using quasi-simplex models applied to longitudinal data. While much of this is straightforward, there are a number of complex technical issues that arise in the application of the theories of measurement and models for reliability estimation discussed in previous chapters. The most basic of these involves the nature of the assumptions one is willing to make about how to model unreliability. In Chapter 5 we argued that the quasi-simplex models for longitudinal data have a number of advantages for reliability estimation that are not present in models that are formulated for cross-sectional data. This is an important advance in the study of survey measurement error, in that virtually all of the current formulations for studying measurement error are developed for cross-sectional data (e.g., Groves, 1991).

In closing, we note that in the application of these longitudinal assessments to reliability the present chapter has addressed several relevant technical issues and our empirical analyses have investigated several aspects of these several estimates: (1) our empirical analyses have revealed that the assumption of equal reliability across waves is a relatively robust one; (2) we find that standard Pearson-based correlation/covariance methods of estimating statistical associations among variables are appropriate for *interval-scale measurement* (when the number of response categories exceeds 15), whereas for *ordinal scale measurement* polychoric-based methods are viewed to have several advantages; and (3) we have addressed the issue of panel attrition in the estimation of reliability, and while we do not consider it to be a major problem, our empirical analyses demonstrated that there was a consistent bias involved in the use of listwise-present approaches to incomplete data.

CHAPTER SEVEN

The Source and Content
of Survey Questions

... for it is the mark of an educated man to look for precision in each class of things just so far as the nature of the subject admits ...

Aristotle[1]

In Chapter 2 I argued that it is possible to better understand the potential errors in survey measurement if we make explicit the assumptions that are often made in the collection of survey data. These included the following key elements: (1) that the question asked is an appropriate and relevant one, which has an answer; (2) that the question is posed in such a way that the respondent or informant understands the information requested; (3) that the respondent or informant has access to the information requested; (4) that the respondent or informant can retrieve the information from memory; (5) that the respondent or informant is motivated to make the effort to provide an accurate account of the information retrieved; and (6) that the respondent or informant can communicate this information using the response categories provided by the survey question.

In this chapter I focus primarily on two issues that relate to the element of *access*—the *source of the information requested* and the *content and topic of the questions*. These are issues that have not been given a great deal of attention in the survey methods literature concerned with assessing measurement error. In my previous discussion I argued that *respondents are better at reporting information about themselves than about others* and that self-reports will on average be more reliable than proxy reports. I also hypothesized from the point of view of ease of information retrieval that the content of the information—whether factual or nonfactual—requested from the respondent would be more reliably reported if it had a factual basis, on the assumption that facts are more accessible than nonfacts. For a number of reasons, variables that involve subjective content are expected to have lower reliabilities because of the difficulties of translating internal cues related to such content into the response framework offered to the respondent (see Kalton and Schuman, 1982).

---

[1]Quoted in McKeon (1992, p. 239).

Table 7.1.   Reliability estimates for organizational and interviewer reports

Source and Content of Survey Measures	n	Reliability Estimates[1]
Organization reports--Facts	9	0.994
Interviewer reports--Facts	4	0.872
Interviewer reports--Beliefs about respondent characteristics	12	0.601
Interviewer reports--Beliefs about respondent reaction to interviewer	10	0.492

[1]Wiley-Wiley polychoric-hybrid estimates based on listwise-present data.

In this chapter I present the results from our study that pertain to the role of these factors in contributing to differences in the reliability of measurement. We have a small amount of data on organizational and interviewer reports of factual material, and we have some data assessing the reliability of interviewer beliefs about respondents and their reactions to the interview. However, the main issue we address with regard to the source of information pertains to the *self-report versus proxy report* issue. We have some information on proxy- versus self-report measures, where we can compare variables that involve the same or similar content reported for self and others obtained *from the same respondents.*

7.1   SOURCE OF INFORMATION

It is possible to distinguish four different sources of information in the panel studies we analyzed: (1) respondent self-reports; (2) respondent reports on other household members (e.g., head of household and/or spouse), hereafter referred to as *proxy reports*; (3) interviewer reports of observed characteristics of the respondent, e.g., the respondent's cooperativeness or apparent intelligence; and (4) information added to the data file by the survey organization, based on US Census information pertaining to the location of the household, e.g., city size. The bulk of the variables in the present study, as in surveys more generally, are respondent self-reports, and self-report questions are quite diverse in content, so it is difficult to draw any strong comparisons about the source of information except by controlling for content.

In Table 7.1 I summarize the information on organizational and interviewer report measures. This information, although limited in certain key respects, does provide a great deal of clarity to some of our major results. As one would expect, organizational reports are almost perfectly reliable. On average they reveal the highest possible reliability. These reports primarily involve information of a factual nature on the characteristics of the location of the sampled household, e.g., size of place and distance from the interview place to the central city of the nearest standard metropolitan statistical area (SMSA). This typically involves a clerical lookup via computer of information from US Census records for the sampled household. This information is independent of the respondent or interviewer with regard to the gathering of the data, but it is subject to change. We should expect, however, that there are very few measurement errors associated with this type of information gathering.

In many ways these estimates provide strong testimony for the validity of the quasi-simplex methods used in this study to estimate reliability. We should expect the measurement of organizational reports to be relatively flawless, but we understand at the same time that for many respondents their residential location changes across waves of the study. Our models allow for the incorporation of change while at the same time separating information on the reliability of measurement from true change. Because the models estimated for organizational reports provide uniformly high levels of reliability as predicted, these results provide a benchmark against which to evaluate the reliability of other sources of survey measures.

Table 7.1 also presents information on reliability of measuring interviewer reports of three types: facts pertaining to the household, beliefs about respondent characteristics, and beliefs about respondent reactions to the interview. There is a fairly large literature on "interviewer effects," which should come to our assistance in evaluating this issue, but that literature has focused on interviewer variance rather than interviewer reliability (see O'Muircheartaigh, 1991; Groves, 1989).

In keeping with our general observation throughout this book, the reliability estimates for interviewer reports of factual information about the household (primarily the number of persons in the household) are very high, around .85 on the average. Again, we should stress that our models allow the true values of these variables to change over time, and we assess measurement reliability within the context of a model that allows for such change.

Estimates of the reliability of interviewer beliefs, on the other hand, tend to be relatively low, among the lowest in the study. The interviewer assigned to a given household at a particular wave of the study is not necessarily the same person across waves of the study. Therefore, these reliability estimates are ideally estimates of interinterview agreement about the respondent's behavior. A case can be made that this is what the reliability estimates for *respondent characteristics* assess, since these all represent latent variables conceptualized at the level of the respondent. Even if the behaviors exhibited by the respondent are highly situational or contextual in the sense that it depends on the conditions under which the interview is administered, including who the interviewer is, one can argue that our models allow for change in such behaviors and therefore that we have a justifiable estimate of the reliability of interviewer beliefs. On the other hand, in the case of what we have here termed *respondent reactions to the interview,* much less of an argument can be made that our design and statistical model fit the data gathered in this way. While it is one thing to argue that respondent characteristics represent the "same object" assessed over time, this can hardly be the case with respondent reactions to the interview. Indeed, one could argue that we should disqualify these latter items on the grounds that they violate one of our basic requirements for the inclusion of question in this study, namely, that they assess the same object over time. On the other hand, as we have argued repeatedly, our model allows for change in respondent behavior and beliefs over time; indeed this is one of the redeeming qualities of the quasi-simplex model.

The quantities measured in this category of *respondent reactions* thus represent just another type of behavior, one that is very situational and that may be very difficult to measure reliably, given its dependence on interviewer judgment. This latter category may therefore represent the other extreme from organizational reports with respect to the content included in the present study, and we may be justified in the conclusion that asking interviewers to rate respondent reactions to the interview is an objective fraught with problems. On this basis it can be hypothesized that interviewer reports of facts to which they have access, e.g., characteristics of the household, are highly reliable, although their reports of beliefs may be in the same range as respondent reports of beliefs.

7.2   PROXY REPORTS

Gathering data by proxy in survey research is commonplace. Respondents to surveys are often asked questions about other people, including their spouses and children, and sometimes, their friends and coworkers. Because of differences in the process of reporting about the characteristics of others and the more common self-report method, one would expect that the nature of measurement errors might be different for proxy vs. self-reports (Blair, Menon and Bickart, 1991). Recent evidence, however, suggests that "for many behaviors and even for some attitudes, proxy reports are not significantly less accurate than self-reports (Sudman, Bradburn and Schwarz, 1996, p. 243). If true, this is an encouraging result because it is seldom possible to obtain self-reports in all cases and the proxy is frequently the only source of information on the person in question, e.g., the head of household. We should note, however, that the broader literature on this topic presents inconsistent results regarding the quality of reporting by proxies (see Moore, 1988).

Our study can address the issue of self- versus proxy reporting and measurement error with respect to the reliability of measurement. We compared six measures involving self-reports with six measures involving proxy reports (in this case reports on spouses) *where the content of the questions was identical*, e.g., respondent's education versus spouse's education, and *the same person is reporting in each case*, usually at a different place in the questionnaire. These results are shown in Table 7.2. In virtually every case the self-report measure is higher in reliability than the proxy-report measures. On average across the six variables, the difference is approximately .90 versus .80 using the listwise-present results, which, if based on a larger sample of measures, would certainly be considered a substantively meaningful difference. The statistical test of this difference (based on six cases) is nonetheless marginally significant (p < .02), which is remarkable given the small number of cases. The difference obtained is slightly smaller using the Allison model, and the systematic nature of the difference is less apparent, but overall the two sets of results are largely consistent. This leads us to conclude that proxy reports are significantly less reliable than are self-report measures, which is consistent with both theory and experience. Further research is necessary, however, to create a broader inferential basis for this conclusion.

Table 7.2.   Comparison of reliability estimates of self- and proxy-report measures of smilar content

| Content of Question | Study | Reliability Estimates[1] | | | | | |
| | | Listwise | | | Allison | | |
		Self	Spouse	N	Self	Spouse	N
Years of schooling	NES 70s	0.971	0.930	815	0.966	0.916	3,216
Years of schooling	SAF Mothers	0.954	0.910	799	0.944	0.908	1,111
Occupational status	NES 70s	0.887	0.806	299	0.812	0.802	2,017
Occupational status	NES 90s	0.859	0.717	197	0.896	0.711	1,197
Hours works/worked per week	NES 70s	0.835	0.710	212	0.728	0.735	1,501
Hours works/worked per week	NES 90s	0.881	0.612	141	0.835	0.580	959
Total		0.898	0.781	6	0.864	0.775	6
		t	p-value		t	p-value	
		3.365	0.020		2.033	0.098	

[1]Wiley-Wiley Pearson-based reliability estimates.

7.3   CONTENT OF QUESTIONS

There is a hypothesis in the survey methods literature that the measurement of *facts*, all other things being equal, ought to be carried out more reliably than the measurement of subjective variables, or *nonfacts* (Kalton and Schuman, 1982; Schuman and Kalton, 1985). Although little research has actually been done on this topic, it seems self-evident that various domains of content may inherently differ in the extent to which they may be measured reliably. A few studies suggest that some types of survey content can be more reliably measured than others. Specifically, in a study of political identities, beliefs, and attitudes, measures of party affiliation and candidate preferences were measured more reliably than attitudes on policy issues (Alwin and Krosnick, 1991a, 1991b; Alwin, 1992). Another study based on reinterview data from the General Social Survey revealed slightly higher estimated reliabilities for questions seeking factual information compared to questions soliciting attitudes or self-appraisals (Alwin, 1989), and a preliminary examination of these issues using a subset of questions studied in the present book reported significantly higher levels of reliability for facts versus nonfacts (Alwin, 1995).

We can see from the way in which we have defined facts versus beliefs, for example, that there is much more room for measurement error in the latter versus the former. We define facts as "objective information" having to do with respondent or household characteristics that can be verified, whereas beliefs (our principal type of nonfact) as "perceptions or subjective assessments of states and or outcomes for self and others." By definition, subjective assessments cannot be verified against an "objective" standard, and there are three key theoretical reasons to expect nonfactual reports to be less accurate than reports of facts: (1) issues of "trait" existence, (2) ambiguity in internal cues, and (3) ambiguity in response scale alternatives (see Alwin and Krosnick, 1991b, pp. 144–147).

In their classic study of questionnaire development, Rugg and Cantril (1944) actually used the term "reliability" in their discussion of what is sometimes lacking in the measurement of nonfactual content. They acknowledged that response errors were often less likely if persons "had standards of judgment resulting from stable frames of reference," as opposed to situations "where people lack *reliable* standards of judgment and *consistent* frames of reference" (Rugg and Cantril, 1944, pp. 48–49). Clearly, if a stable frame of reference surrounding the topic of the question, e.g., beliefs about government performance or attitudes toward government policy, is lacking, respondents may have no firm basis for providing a response. This often results from respondents having little knowledge of the subject matter or may not have thought about it.

We have already mentioned Converse's (1964) important concern that in attitude studies respondents may not always have an attitude, but feel pressure during survey interviews to offer opinions in response to questions even when they have none. By concocting attitude reports when they have none, as Converse argued, respondents often produce what are essentially random choices from among the offered response options. The production of a random selection of answers is considered to be much more likely in the case of nonfactual questions, given their failure in many cases to meet the assumptions mentioned above.

While we can almost always assume that respondents have knowledge and access to information sought by the researcher when it is of a factual nature, there may be considerable ambiguity in some cases in the nature of the information sought and the internal cues that should be tapped to provide a response. Many questions assume that the response called for is associated with univocal, unambiguous internal cues that come to mind quickly and effortlessly when a person simply thinks about the content sought. In fact, in many cases the subjective content that is the focus of the question may be associated with ambiguous or conflicting internal cues or with cues that are relatively inaccessible and come to mind only as the result of considerable cognitive effort (Fazio, Herr, and Olney, 1984).

In the case of measuring attitudes, for example, theory would suggest that attitudes are not represented by clear points on a latent attitude continuum, but rather people have what Sherif and his colleagues (see Sherif and Hovland, 1961; Sherif, Sherif and Nebergall, 1965) called "latitudes of acceptance" and "latitudes of rejection," regions of the attitude dimension they find acceptable or unacceptable, and in which their attitude toward the object involves "acceptance" or "rejection." In other words, at any given time, there may be a range of possible responses that could be produced, and given the ambiguity of the cues the respondent may provide any one of a number of possible responses. In order to report one's attitudes when faced with an interviewer's query, the respondent may be forced to choose from within this range in order to satisfy the demands of the reporting task, and the resolution of the choice within the range of acceptability and unacceptability may involve a component of random choice that may not be present if the situational demands were different. According to Sherif, it may be a more effective approach to measurement if the respondent is asked to indicate the regions of acceptability and unacceptability that corresponds to her/his attitude rather than choose a single point on an attitude continuum.

Finally, given the presence of the latent variable of interest, that is, its "existence" insofar as the respondent is concerned, and some clear internal representation of the concept of interest, some ambiguities may still remain. Given the need for the survey participant to respond within the framework of the response options provided, there may still be room for random error in matching the internal response to these response options. This process of responding involves one of "mapping" the internal cues on to the most appropriate response alternative. Some random variation in respondent reports may result from the ambiguity in the translation of respondent levels to the response continua offered by the survey questions. This can happen in at least two different ways. First, the respondent might find that his or her internal cues do not correspond to any of the offered response categories. For example, a respondent who feels that abortion should be legal only if the mother's life is endangered, and not under any other circumstances, might have difficulty in mapping that view on to response choices of a question that simply asks whether he or she favors or opposes legalized abortion. Second, the meaning of the response choices may not be completely clear. There is a literature, for example, that has developed around what are known as "vague quantifiers"—words like "constantly, frequently, sometimes, rarely and never" (Bradburn and Miles, 1979; Schaeffer, 1991a; O'Muircheartaigh, Gaskell, and Wright, 1993). The meanings of these response quantities are clearly ambiguous, as are many others, and there may be a high degree of randomness involved in the selection from among them by the respondent.

These sources of ambiguity in measurement are much more likely to exist in the measurement of subjective variables than in the case of the measurement of facts, and consistent with past research, then, we expect that factual information will be more reliably gathered. Our results, shown in Table 7.3, support this hypothesis.

Table 7.3. Comparison of reliability estimates for survey measures of facts and nonfacts

	Reliability Estimates							
	Listwise				Allison			
Panel	Nonfacts	Facts	F-ratio	p-value	Nonfacts	Facts	F-ratio	p-value
NES 50s	0.610	0.887	30.27	0.000	0.462	0.824	45.49	0.000
	(29)	(12)			(29)	(11)		
NES 70s	0.592	0.867	32.34	0.000	0.490	0.829	66.04	0.000
	(87)	(10)			(87)	(10)		
NES 90s	0.651	0.831	12.53	0.001	0.539	0.787	34.67	0.000
	(85)	(11)			(85)	(10)		
ACL	0.654	0.743	6.16	0.015	0.484	0.596	8.93	0.004
	(57)	(29)			(56)	(24)		
SAF (combined)	0.656	0.804	7.90	0.006	0.514	0.790	33.90	0.000
	(89)	(10)			(88)	(7)		
Total	0.634	0.806	71.79	0.000	0.505	0.727	132.46	0.000
	(347)	(72)			(345)	(62)		

Note: The number of questions on which reliability estimates are based is given in parentheses.

On the basis of the measures in our database (n = 426), we examined the variation in our estimates of reliability due to the content of the question. Questions seeking factual content are more reliably measured than those seeking subjective information—.80 versus .63 in the listwise results and .73 versus .50 in the results incorporating attrition patterns. These findings are present in all of the major panel studies, and while there is considerable variability around these averages, with some variables measured with near-perfect reliability, and others being measured quite poorly, *we feel justified in drawing the strong conclusion that self-reported factual content can be quite reliably measured.*

7.3.1 Reliability in Measuring Self-Reported Facts

In Table 7.4 we present a breakdown of the results for 72 factual self-report measures. Here we present only the Wiley-Wiley "hybrid" estimate, given that there is a high degree of agreement among various reliability approaches. The reliability of measuring factual material is routinely found to be quite high. This is especially the case for variables like age, weight, number of children, and the dating of major life events in time, where the reliability estimate approximates perfect reliability—upward from .96! These results are reassuring in the sense that our confidence in survey methods (or, for that matter, our reliability estimation strategies) would be entirely undermined if such high numbers were not attained for measures of such *factual* variables.

Moving down the list in Table 7.4, we see that the estimates of socioeconomic characteristics, such as education, occupation, and income, aspects of the respondent's place of residence, and church attendance, measures of other religious activities, and voting frequency are measured with reliability ranging from .80 to .90. While not perfect, these reflect high levels of reliability. Measures of other economic characteristics, hours worked per week and other employment-related variables are measured with reliabilities between .70 and .80. Finally, measures of health behaviors, (nonchurch) activities, other political behaviors, and social interactions are measured with reliabilities ranging from .60 to .70.

Although it is in general the case that factual content is measured quite reliably, those fact measures resulting in lower levels of assessed reliability are more often than not retrospective assessments of the frequency of behavior or are variables measured on scales labeled with vague quantifiers. For example, in the measurement of health behaviors respondents are asked to report "the average number of cigarettes the respondent smokes in a day," or "the number of days during the last month the respondent drank beer, wine or liquor," and "on days respondent drinks, how many cans of beer, glasses of wine, or drinks of liquor does the respondent usually have?" Or, for example, in the case of the questions assessing "church attendance" in the NES50s the categories were "regularly," "often," "seldom," and "never," whereas in the NES70s the categories were "every week," "almost every week," "once or twice a month," "a few times a year," and "never." While the options "never" and "every week" may be unambiguous, the response categories in between may be less than completely clear. Or consider the measurement of "other religious activities" in the

Table 7.4. Estimates of reliability of self-reported measures of facts by topic of question

| | Content | Reliability Estimates[1] | | | | | |
		Listwise	n	Allison	n	Studies
1	Age	0.997	5	0.987	3	ACL, NES 50s, NES 70s, NES 90s
2	Year or month of events	0.962	4	0.914	4	ACL, NES 90s
3	Weight	0.956	1	---	0	ACL
4	Number of children	1.000	2	1.000	2	NES 50s, SAF mothers
5	Education	0.884	4	0.897	4	NES 50s, NES 70s, SAF mothers, SAF children
6	Aspects of place of residence	0.839	5	0.802	5	NES 50s, NES 70s, NES 90s
7	Occupation	0.813	2	0.819	2	NES 70s, NES 90s
8	Income	0.893	6	0.868	2	ACL, NES 50s, NES 70s, NES 90s, SAF mothers
9	Other economic characteristics	0.800	4	0.630	1	ACL, SAF mothers
10	Hours worked per week	0.765	4	0.766	4	ACL, NES 70s, NES 90s, SAF mothers
11	Other aspects of employment	0.809	4	0.702	4	ACL, NES 50s
12	Social interactions	0.600	7	0.493	7	ACL
13	Activities (non-church)	0.680	10	0.555	10	ACL, NES 90s
14	Church attendance	0.827	5	0.776	5	ACL, NES 50s, NES 70s, SAF mothers, SAF children
15	Other religious activities	0.863	2	0.789	2	NES 90s
16	Voting frequencies	0.914	2	0.864	2	NES 50s, NES 70s
17	Other political behaviors	0.596	2	0.484	2	NES 50s, NES 70s
18	Health behaviors	0.719	3	0.600	3	ACL
	Total	0.806	72	0.727	62	

[1]Reliability estimates presented here are based on the average Wiley estimate for polychoric-based covariances, supplemented by average Wiley estimates for Pearson-based covariances where the number of categories is greater than 15.

NES90s, where respondents were asked to report how often they "pray, outside of attending religious services," and how often they "read the Bible, outside of attending religious services," the scale used ranged across the following categories: "several times a day," "once a day," "few times a week," "once a week or less," or "never."

Thus, there is an inherent confounding between substantive content and the approach used to measure it, as we have emphasized throughout this discussion. It is not the factual nature of these variables that contributes to their lower level of estimated reliabilities, but more than likely the vagueness of the response categories used to assess frequencies of past behavior. It would be interesting to see if reliability in the measurement of these phenomena would improve if the questions were open-ended rather than fixed-category rating scales such as these. There are no studies to date that have addressed this issue, but given the technology introduced by modern computer-based methods that can code free-format responses, there may be several advantages to abandoning the use of vague quantifiers.

It would be theoretically possible to compare some of these retrospective reports with concurrent reports, in order to examine the comparative utility of retrospective measurement, but we have not as yet explored these possibilities. Reports of past events, however, must be interpreted with caution, based on what is known about memory processes. Psychologists argue that memory is often reconstructive, and thus reliability estimates of retrospective reports may be falsely inflated due to the respondent's use of schemas for what must have been true at a previous time (Ross, 1988; Dawes and Smith, 1985). This complicates the interpretation of reliability estimates of retrospective reports.

7.3.2   The Reliability of Measuring Self-Reported Nonfacts

We extend our analysis of the content of the questions to an examination of the variation among the nonfacts. We invested a great deal of effort in classifying our nonfactual questions into beliefs, values, attitudes, self-perceptions, and self-assessments (see Chapter 6). The major comparison we focus on here concerns whether the self-reports deal with self-evaluation or self-assessment on one hand and evaluation of external objects on the other. The latter involves beliefs about external objects and attitudes toward external objects (recall that values are a type of belief). Table 7.5 presents the results by panel study of the differences between these two major types of subjective variables.

These results indicate that regardless of the type of subjective content assessed there is a great deal of uniformity in the level of reliability. There are some panel studies in which there is a difference in the reliability of measuring self-objects versus external objects, but the differences are not consistent across studies. In the ACL and SAF mothers studies the external-object measures had higher reliabilities than the self-object measures, whereas in the NES70s study the observed pattern was reversed. Overall, there was no systematic difference across studies that afforded a basis for interpretation, and we conclude that self-assessments/perceptions can be measured no more or less reliably as measures of beliefs, attitudes, and values.

Table 7.5. Comparison of reliability estimates among nonfactual measures by panel study

| | Reliability Estimates | | | | | | | |
| | Listwise | | | | Allison | | | |
Panel	Self-Assessments/ Perceptions	Beliefs Attitudes Values	F-ratio	p-value	Self-Assessments/ Perceptions	Beliefs Attitudes Values	F-ratio	p-value
NES 50s	0.625 (14)	0.596 (15)	0.266	0.610	0.486 (14)	0.440 (15)	0.760	0.391
NES 70s	0.687 (12)	0.577 (75)	5.910	0.017	0.547 (12)	0.480 (75)	2.956	0.089
NES 90s	0.689 (13)	0.644 (72)	0.854	0.358	0.576 (13)	0.532 (72)	1.366	0.246
ACL	0.626 (42)	0.731 (15)	6.186	0.016	0.463 (42)	0.547 (14)	5.448	0.023
SAF Child	0.649 (14)	0.680 (30)	0.439	0.511	0.485 (14)	0.528 (30)	1.126	0.295
SAF Mother	0.559 (17)	0.692 (28)	6.524	0.014	0.488 (17)	0.529 (27)	1.317	0.258
Total	0.633 (112)	0.635 (235)	0.022	0.882	0.495 (112)	0.51 (233)	1.048	0.307

Note: The number of questions on which reliability estimates are based is given in parentheses.

In order to more thoroughly exhaust the question of "topic effects" on reliability of measurement, we explore this issue through a more detailed examination of differences among nonfactual questions with respect to the domain of measurement. Here we partition the 347 self-reports of nonfactual content into a set of 21 categories corresponding to the topic of the question. These categories conform to those reported in the Appendix. In order to present as much detail here as possible we also subcategorize content within a given topic area into the content categories used above (self-perceptions, beliefs, attitudes, etc.).

We do not perform any tests of significance across these various categories because they include quite diverse numbers of questions and a variety of different types of content. It is not clear what one would learn from such tests. On the other hand, by inspecting the differences in the reliability estimates, one gets a sense of where the variation exists both between and within categories. For this purpose we have constructed two measures of variability about the mean reliability for a given category—the range and the standard deviation. The typical nonfactual question has a reliability of .63, but this includes a range from estimates of .12 for some self-perception questions from the Study of American Families to estimates of 1.0 for some measures of attitudes toward public officials in the National Election Studies. Even if we take out the items at the extremes, the empirical range reflected in these results extends from .23 to .98! This spans nearly the entire theoretical range for reliability—from 0 to 1.0—and while it is reassuring that the reliability models work in the sense that they are sensitive to differences in levels of measurement error, it is disconcerting that despite efforts to reduce errors of measurement, some phenomena are difficult to measure at even modest levels of reliability.

We have ranked the topic-specific reliability estimates in Table 7.6, presenting the estimates for the nonfactual categories with the highest reliabilities at the top and those with the lowest at the bottom of the table. It is obvious that survey measurement of some nonfactual content can be quite high. The best examples of this are self-perceptions of functional status in the ACL (.82), perceptions of political self-identification in the NES studies (.79), and attitudes toward political candidates (.77). The reliability of measurement for these variables approach or exceed the typical level of reliability that we estimate for measures of facts (.81)—see Table 7.4. There is clearly overlap in the distributions of estimated reliability for facts and nonfacts, even though on average facts are measured more reliably.

Moving down the list of questions gathering information on nonfactual variables in Table 7.6 we see that most of the average reliability estimates there fall within the range from .50 to .60. There are some exceptions—e.g., the NES measures of attitudes toward public officials (.804) and political parties (.683), and several studies of attitudes and values with respect to family topics (.72)—but overall, the typical measures of attitudes, beliefs and values fall within the .50–.60 range. As we point out in Chapter 12, this does not bode well for use of single indicators of subjective concepts, and, as is well known in the psychometric literature, reliability in such cases needs to be bolstered through the combination of measures.

Table 7.6. Estimates of reliability of self-reports of nonfacts by topic and content

Topic / Content	n	Mean	Range	St.Dev	Studies
			Reliability Estimates		
1 Functional status					
Self-perceptions	**5**	**0.820**	**.66 - .95**	**0.115**	ACL
2 Political identification					
Self-perceptions	**13**	**0.793**	**.63 - .95**	**0.109**	NES 50s, NES 70s, NES 90s
3 Political candidates					
Attitudes	**10**	**0.769**	**.54 - .90**	**0.096**	NES 70s, NES 90s
4 Morality and spirituality					
Values	**4**	**0.731**	**.69 - .83**	**0.065**	ACL, NES 90s
5 Relationships					
Beliefs	**26**	**0.692**	**.50 - .96**	**0.119**	ACL, SAFMO, SAFCH
6 Economy and society					
Beliefs	**12**	**0.675**	**.43 - .97**	**0.145**	ACL, NES 50s, NES 90s, SAFMO
7 Quality of life					
Self-perceptions	11	0.680	.48 - .91	0.122	ACL, SAFMO, SAFCH
Self-assessments	8	0.659	.50 - .80	0.116	ACL, NES 50s, SAFMO
Total	**19**	**0.671**	**.48 - .91**	**0.116**	
8 Employment					
Self-assessments	**4**	**0.662**	**.59 - .81**	**0.099**	ACL, NES 90s
9 Family					
Attitudes	15	0.717	.36 - .86	0.137	ACL, SAFMO, SAFCH
Values	18	0.724	.50 - .98	0.110	NES 70s, NES 90s, SAFMO, SAFCH
Beliefs	10	0.580	.45 - .70	0.082	SAFMO, SAFCH
Self-perceptions /					
self-assessments	20	0.600	.12 - .96	0.235	ACL, SAFMO, SAFCH
Total	**63**	**0.660**	**.12 - .98**	**0.172**	
10 Public officials					
Attitudes	7	0.804	.70 - 1.0	0.100	NES 90s
Beliefs	18	0.572	.42 - .84	0.122	NES 70s, NES 90s
Total	**25**	**0.651**	**.42 - 1.0**	**0.150**	
11 Self-efficacy					
Self-perceptions	**8**	**0.625**	**.40 - .76**	**0.135**	ACL, NES 50s, NES 70s
12 Social groups					
Attitudes	29	0.585	.28 - .83	0.142	NES 70s, NES 90s
Values	6	0.688	.60 - .84	0.104	NES 50s, NES 70s, NES 90s
Total	**35**	**0.603**	**.28 - .84**	**0.140**	
13 Self-esteem					
Self-assessments	**10**	**0.594**	**.36 - .74**	**0.107**	ACL, SAFCH
14 Government					
Attitudes	5	0.649	.39 - .91	0.237	NES 50s, NES 70s, NES 90s
Values	15	0.570	.40 - .74	0.109	NES 50s, NES 70s, NES 90s
Beliefs	13	0.579	.28 - .86	0.179	NES 50s, NES 70s, NES 90s
Total	**33**	**0.585**	**.28 - .91**	**0.158**	
15 Financial situation					
Self-perceptions	**10**	**0.592**	**.40 - .85**	**0.130**	ACL, NES 50s, NES 70s, NES 90s,SAFMO
16 Relationship to social groups					
Self-perceptions	**6**	**0.584**	**.23 - .81**	**0.220**	NES 50s, NES 70s, NES 90s
17 Civil rights					
Attitudes	**6**	**0.580**	**.34 - .94**	**0.206**	NES 90s
18 Influence of social groups					
Beliefs	**10**	**0.571**	**.29 - .77**	**0.145**	NES70s

Table 7.6 *Continued.*   **Estimates of reliability of self-reports of nonfacts by topic and content**

19	Political parties				
	Attitudes	4	0.683	.57 - .80	0.104 NES 70s, NES 90s
	Beliefs	27	0.533	.25 - .78	0.151 NES 50s, NES 70s, NES 90s
	Total	**31**	**0.553**	**.25 - .80**	**0.153**
20	Political interest				
	Self-perceptions	**4**	**0.515**	**.39 - .74**	**0.156** NES 50s, NES 90s
21	Stress and mental health				
	Self-perceptions	**13**	**0.511**	**.33 - .64**	**0.096** ACL
	Total	**347**	**0.634**	**.12 - 1.0**	**0.157**

7.4   SUMMARY AND CONCLUSIONS

The results reported in this chapter make several things clear. The first is that the reliability of measuring factual information in surveys is generally quite high. In this regard we have confirmed the commonly held view that facts are reported more reliably than are nonfacts. Facts reported by the survey organization are perhaps the most reliable. Interviewers are also highly reliable in reporting factual aspects of the household, but with respect to their judgments regarding respondent characteristics, their reports are substantially less reliable.

In general, one would expect that self-reports are likely to be more reliable than information that can be provided secondhand by a proxy informant. Proxy informa-tion, however, can be used to gain some insight into the reliability of self-reports. We drew on a controlled comparison of self- and proxy reports by using only comparable self-report/proxy-report information, which reinforced this conclusion.

Finally, we reported major differences in the reliability of reporting factual con-tent, consistent with both theory and past research. While factual material is the most reliable in the aggregate, it is also clear that there is variation both within and between content categories. Indeed there are some facts that are measured no more reliably than the typical nonfact (or subjective) variable.

The factual variables with the lowest reliabilities were those in which the ques-tion used "vague quantifiers" to assess frequencies of past behavior. To our knowl-edge there have been no systematic studies of the reliability of measuring factual content with such rating scales using vague quantifiers, versus using open-ended questions. In the following chapter we show that questions using an open-ended for-mat generally obtain higher levels of reliability, but as we note there, more research is necessary to establish the utility of open-ended questions that employ modern computer-based approaches to coding free-format responses.

These results raise questions, not only concerning what can be done about low levels of reporting reliability in the measurement of subjective or nonfactual vari-ables, as well as some factual variables, but also questions about what accounts for the lower levels of estimated reliability of questions gathering survey information. One possible explanation is that some types of content, specifically "subjective" con-tent, are more difficult to measure. Subjective content (attitudes, beliefs, perceptions, etc.) is often inherently less easily clarified to the respondent and interviewers, and it is difficult for the investigator to phrase a question that is utterly clear with respect to what information is desired (Schaeffer and Presser, 2003). An opposing viewpoint is that it is not the content that produces the high levels of unreliability, but how we measure it that contributes to poor measurement. There may be a number of features of survey questionnaires that contribute to measurement errors, such as ambiguities in response options, the use of long series or batteries of questions, the use of overly long questions that make comprehension more difficult, the use of questions that place cognitive demands on respondents that make it difficult or impossible for them to produce reliable answers.

Although I cannot resolve these matters here, we can explore the possibility that some of the measurable features of survey questionnaires and survey questions contribute to unreliability of measurement. In the following two chapters I explore some of these issues. In Chapter 8, I address the question of *question context*, whether reliability of measurement is affected by the context in which the questions are framed. Specifically, I investigate whether the location of the question in the questionnaire, or the uses of series or batteries of questions produce detectable differences in levels of measurement error. In Chapter 9 I address the question of whether the formal *properties of survey questions* affect the quality of data that results from their use. I examine how reliability is affected, if at all, by the form of the question, the length of the question, the number of response categories, the extent of verbal labeling, and the provision of an explicit Don't Know response option.

CHAPTER EIGHT

Survey Question Context

It is a wise experimenter who knows his artifact from his main effect; and wiser still is the researcher who realizes that today's artifact may be tomorrow's independent variable.

William J. McGuire (1969)[1]

Virtually anyone with even a modest amount of experience in conducting surveys has some firm ideas about *how to organize a good questionnaire.* There are several available sources about questionnaire design providing practical wisdom about "best practices" for putting questions together into questionnaires (e.g., Dillman, 1978; Krosnick and Fabrigar, 1997; Schaeffer and Presser, 2003). The choices that survey researchers make when it comes to the nature of questions and questionnaires are often perceived as much a matter of art as they are a science. While we do not dispute this observation, we would argue that much more can be done to bring empirical evidence to bear on the general issue of the desirable attributes of survey questionnaire design.

In keeping with this principle, in the mid-1970s Sudman and Bradburn's (1974) path-breaking *Response effects in surveys* charted the early methodological terrain with the goal of assessing the impact of various methodological features of survey questionnaires on the quality of survey data, i.e., what they referred to as "response effects." Their research set the stage for nearly three decades of methodological research into various aspects of survey methods and increasingly many of these factors have been isolated as sources of measurement error. Much of this research is reviewed in the present and subsequent chapters as it bears on the relationship of features of questions and questionnaires to measurement errors.

One feature of much of the early methodological research on questionnaire design was it's "question-centric" approach, focusing almost exclusively on the form of survey questions. For example, Schuman and Kalton (1985, p. 642) noted, "the fundamental unit of meaning in a survey is the single question . . . (playing a role)

---

[1]From William R. McGuire (1969), Suspiciousness of experimenter's intent. In R. Rosenthal and R.L. Rosnow (Eds.), *Artifact in behavioral research* (p. 13). New York: Academic Press [quoted in Seymour Sudman and Norman Bradburn, *Response effects in surveys* (1974, p. 1)].

analogous to that of atoms in chemistry . . . (with) words, phrases, and similar constituents of questions . . . regarded as the subatomic particles in the survey process." This view is compatible with the one adopted in the present research, in that for the most part the survey question is our *unit of analysis*, but we should note that, in addition to the question, we can broaden this conceptualization to include other elements of the survey data collection process, e.g., the design of the questionnaire and the motivational context of the setting. Schaeffer (1991b) expanded the framework even further, arguing that three general sets of factors interact to shape the quality of survey data—questions, respondents, and interviewers.

In addition to the formal properties of questions, which we discuss in the next chapter, there are several formal characteristics of survey questionnaires that are believed to affect the quality of measurement. Perhaps the most severe limitation of previous research on questionnaire design is its lack of any rigorous and comprehensible standard for evaluation. In order to assess the issue of questionnaire design and to evaluate the linkage between question and questionnaire characteristics and the quality of data, one needs a set of criteria that can be used in deciding which questionnaire choices are better than others. One criterion—one that we employ here—is the amount of measurement error or level of reliability associated with a particular question, question type, or questionnaire organization. The choice of this criterion is not meant to deny the relevance of many other pertinent criteria, but this choice is consistent with an emerging literature that evaluates the properties of questions in terms of levels of measurement error (e.g., Alwin and Krosnick, 1991b; Andrews, 1984; Biemer and Stokes, 1991; Saris and Andrews, 1991; Scherpenzeel and Saris, 1997).

Unfortunately, little of the research focusing on the extent of measurement error has looked at the impact of the formal attributes of the context of questions in questionnaires. Sudman and Bradburn's (1974) research provides an early exception, in that their research explored such issues as position of the question in the questionnaire (p. 60). Another exception was the work of Andrews, Saris, and their collaborators (Andrews, 1984; Saris and van Meurs, 1990; Saris and Andrews, 1991; Scherpenzeel, 1995; Scherpenzeel and Saris, 1997) who focused on a number of issues linked to question context, namely, effects associated with position in the questionnaire, the presence of the question in a battery of questions, introductions to batteries, and battery length. Following in this research tradition, in this chapter we focus on several features of the context in which survey questions are presented to respondents and their potential impact on measurement reliability. We begin our examination of these issues with a discussion of the architecture of survey questionnaires, focusing on question context, including the position of the question in the questionnaire, whether it is part of a series or battery of questions, the position of the question in the series or battery, and the nature of the introduction to the series or battery. In the next chapter we turn to a discussion of question form, including the consideration of question length, the number of response options provided in closed-form questions, the use of visual aids and labeling of response options, and the provision of a Don't Know option.

We have chosen to deal with the issue of *question context*, i.e., the organizational context of the question, prior to the consideration of *question form*, although we could just as easily have reversed the order of topics. As we point out in subsequent sections, it is a very tricky proposition to isolate the "effects" of formal attributes of questions, due largely to the confounding of question content, context, and question characteristics. There is a certain logic in moving, as we have here, from question content to question context and then to question form, as this sequence parallels the development of issues considered in the construction of questionnaires. On the other hand, ultimately these issues must be considered simultaneously, and we make an effort to systematically control for other relevant features of survey questions when we attempt to isolate effects attributable to specific aspects of question context and question form.

8.1   THE ARCHITECTURE OF SURVEY QUESTIONNAIRES

When survey researchers put questions together to build questionnaires, there are considerations that affect the measurement errors associated with questions other than the formal properties of the questions themselves. In short, the development of survey measurement involves more than writing the questions; it also involves the organization of questions into questionnaires. The inspection of any survey questionnaire reveals that it is organized into subunits larger than the question. Ordinarily it is divided into *sections*, which are subquestionnaires focusing on similar general content, and within sections there are often *series* of questions that pertain to the same specific topic. Sometimes the series of questions will include *batteries* of questions that not only deal with the same topic but also involve the exact same response formats.

Within a given section of the questionnaire, then, it is possible that questions fall into one of three basic organizational categories: (1) stand-alone questions, which do not bear any particular topical relationship to adjacent questions, except perhaps at a very general level; (2) questions that are a part of a series, all of which focus on the same specific topical content; and (3) questions that appear in batteries focusing on the same or similar content, using the identical response format. What distinguishes a battery from a series is the use of the same response format. In many (but by no means all) cases the set of questions included in series and batteries is preceded by an introduction that is read by the interviewer to the respondent in order to clarify the nature of the topic on which the questions will focus, or in the case of batteries specifically, give some directions about the characteristics of the response scales or other information peculiar to the questions that follow. Series and (especially) batteries are constructed in part to streamline the questionnaire and economize with respect to the time it takes to gather responses to questions covering similar content and/or using similar response categories.

We use the term "question context" to refer to the following five categories reflecting organizational placement of questions in the questionnaire:

Table 8.1.   Cross-classification of question content and question context

Question Content	Alone		Series				Battery				Total
			Introduction		No Introduction		Introduction		No Introduction		n
Facts	21	(27%)	11	(14%)	45	(57%)	1	(1%)	1	(1%)	79	(100%)
Beliefs	2	(2%)	22	(19%)	8	(7%)	73	(63%)	11	(9%)	116	(100%)
Values	2	(5%)	9	(21%)	3	(7%)	29	(67%)	0	(0%)	43	(100%)
Attitudes	1	(1%)	3	(4%)	8	(11%)	64	(84%)	0	(0%)	76	(100%)
Self-assessments	1	(4%)	6	(24%)	8	(32%)	10	(40%)	0	(0%)	25	(100%)
Self-perceptions	3	(3%)	21	(24%)	33	(38%)	30	(34%)	0	(0%)	87	(100%)
Total n	30	(7%)	72	(17%)	105	(25%)	207	(49%)	12	(3%)	426	(100%)

(Question Context spans Alone, Series [Introduction / No Introduction], and Battery [Introduction / No Introduction])

- Stand-alone question (not in topical series or in batteries)
- Questions that are part of a topical series, which has an introduction
- Questions that are part of a topical series, which has *no* introduction
- Questions that are part of a battery—a battery is a type of series where the response scale is identical for all questions in the battery—which has an introduction
- Questions that are part of a battery that has *no* introduction

It is possible in some instances for batteries to be embedded within a larger series of questions, not all of which employ the same response format. However, in order to provide a workable definition of what we mean by a "topical series" or "battery," we employ the idea here that series and batteries are nonoverlapping subunits of survey questionnaire organization. There are other ways in which the term "question context" is used, namely, to refer to the sequence or order of questions within a particular part of a questionnaire (e.g., Schuman and Presser, 1981), and while we consider this to be an example of question context, it is just one part of the broader set of issues that can be addressed. Given the *nonexperimental* nature of our approach, we do not consider this aspect of question context here.

One of the issues we consider here is the nature and extent of *introductory* text and whether this bears on the extent of measurement error. Generally speaking, individual questions *do not* have introductions—such "prequestion" text is typically introduced at the beginning of the interview in many cases, or at the beginning of sections of the questionnaire, or as we have indicated in the above classification, they are typically used to introduce a series of questions of questions in a battery. We do not consider stand-alone questions with introductions, although there is an occasional question that is preceded by its own introduction.

Table 8.1 presents the cross-classification of the 426 self- and proxy-report questions analyzed in this study in terms of their organizational attributes, or *question context*, and the content of the questions. As these results show, there is an association between the nature of the organizational placement of questions in the questionnaire and their content. Specifically, these results reveal several things about

Table 8.2.   Descriptive characteristics of series and batteries in study
questionnaires

	Question Context	
Characteristic	Series	Battery
Number	77.00	45.00
Number with introductions	23.00	41.00
Mean introduction length (number of words)	20.52	43.34
Median introduction length (number of words)	13.00	37.00
Minimum introduction length (number of words)	5.00	4.00
Maximum introduction length (number of words)	97.00	174.00
Mean number of questions	5.68	7.49
Median number of questions	5.00	7.00
Minimum number of questions	2.00	2.00
Maximum number of questions	24.00	23.00

the context of survey questions in the questionnaires used in the present study:
(1) factual questions (both self- and proxy reports) are on the whole not found in
batteries; (2) the overwhelming majority of self-reported beliefs, values, and attitudes are found in batteries; (3) virtually all self-reported beliefs, values, and attitudes involve placement in a series or battery that uses an introduction; (4) questions
involving self-assessments and self-perceptions tend to be evenly distributed between
series and batteries, with few included as stand-alone questions; (5) self-assessments
and self-perceptions appear largely in series and batteries that use introductions; and
(6) there are virtually no batteries that do not have an introduction, whereas more
than a third of questions in our study sample are in series without an introduction.

These results reveal the natural confounding of question content and question
context that presents certain challenges in disentangling the contribution of either to
our assessment of their effects of the level of measurement error. The extent to which
their effects can be separated may be somewhat limited, but some comparisons can
be made. We can, for example, draw comparisons between reliability estimates for
measures of *factual content* within series with fact measures included as stand-alone
questions; and among measures of *nonfactual content*, we can compare the reliability estimates for stand-alone questions versus those in series and batteries, but even
here these qualities may be confounded with the formal properties of questions. We
return to an analysis of the role of question characteristics in the next chapter.

Table 8.2 presents some descriptive information on the nature of the introductions used in the *series* and *batteries* present in the six surveys in which our 426
questions are embedded. Here the information presented is at the level of the questionnaire subunit, i.e., at the series and battery level. Consistent with the information
given at the question level in Table 8.1, virtually all batteries include an introduction,
whereas roughly one-third of series do. The typical series introduction is relatively
short—a median length of 13 words—while the typical battery introduction is nearly
3 times that—a median length of 37 words. As indicated earlier, the introductions

Table 8.3.   Comparison of reliability estimates by question content and question context

| Question Content | Question Context | | | | F-ratio[1] | p-value |
	Alone	Series	Battery	Total		
Respondent self-report and proxy facts	**0.91**	**0.77**	0.76	0.81	16.26	0.000
	(21)	(56)	(2)	(79)		
Respondent self-report nonfacts	0.66	**0.67**	**0.61**	0.63	12.96	0.000
	(9)	(121)	(217)	(347)		
Total	0.84	0.70	0.61	0.67		
Total n	(30)	(177)	(219)	(426)		
F-ratio[2]	27.18	15.00				
p-value	0.000	0.000				

[1]Test within facts excludes 2 battery fact triads; test within nonfacts excludes the 9 stand-alone items.
[2]Test within batteries was not carried out.
Note: The number of questions on which reliability estimates are based is given in parentheses.

to series tend to be transitional sentences, e.g., "Now I would like to ask you some questions about your relationship with your parents," which serve the function of getting from one topic to another. Introductions to batteries, on the other hand, tend to be instructive in purpose, clarifying the nature of the task and/or to give some directions about how the response framework is constructed. Finally, we note that with respect to series and battery length, they typically contain the same number of questions. We return to a consideration of the possible effect of "introduction length" and "unit length" (i.e., series and/or battery length) to the reliability of measurement for questions in series and batteries for those sets of questions that employ them.

Table 8.3 presents our reliability estimates by categories of the cross-classification of question context and content using a collapsed set of categories for both variables. For purposes of this presentation we arrange the data on question *content* into facts versus nonfacts, and arrange the data for question *context* into three categories: stand-alone questions, questions in series, and questions in batteries, ignoring for the moment the presence or absence of an introduction. The results here indicate that there are significant differences in the reliabilities of facts measured using stand-alone versus series formats. Also, among nonfacts there are significant differences among the three formats, with those in batteries showing the lowest estimated reliabilities. These results suggest that, net of question content, stand-alone questions have the highest level of reliability, followed by questions in series and batteries, with questions in batteries having the lowest level of reliability. These results are completely consistent with what was reported by Andrews (1984) where he found that, net of content, questions not in batteries had the lowest levels of measurement error and questions in batteries containing five or more questions had the highest levels of measurement error. We return to our interpretation of these results after we consider a comparison of questions in series and questions in batteries (see Table 8.4).

Table 8.4. Comparison of reliability estimates for questions in batteries and questions in series by type of nonfactual content

Question Content	Question Context			F-ratio	p-value
	Series	Battery	Total		
Beliefs	0.67	0.58	0.61	7.73	0.006
	(30)	(84)	(114)		
Values	0.62	0.69	0.67	2.30	0.137
	(12)	(29)	(41)		
Attitudes	0.74	0.65	0.66	3.38	0.070
	(11)	(64)	(75)		
Self-assessments	0.66	0.59	0.63	1.81	0.193
	(14)	(10)	(24)		
Self-perceptions	0.68	0.54	0.63	12.23	0.001
	(54)	(30)	(84)		
Total	0.67	0.61	0.63	12.96	0.000
Total n	(121)	(217)	(338)		

Note: The number of questions on which reliability estimates are based is given in parentheses.

8.2 QUESTIONS IN SERIES VERSUS QUESTIONS IN BATTERIES

In order to examine the origin of the difference observed in estimated reliability among nonfacts, we also compare the differences between the reliabilities of questions in series and in batteries *within categories of nonfactual questions*: beliefs, values, attitudes, self-assessments, and self-perceptions. Results indicate that with few exceptions, the conclusion reached above can be generalized across categories of content of nonfactual questions. Except for self-reported values, where we have relatively few measures in our sample, questions in batteries have significantly lower levels of reliability, controlling for the content of the question.

Our tentative conclusion regarding the effects of *question context*, then, is that net of content of questions (i.e., facts versus nonfacts) stand-alone questions produce the highest level of reliability, questions in series have somewhat less reliability, and questions in batteries have the lowest relative reliability. The results cannot be completely generalized across content, in that factual questions are hardly ever included in batteries, and nonfactual questions are relatively less often asked as stand-alone questions. Within the limitations of the data, however, it appears that as one adds *contextual similarity* to questions, reliability decreases. Moving, for example, from stand-alone questions to series where questions are homogeneous with respect to content, more measurement errors appear to be produced, and as one moves to the situation of questions in batteries, where questions are homogeneous not only with respect to content but to response format as well, the estimated reliability is lowest.

Thus, while placing questions in series and batteries increases the efficiency of questionnaire construction, it may reduce the quality of the data. If true, this has serious implications for the ways in which survey questionnaires are organized.

8.3   LOCATION IN THE QUESTIONNAIRE

There is considerable folklore among survey researchers about where to place certain types of questions within the questionnaire. One school of thought regarding such placement suggests that questions with factual content should be placed at the beginning and end of the questionnaire with nonfactual content sandwiched in between. Another view is that questions that require a great deal of effort and thought on the respondent's part should be placed in the middle of the questionnaire to avoid turning the respondent off early in the interview and to avoid any of the possible effects of fatigue that may result from being placed at the end of the interview. There is very little actual research that has been conducted to assess this issue, although Sudman and Bradburn (1974, p. 33) indicate that position in the questionnaire "has by itself little biasing effect for behavioral items and a negligible effect for attitudinal items." By contrast, Andrews' (1984) research suggested that, regardless of content, questions in the first and last of the questionnaire had significantly greater amounts of measurement error. He concluded that "better data quality comes from items that fell in the 26th to 100th positions" (p. 432), reasoning that data gathered early in the interview were of poorer quality because respondents had not warmed up to the task and that data gathered later suffered in quality because of respondent or interviewer fatigue (Andrews, 1984, p. 432).

Such findings may have little practical utility—it would not be feasible to advise researchers to place all questions in the middle of the questionnaire. That would be a bit like the pediatrician's advice to parents to have their first child second, in order to avoid the effects of being first born. It could be, however, that some types of survey questions may be less vulnerable to position in the questionnaire. On the basis of the assumption that there may be validity to some of these claims, we examined the role of questionnaire location on the reliability of survey self-report measures. Our hypotheses were stimulated primarily by an interest in whether there may be motivational factors at work. We hypothesized that, controlling for question content, responses to questions asked in the middle of the interview might in fact be more reliably reported than those asked either earlier or later on. We reasoned that respondents may be more highly motivated to produce accurate responses earlier in the questionnaire, and thus, reliability of measurement would be affected by a question's location. Of course, there is a confounding of question content with position location, where it appears in the questionnaire—factual material is often gathered earlier or later in the interview, for example; and for this reason we have partitioned the data to analyze the potential contribution of questionnaire position separately for facts and nonfacts. There may also be a confounding of the effect of position with question context, in that lengthy batteries may typically not be included early in the questionnaire, lest the respondent prematurely break off the interview.

In Table 8.5 we present information that is relevant to evaluating the *position of the question* in the questionnaire for the reliability of measurement. Questionnaires vary in length, and it is possible to compare absolute position in the questionnaire only insofar as questionnaires are similar in length. Measuring relative position (e.g., question number divided by the number of questions in the questionnaire) makes little sense because normally respondents have no idea how long the questionnaire will be. Thus, we measure the *cardinal position* in the questionnaire from the beginning, by arbitrarily categorizing questions according to whether they appear in (1) the first 25 questions, (2) the second 25 questions, (3) the next 50 questions, and so forth (see Table 8.5).

The information presented in Table 8.5 examines the conventional assumptions about the confounding of question location and question content and context. First, it is important to note that question position in the questionnaire is *not* statistically independent of either question context or question content.[2] Interestingly, what these results show is that facts and nonfacts are proportionally represented in the first 25 questions, but disproportionately represented thereafter. Nonfacts are much more likely to be placed in the middle sections of the questionnaire and facts at the end. Somewhat surprisingly, with respect to question context, questions in batteries are much more likely to be placed in the early parts of the questionnaire, questions in series are significantly more likely to be placed in the middle and end, and stand-alone questions are more likely to be placed toward the end of the questionnaire. Whether the six questionnaires used in this study reflect survey practice in general is difficult to assess, but these results provide a basis for the assumption that in order to assess the influence of question position it is important to control for question *content* and question *context*.

Using the partitions of question position given above, we present estimates of reliability by position in the questionnaire controlling separately for (1) study, (2) question context, and (3) question content in Table 8.6. The results in this table provide relatively uniform support for the null hypothesis of *no effect of question position* on estimates of reliability controlling (separately) for question context and content. There is hardly any difference in the estimated reliability for questions that appear early in the questionnaire compared to those appearing later. For neither facts nor nonfacts does position in the questionnaire appear to make a difference. However, there does appear to be a significant interaction between position in the questionnaire and question context—specifically, batteries placed later in the questionnaire produce lower data quality. Our tentative conclusion with regard to position in the questionnaire is that while there is no systematic main effect of position, there is an interaction with question context such that position seems to matter primarily for questions in batteries.

Why isn't position in the questionnaire a more general predictor of measurement reliability? Perhaps the best answer to this question is that most survey interviews

---

[2]The χ^2 (with 12 df) for the relationship between question position and question context is 41.05 (p < .0001). The χ^2 (with 6 df) for the relationship between question position and question content is 31.41 (p < .0001).

Table 8.5. Distribution of study questions by position in the questionnaire for six panel studies by question content and question context

Panel	Length[1]	Position in Questionnaire													Total n	
		1-25		26-50		51-100		101-150		151-200		201-250		251+			
NES 50s	222	5	(12%)	12	(29%)	6	(14%)	1	(2%)	8	(19%)	10	(24%)	0	(0%)	42	(100%)
NES 70s	339	1	(1%)	0	(0%)	7	(7%)	20	(20%)	18	(18%)	23	(23%)	31	(31%)	100	(100%)
NES 90s	322	4	(4%)	22	(22%)	4	(4%)	9	(9%)	23	(23%)	16	(16%)	20	(20%)	98	(100%)
ACL	378	10	(12%)	9	(10%)	12	(14%)	9	(10%)	20	(23%)	11	(13%)	15	(17%)	86	(100%)
SAF Children	212	18	(39%)	0	(0%)	6	(13%)	5	(11%)	17	(37%)	0	(0%)	0	(0%)	46	(100%)
SAF Mothers	58-151	25	(46%)	19	(35%)	9	(17%)	1	(2%)	0	(0%)	0	(0%)	0	(0%)	54	(100%)
Total n (%)		63	(15%)	62	(15%)	44	(10%)	45	(11%)	86	(20%)	60	(14%)	66	(15%)	426	(100%)

Question Context	Position in Questionnaire													Total n	
	1-25		26-50		51-100		101-150		151-200		201-250		251+			
Stand-alone	3	(10%)	1	(3%)	2	(7%)	5	(17%)	8	(27%)	6	(20%)	5	(17%)	30	(100%)
Series	17	(10%)	25	(14%)	25	(14%)	19	(11%)	25	(14%)	22	(12%)	44	(25%)	177	(100%)
Battery	43	(20%)	36	(16%)	17	(8%)	21	(10%)	53	(24%)	32	(15%)	17	(8%)	219	(100%)
Total n (%)	63	(15%)	62	(15%)	44	(10%)	45	(11%)	86	(20%)	60	(14%)	66	(15%)	426	(100%)

Question Content	Position in Questionnaire													Total n	
	1-25		26-50		51-100		101-150		151-200		201-250		251+			
Facts[2]	12	(15%)	6	(8%)	6	(8%)	6	(8%)	13	(16%)	8	(10%)	28	(35%)	79	(100%)
Nonfacts	51	(15%)	56	(16%)	38	(11%)	39	(11%)	73	(21%)	52	(15%)	38	(11%)	347	(100%)
Total n (%)	63	(15%)	62	(15%)	44	(10%)	45	(11%)	86	(20%)	60	(14%)	66	(15%)	426	(100%)

[1]Number of questions in the wave 2 questionnaire.
[2]Includes both self-report and proxy facts.

174

Table 8.6.   Comparison of reliability estimates for questions differing in position in the questionnaire by study sample, question content and question context

Panel	1-25	26-50	51-100	101-150	151-200	201-250	251+	Total	F-ratio	p-value
			Position in Questionnaire							
NES 50s	0.61	0.58	0.73	0.76	0.95	0.66	---	0.70	---	---
NES 70s	0.69	---	0.75	0.57	0.51	0.61	0.71	0.63	---	---
NES 90s	0.57	0.68	0.82	0.77	0.61	0.58	0.76	0.67	---	---
ACL	0.67	0.69	0.68	0.75	0.66	0.69	0.69	0.68	---	---
SAF Children	0.67	---	0.66	0.63	0.69	---	---	0.67	---	---
SAF Mothers	0.68	0.64	0.70	0.95	---	---	---	0.67	---	---
Total	0.66	0.65	0.71	0.66	0.65	0.63	0.72	0.67	2.50	0.022

Question Context	1-25	26-50	51-100	101-150	151-200	201-250	251+	Total n	F-ratio	p-value
			Position in Questionnaire							
Stand-alone	0.66	0.98	0.78	0.78	0.87	0.84	0.92	0.84	1.09	0.399
Series	0.65	0.66	0.74	0.73	0.74	0.62	0.74	0.70	2.74	0.014
Battery	0.67	0.63	0.66	0.57	0.57	0.59	0.60	0.61	2.21	0.044
Total	0.66	0.65	0.71	0.66	0.65	0.63	0.72	0.67	2.50	0.022
F-ratio	0.07	2.08	1.54	9.28	19.00	5.94	11.82	35.25		
p-value	0.932	0.134	0.226	0.000	0.000	0.005	0.000	0.000		

Question Content	1-25	26-50	51-100	101-150	151-200	201-250	251+	Total n	F-ratio	p-value
			Position in Questionnaire							
Facts[1]	0.71	0.81	0.76	0.80	0.88	0.83	0.82	0.81	1.40	0.226
Nonfacts	0.65	0.63	0.70	0.64	0.61	0.60	0.65	0.63	2.28	0.036
Total	0.66	0.65	0.71	0.66	0.65	0.63	0.72	0.67	2.50	0.022
F-ratio	1.74	5.83	0.73	5.98	33.11	14.86	26.48	79.32		
p-value	0.192	0.019	0.398	0.019	0.000	0.000	0.000	0.000		

[1]Includes both self-report and proxy facts.

are relatively short, and this would argue against the claim that respondent fatigue over the course of the interview should affect the reliability of responses. Indeed, most researchers report a high degree of interest and motivation among survey respondents, except when respondent burden is excessively high, as in the case of questions in batteries. The present results appear to indicate that it is only in the case of batteries that the advice often given in other spheres of human activity—that of "location, location, location"—really matters for survey design.

8.4   UNIT LENGTH AND POSITION IN SERIES AND BATTERIES

We reported above that position in the questionnaire had little, if any, bearing on the reliability of measurement (except in the case of questions in batteries), a finding that should be highly reassuring to those who design questionnaires. In the present study the majority of questions are placed either in series or in batteries. To review, specifically there are 177 questions in 77 different series and 219 questions in 45 different batteries. Two issues regarding questionnaire context that remain to

Table 8.7. Comparison of reliability estimates of questions in series and batteries of different lengths by question content

Battery/Series Length[1]	Series			Battery			F-ratio[2]	p-value
	Facts	Nonfacts	Total	Facts	Nonfacts	Total		
2 - 5	0.81	0.68	0.72	0.96	0.60	0.61	4.80	0.031
	(25)	(62)	(87)	(1)	(39)	(40)		
6 - 10	0.73	0.65	0.69	0.57	0.61	0.61	2.45	0.120
	(23)	(32)	(55)	(1)	(107)	(108)		
11 +	0.76	0.69	0.70	---	0.62	0.62	3.71	0.057
	(8)	(27)	(35)	(0)	(71)	(71)		
Total	0.77	0.67	0.70	0.76	0.61	0.61	12.96	0.000
Total n	(56)	(121)	(177)	(2)	(217)	(219)		
F-ratio	1.90	0.40	0.68	---	0.29	0.25		
p-value	0.159	0.674	0.506	---	0.746	0.781		

[1]Number of questions in battery or series.
[2]Test for difference between Series nonfacts and Battery nonfacts.
Note: The number of questions on which reliability estimates are based is given in parentheses.

be addressed are whether (1) the length of a series or battery affects the reliability of measurement and (2) the position of a question within a series or battery has any effect on the reliability of measurement. The only previous research carried out on this topic was that of Andrews (1984, pp. 430–431), who found that questions in longer batteries had higher levels of measurement error. His research did not employ the distinction we have drawn here between series and batteries, and his results presumably pertain to the measurement of attributes of questions in a series as well as those of questions in batteries.

The typical series in this sample of questions contains five questions, while the typical battery contains seven questions (see Table 8.2). In both cases the shortest series or battery contains just two questions and the longest 23 or 24. Table 8.7 presents estimates of reliability by the length of series and batteries. Somewhat surprisingly, these results indicate that there is no net difference in the reliability of questions attributable to the length of the battery or series in which the questions are located. These results differ from those reported by Andrews (1984), who found that the longer the battery, the lower the data quality. This is true for both factual and nonfactual questions in series and nonfactual questions in batteries (recall that there are virtually no factual questions appearing in batteries). We continue to observe in this table that among nonfactual questions, those situated in batteries are significantly less reliable, even when we control for the length of the series or battery.

In order to address the issue of the effect of the position of the question within a battery or series on the reliability of measurement, we partitioned our sample of survey questions into categories defined by the *cardinal position* within the two types of questionnaire subunits (see Table 8.8). We analyze the effect of position separately for questions in series and questions in batteries. These results indicate

Table 8.8.   Comparison of reliability estimates for questions in different positions in series and batteries by question content

	Series				Battery		
Position in Series[1]	Facts	Nonfacts	Total	Position in Battery[2]	Facts	Nonfacts	Total
1	0.78	0.70	0.72	1	0.57	0.64	0.64
	(13)	(29)	(42)		(1)	(30)	(31)
2	0.79	0.66	0.70	2	---	0.57	0.57
	(15)	(25)	(40)		(0)	(18)	(18)
3	0.78	0.69	0.72	3	---	0.64	0.64
	(9)	(22)	(31)		(0)	(22)	(22)
4	0.82	0.67	0.72	4	---	0.60	0.60
	(6)	(11)	(17)		(0)	(16)	(16)
5	0.63	0.67	0.66	5	---	0.55	0.55
	(4)	(9)	(13)		(0)	(20)	(20)
6+	0.74	0.65	0.68	6	---	0.59	0.59
	(9)	(25)	(34)		(0)	(16)	(16)
Total	0.77	0.67	0.70	7	---	0.67	0.67
Total n	(56)	(121)	(177)		(0)	(9)	(9)
F-ratio	0.99	0.32	0.59	8+	---	0.61	0.61
p-value	0.434	0.900	0.708		(0)	(34)	(34)
				Total	0.57	0.61	0.61
				Total n	(1)	(165)	(166)
				F-ratio	---	1.03	1.02
				p-value	---	0.409	0.420

[1]Cardinal number of question position in series.
[2]Cardinal number of question position in battery.
Note: The number of questions on which reliability estimates are based is given in parentheses.

there is no systematic relationship between the position of a question within a series or battery and the estimated level of measurement error in the questions. These findings are consistent with those of Andrews (1984) and Scherpenzeel and Saris (1997), which indicate only trivial differences in measurement error attributable to position within the questionnaire subunit. Thus, based on the present results it is not the length of the series or battery that matters, nor the position of the question within the series or battery, but *whether* the question is included within a battery *or* a series that affects measurement reliability.

8.5   LENGTH OF INTRODUCTIONS TO SERIES AND BATTERIES

One of the key differences between series and batteries, as units of questionnaire organization, involves the nature and extent of the introductory text that is read to the respondent concerning the questions to follow. We indicated earlier that one of the characteristics of the batteries in our sample of questionnaires is that they nearly always have an introduction (see Table 8.2) that averages some 37 or more words in length. Series are much less likely to have an introduction—most do not—and the length of the typical introduction to a series of questions is much shorter—some 13 or more words. We suggested further that the function of introductions is very different for sets of questions in series compared to those in batteries. In the former an introduction serves primarily as a transition from one set of questions to another, or from one topic to another, whereas in the case of batteries the purpose of the introduction is to explain the nature of the questions and/or response categories to the respondent. In some instances an introduction to the battery will actually contain the question or part of the question as well (see Chapter 9)

One of the types of questions used in the 1970s and 1990s National Election Studies (NES) which illustrates the use of long introductions is the famous Michigan "feeling thermometer," a measuring device typically used to assess attitudes toward social groups, political figures, and political candidates. These questions begin with a long and tedious introduction, such as the following:

> We'd also like to get your feelings about some groups in American society. When I read the name of a group, we'd like you to rate it with what we call a feeling thermometer. It is on page 19 of your booklet. Ratings between 50 and 100 degrees mean that you feel favorably and warm toward the group. Ratings between 0 and 50 degrees mean that you don't feel favorably toward the group and that you don't care too much for that group. If you don't feel particularly warm or cold toward a group, you would rate them at 50 degrees. If we come to a group you don't know much about, just tell me and we'll move on to the next one.

The interviewer then reads the first social group for which a rating is desired, "black militants," or "big business," and so on. After the introduction, the questions are very brief—typically just a few words. This may be a situation where lengthy introductions (in this case 123 words worth of introduction) may have a certain degree of payoff with respect to increasing the clarity of the task.

Here we examine whether the length of the introductions to the series and batteries in which questions are placed have any effect on the reliability of measurement. In order to examine this issue, we partitioned the sample of questions into several categories representing differences in the *length of introduction* of questions in series and batteries. As noted earlier, there is a great deal more variation in the length of introductions in the case of batteries. Moreover, the main type of content for which batteries are used is nonfactual content, so our main focus here is in examining the effect of variation in battery introduction length for nonfactual questions. These results are presented in Table 8.9.

These results indicate that in our sample of questions there is a detectable effect on reliability of measurement that is attributable to the length of the introduction, with those questions in series having either no introductions or short introductions revealing significantly higher reliabilities than questions in series with long introductions. Among questions in batteries, those having no introductions are somewhat

Table 8.9. Comparison of reliability estimates for questions in series and batteries with differing lengths of introduction by question content

Series Introduction Length[1]	Series			Battery Introduction Length[1]	Battery		
	Facts	Nonfacts	Total		Facts	Nonfacts	Total
0	0.78	0.69	0.73	0	0.96	0.80	0.82
	(45)	(60)	(105)		(1)	(6)	(7)
1 - 15	0.74	0.72	0.72	1 - 15	---	0.63	0.63
	(8)	(28)	(36)		(0)	(41)	(41)
16+	0.66	0.60	0.61	16 - 40	0.57	0.64	0.64
	(3)	(33)	(36)		(1)	(76)	(77)
Total	0.77	0.67	0.70	41 - 80	---	0.54	0.54
Total n	(56)	(121)	(177)		(0)	(70)	(70)
F-ratio	1.24	5.20	9.06	81+	---	0.66	0.66
p-value	0.299	0.007	0.000		(0)	(24)	(24)
				Total	0.76	0.61	0.61
				Total n	(2)	(217)	(219)
				F-ratio	---	8.39	9.56
				p-value	---	0.000	0.000

[1]Number of words in introduction to series/battery.

Note: The number of questions on which reliability estimates are based is given in parentheses.

more reliable. Both sets of results are statistically significant. We should offer a caveat here, however, lest the wrong conclusion be drawn. Recall that we have arbitrarily labeled any set of questions on the same specific topical area that share the same set of response options a "battery." If a battery has no introduction, this means that the investigators no doubt believed one was not necessary and that the questions could be answered without any special efforts undertaken to understand the response framework. If so, it is not the absence of an introduction per se that results in higher reliabilities for questions in batteries, but that the content of the questions and the nature of the response scales were such that an introduction was unnecessary.

8.6 CONCLUSIONS

Our evidence to this point is that among factual questions those included as standalone questions are more reliable than questions included in a series of questions on the same topic. For nonfactual questions, those included in batteries are less reliable than those in series. We find little or no effect on reliability of location of the question in the questionnaire or of the position of the question within a series or battery. Question context interacts to a slight degree with position in the questionnaire, in that those questions in batteries that appear later in the questionnaire are somewhat less reliable. In addition, the length of the introduction in both series and batteries seems to matter for reliability of measurement—although it operates some-what differently in the two cases. Among series, those having long introductions (16+ words) appear to have lower reliability, whereas questions in batteries having *any* introduction appear to have lower reliability, with length of introduction matter-ing little for batteries that have an introduction.

It is not clear to us at this stage in our analysis what accounts for these patterns. The "stand-alone > series > battery" ordering with respect to reliability is certainly intriguing, which raises the question of how it can best be explained. Perhaps the best answer is that the same factors that motivate the researcher to group questions together—contextual similarity—are the same factors that promote measurement errors. Similarity of question content and response format may actually distract the respondent from giving full consideration to what information is being asked for, and it may reduce the respondent's attention to the specificity of questions. Mea-surement errors may, thus, be generated in response to the features of the question-naire that make it more efficient and streamlined. Andrews' (1984, p. 431) words are instructive here: "It is possible that respondents and/or interviewers recognize the 'production line' character of this survey strategy and that it promotes carelessness in the way questions are asked and answered." The problem appears to be that the respondents are also more likely to "streamline" their answers when the investigator "streamlines" the questionnaire.

On the other hand, one might argue that it is the type of content measured using various types of questionnaire organization that accounts for these results—that is, the measurement of types of survey content for which batteries are used are simply measured less reliably than are those types for which series are used. Or conversely, that the types of content assessed in stand-alone questions are the kinds of things that are measured more reliably in general. Of course, we have anticipated this argument and have been careful to control for survey content—fact versus nonfact—in all of our analyses of survey context. We noted the limitations inherent in doing this, given that batteries are hardly ever used to measure facts and nonfactual questions are much less likely to be measured using stand-alone questions than with series and bat-teries. However, to the extent we can control on survey content, we believe that this argument cannot be used to successfully account for the effects of question context.

Another possibility is that the *context effects* discovered here result from other aspects of the questions, particularly question form, e.g., type of rating scale used, the length of the question, the number of response categories, the labeling of response categories, or the explicit provision of a Don't Know option. In other words, reliabil-ity of measurement might have little to do with question context (stand-alone, series or batteries) per se, and the effects may instead be due to the formal properties of questions appearing in one or the other type of questionnaire unit. If we can account for the observed differences in reliability by reference to the formal attributes of questions rather than the context in which they are embedded, then the focus of our attention can be simplified; that is, we can pay less attention to the way in which the survey questionnaire is organized and more attention to the intrinsic characteristics of the questions themselves. We turn now to a discussion of the formal attributes of survey questions and their role in measurement reliability.

CHAPTER NINE

Formal Properties of Survey Questions

[Survey questions should meet four requirements. They should] . . . (a) be quickly and easily understood; (b) admit of easy reply; (c) cover the ground of enquiry; (d) tempt the co-respondents to write freely in fuller explanation of their replies and cognate topics . . .

Sir Francis Galton, Inquiries into the Human Faculty, 1893[1]

There is an abundance of available advice about *what constitutes a good question.* Vast amounts have been written on this topic from the earliest uses of surveys down to the present. As with questionnaire design, there is a prevalence of "expert opinion" on how to write good questions (e.g., Belson, 1981; Krosnick and Fabrigar, 1997; Schaeffer and Presser, 2003). Still, there is very little consensus on the specifics of question design, and many view the development of survey questions as more of an art than a science. Many guides to questionnaire development reiterate the kinds of general ideas expressed in the quote from Galton given above. Except for possibly the last item on his list—"to tempt co-respondents to write freely in fuller explanation of their answers"—hardly anyone would disagree with Galton's desiderata. Survey questions should be, in modern vernacular, "short, sweet, and to the point." These early suggestions have been echoed by similar efforts to codify what attributes are possessed by "good questions" and/or "good questionnaires," and on some points there is a modest degree of agreement. On others there is little at all.

Many of the issues discussed in the survey methodology literature and at conferences on questionnaire design hark back to some of the earliest writings on the subject. For example, in their pioneering book, *Gauging public opinion,* Cantril and Fried (1944) developed a taxonomy of chronic problems in writing good questions, including such things as the vagueness of questions, the use of unfamiliar or technical words, or the use of response categories that are incomplete. In this same tradition, Payne's (1951) early book on survey question design, *The art of asking questions,* codified a list of "100 considerations" in the development of good survey questions. Payne's "handbook" discussed the advantages and disadvantages of

---

[1]Cited in O'Muircheartaigh (1997) from Ruckmick (1930).

"open-" versus "closed-" form questions, the problems with using vague quantifiers, the perils of using "loaded" questions, and many others.

Guidelines for writing good survey questions often represent the "tried and true" and are aimed at codifying a workable set of "common sense" rules for question construction. Beyond some of the general considerations advanced in the earliest writings on this subject, more recent treatments have developed guidelines that are quite specific with respect to the form and content of questions. Writing specifically about questions for cross-national or cross-cultural surveys, for example, Brislin (1986) listed several *detailed* guidelines for writing survey questions, although his recommendations are quite general and can be considered to apply equally to mono-cultural research. He emphasized such things as the use of short simple sentences of less than 16 words, the use of the active rather than the passive voice, avoiding metaphors and colloquialisms, adding sentences to provide context to key items, and to reword key phrases to provide redundancy. He also suggested that good questions avoid the subjunctive mood, conditional or hypothetical verb forms, repeat nouns instead of using pronouns, avoid adverbs and prepositions telling "where" or "when," avoid possessive forms where possible, and use specific rather than general terms. Finally, he encouraged question writers to use wording familiar to the respondent whenever possible and to avoid words indicating vagueness regarding some event or thing (e.g., probably, maybe, perhaps).

Although it is hard to argue against any of these criteria for good questions, as is true with most available advice on how to design survey questionnaires, there was little hard evidence to support the idea that following these rules contributes to improved quality of data. In order to assess these issues more broadly and to evaluate the linkage between question/questionnaire characteristics and the quality of data, one needs a set of criteria deciding which questions are better than others, other than an appeal to common sense. Modern empirical survey researchers demanded, not simply the good advice of experts, but empirical evidence on the advantages of particular question forms and approaches to interviewing. Several pioneering efforts were undertaken in the mid-1970s to advance an "experimental approach" to the evaluation of survey questions—by experimentally varying various forms of a question and evaluating the differences empirically (Sudman and Bradburn, 1974; Bradburn and Sudman, 1979; Hyman, 1975; Schuman and Presser, 1981; Cannell, Miller, and Oksenberg, 1981).

By requiring empirical evidence for the differences in survey results and the advantages or disadvantages of various questioning strategies, the standards for evaluating questions were substantially raised. This approach to the evaluation of the form and context of survey questions, along with several efforts to bring the principles of "cognitive science" to bear on the development of survey methods stimulated a tradition of research focusing on the effects of formal attributes of survey questions that has become a virtual cottage industry (see Jabine, Straf, Tanur, and Tourangeau, 1984; Hippler, Schwarz and Sudman, 1987; Schwarz and Sudman, 1994; Turner and Martin, 1984; Sirken, Herrmann, Schechter, Schwarz, Tanur, and Tourangeau, 1999; Tourangeau, 1984; Tourangeau, Rips and Rasinski, 2000; Schaeffer and Presser, 2003; Krosnick and Fabrigar, 1997). These early experiments ushered in a renewed concentration on the tools of the trade within the field of survey methods, focusing specifically on the nature of survey questions and responses. Much of the early experimental work on "question wording" effects, however, was not guided by any strong theoretical concerns, focusing instead on manipulations of common forms to see what differences might result (Hippler et al., 1987). Instead, most of this research has focused on differences in the marginal distributions of different forms of question wording or other experimental variations in question form. What is perhaps most striking about these *experimental* developments is that they did not employ a model for the evaluation of measurement error—very little of this work focused on the linkage between question form and the extent of measurement error (see, e.g., Schaeffer and Presser, 2003).

Another criterion for the evaluation of questions—a *non-experimental approach*, one that I employ throughout this book—focuses on the amount of measurement error associated with a particular question or question type. A preference for this criterion is not meant to deny the relevance of many other pertinent criteria, but this choice is consistent with the view that survey questions must be evaluated not simply with respect to their univariate characteristics, but also in terms of levels of measurement error (e.g., Biemer, Groves, Lyberg, Mathiowetz, and Sudman, 1991; Andrews, 1984; Saris and Andrews, 1991). Fortunately, although little of the research on the impact of the formal attributes of questions has focused on assessing the extent of measurement error, there are several key exceptions on which we build here. Principal among these is the work of Andrews (1984) and his collaborators (Saris and Andrews, 1991; Scherpenzeel and Saris, 1997). Employing the multitrait-multimethod approach to designing replicate measures of questions (described in Chapter 3), this research has contributed to the debate about question length, open- vs. closed-form question, number of response options, and the explicit offering of a Don't Know option.

Using measurement reliability as a criterion for evaluating differences among questions, in this chapter we focus on several features of survey questions that may have a potential impact on the quality of data. We begin our investigation with a focus on the topic of *question form*, and move from there to a consideration of several additional attributes of questions, specifically the *number of response options* provided in closed-form questions, the use of *unipolar* versus *bipolar* response formats, the use of *verbal labeling* of response categories, the provision of explicit *Don't Know options*, and the *length of questions* used in survey questionnaires. As I have pointed out in the previous chapter, it is challenging to isolate the "effects" of formal attributes of questions, due to the confounding of question content, context and question characteristics. However, I argue that in most cases it is possible to isolate critical comparisons of specific forms of questions by carefully controlling for other pertinent features of question content and context.

9.1   QUESTION FORM

One of the earliest areas of debate in the design of questionnaires originated during World War II involving whether "open" versus "closed" question forms should be used (see Converse, 1987, pp. 195–202). Rensis Likert was an advocate of the "fixed question/free answer" approach in which a standardized question was used and responses were transcribed verbatim, even for nonfactual questions. This may seem somewhat ironic, given that Likert's name has come to be associated with the *Likert scale,* a well-known fixed-format approach to measuring attitudes (see below). This position was countered by the "closed question" approach (advocated by Paul Lazarsfeld, among others) in which response options were fixed beforehand and read to the respondent as part of the question (see Converse, 1987, pp. 195–202). Eventually the "closed question" or "forced choice" approach won out for most survey purposes, although even in the early 1980s there was concern that such questions might be suggesting unconsidered options to respondents (Schuman and Presser, 1981). There are clearly many uses for open-ended questions in survey research, but the more or less standard approach in modern survey research is to formulate questions within the framework of a fixed-format set of response categories. In the words of experts Converse and Presser (1986, p. 35) " . . . in most instances, a carefully pretested closed form is to be preferred for its greater specificity."

It seems from this discussion that information sought through the use of *open-ended* response formats will *not necessarily* have systematically higher levels of reporting reliability than questions with *closed-form* response formats. On the other hand, the results of Andrews' (1984) study suggest that questions employing an open-ended format have higher reliabilities. This is probably due to some extent to the nature of the content being measured, in that researchers are more likely to use open-ended questions when the categories are both more clear and more concrete (e.g., occupational information), whereas closed-form response formats are more likely to be used when the categories are less clear and must be interpreted by the respondent.

In Table 9.1 we present the results of our comparison of reliability estimates for open-ended and closed-form questions by question content and question context. If our results are at all representative—and we believe they are—it illustrates the strong tendency for open-ended questions to be employed in a relatively selective way. Our results show that there are hardly any nonfactual questions that are open-ended—only six of 347 nonfactual questions. Examples of this type of open-ended nonfactual question include "Can you tell me about what the present value of your house is, that is, about what it would bring if you sold it today?" "If you sold this (house/apartment/farm) today, how much money would you get for it (after paying off the mortgage)?" and "What do you think is the ideal number of children for the average American family?"

There is also a relationship between question form and the context in which the question is presented. Of the 48 factual open-ended questions in our sample of questions, all but one are stand-alone questions or questions that are a part of a series. The exception was a series of factual questions about the possession of various forms of assets, asked in an open-ended format. These are the only open-ended questions in our sample coded as being included in a battery. By contrast, closed-form questions are almost exclusively contained within series and batteries—questions in batteries

Table 9.1. Comparison of reliability estimates for questions using open-ended and closed-form response formats by question content and question context

| Response Format | Stand-alone | | | Series | | | Battery | | | Total |
	Facts	Nonfacts	Total	Facts	Nonfacts	Total	Facts	Nonfacts	Total	Total
Open-ended	0.92	---	0.92	0.79	0.83	0.80	0.57	---	0.57	0.834
	(18)	(0)	(18)	(29)	(6)	(35)	(1)	(0)	(1)	(54)
Closed-form	0.88	0.66	0.71	0.75	0.67	0.68	0.96	0.61	0.61	0.642
	(3)	(9)	(12)	(27)	(115)	(142)	(1)	(217)	(218)	(372)
Total	0.91	0.66	0.84	0.77	0.67	0.70	0.76	0.61	0.61	0.67
Total n	(21)	(9)	(30)	(56)	(121)	(177)	(2)	(217)	(219)	(426)
F-ratio	0.20	---	15.49	1.26	6.71	16.42	---	---	0.06	69.39
p-value	0.664	---	0.000	0.266	0.011	0.000	---	---	0.800	0.000

Note: The number of questions on which reliability estimates are based is given in parentheses.

are almost always closed form (218 out of 219 cases), and this is also the tendency for questions in series as well (see Table 9.1).[2]

Our results with respect to reliability shown in Table 9.1 reinforce the conclusion often drawn that information sought through the use of *open-ended* response formats tends to have systematically higher levels of reporting reliability than questions with *closed-form* response formats. In anticipation of the concern that this may be due in part to the nature of the content typically assessed by particular question forms, we should note that these conclusions apply even when controlling for the fact versus nonfact distinction. We find substantial differences favoring the open-ended format within categories of facts and nonfacts, although, as we already mentioned, there are too few cases in which the measurement of nonfacts employs the open-ended format, so we do not have the data to establish this empirically. This comparison is especially meaningful in the context of factual questions, however, as it suggests that researchers might more profitably exploit the uses of the open-ended approach, relying more on *open-ended* formats when measuring factual content.

9.2 TYPES OF CLOSED-FORM QUESTIONS

Within the set of closed-form questions there are several discrete types: Likert-type or "agree-disagree" questions, "forced-choice" questions, "feeling thermometers," and various kinds of *other* rating scales. The key distinction we make here among the types of *other* rating scales is that between "unipolar" versus "bipolar" rating

---

[2]The single closed-form factual question coded here as occurring in a battery was the following: "Now we are interested in the income that you yourself received in 1993, not including any of the income received by (your spouse and) the rest of your family. Please look at this page and tell me the income you yourself had in 1993 before taxes. This figure should include salaries, wages, pensions, dividends, interest, and all other income." This measure was coded as being included in a battery because the previous question used the same scale but asked about household income instead of personal income.

scales, but this difference refers not to the form of the question so much as to the form of the content being measured. Typically, the ends of a bipolar scale are of the same intensity but opposite valence, while the ends of a unipolar scale tend to differ in amount or intensity, but not valence. The differences among the estimated reliabilities of these forms of questions are presented in Table 9.2. We group these questions within categories of question content (fact versus nonfact) cross-classified by question context (stand-alone, series, and/or battery), although we should note that the distribution of our sample of questions across these categories (for reasons that will become apparent) is quite lumpy.

As these results indicate, there are few apparent differences in the reliabilities associated with these several types of closed-form question formats. Their estimated reliabilities do not vary a great deal around the average reliability of .64 for closed-form questions generally (in contrast to .83 for the typical open-ended questions), although there do appear to be some small significant differences. Although we present these here, we are not in a position at this point in the analysis to draw significant import from these results without further consideration of all relevant attributes of the question forms. Thus, before ascribing these differences to question form per se, we discuss the key features of these different question forms and examine the key differences among them in the measurement of nonfacts before proceeding on to an analysis of the differences among them with respect to levels of measurement error (see Sections 9.3 and 9.4).

9.2.1   Agree-Disagree Questions

In the publication of his doctoral dissertation, Rensis Likert (1932) suggested that attitudes could be measured relatively easily be presenting respondents with a five-category scale that included the measurement of three elements relating to the attitude concept: the direction of the attitude (agree vs. disagree, approval vs. disapproval etc.), the strength of the attitude (e.g., agree vs. strongly agree and disagree vs. strongly disagree), and a neutral point (neither agree nor disagree) for those respondents who could not choose between these alternatives. Likert also suggested offering an explicit Don't Know to distinguish between those people who had no opinion and those who were truly neutral. (We return to the issue of offering a Don't Know response option and its effect on reliability in a later section.)

Such question forms have become known as *Likert scales*, although the term is often used more broadly to refer to any bipolar survey question (regardless of the number of categories) that attempts to assess the direction and strength of attitudes, using labeled response categories. In the present sample of questions we have 64 such questions, using either four or five response categories. Although historically such scales were developed for the measurement of attitudes, in the present set of studies we should note that this form is used to measure a range of nonfactual content (attitudes, beliefs, values, and self-assessments). As indicated by the results in Table 9.2, *Likert scales* tend to produce an average level of reliability among all closed-form measures.

Table 9.2. Comparison of reliability estimates for questions with closed-form response formats by question content and question context

Question Form	Question Context									Question Content		
	Stand-alone			Series			Battery					
	Facts	Non-Facts	Total	Facts	Non-Facts	Total	Facts	Non-Facts	Total	Facts	Non-Facts	Total
Agree disagree	--	--	--	--	0.57 (8)	0.57 (8)	--	0.64 (56)	0.64 (56)	--	0.63 (64)	0.632 (64)
Forced choice	--	0.66 (2)	0.66 (2)	1.00 (1)	0.70 (23)	0.71 (24)	--	0.69 (6)	0.69 (6)	1.00 (1)	0.70 (31)	0.71 (32)
Feeling thermometer	--	--	--	--	--	--	--	0.65 (6)	0.65 (6)	--	0.65 (31)	0.65 (32)
Other ratings/unipolar	0.89 (2)	0.69 (4)	0.75 (6)	0.74 (24)	0.65 (53)	0.68 (77)	0.96 (1)	0.62 (57)	0.62 (58)	0.76 (27)	0.63 (114)	0.66 (141)
Other ratings/bipolar	--	0.61 (3)	0.61 (3)	0.50 (1)	0.69 (31)	0.69 (32)	--	0.53 (51)	0.53 (51)	0.50 (1)	0.59 (85)	0.592 (86)
Total	0.89 (2)	0.66 (9)	0.70 (11)	0.74 (26)	0.67 (115)	0.68 (141)	0.96 (1)	0.61 (217)	0.61 (218)	0.76 (29)	0.63 (341)	0.64 (370)
Total n										F-ratio	2.91	4.03
										p-value	0.022	0.003

Note: The number of questions on which reliability estimates are based is given in parentheses.

187

9.2.2 Forced-Choice Questions

It is often the case that attitude researchers do not wish to assess magnitudes of attitude strength, as is the case in the Likert scale, and choose questions with fewer response options. The choice is often between two or three response categories, although we include only two category scales under this "forced choice" category in the present analysis. In the present sample there are 30 such questions covering a range of content, including facts, beliefs, attitudes, and self-perceptions. Forced-choice questions may be bipolar in the sense of "agree vs. disagree" questions, or they may be unipolar, meaning that the question assesses the presence or absence of some quantity.

There has been a substantial amount of research on the relative advantages and disadvantages of forced-choice questions with a neutral point or middle alternative. Schuman and Presser (1981) found that middle alternatives were more often chosen when they are explicitly offered than when they are not, suggesting that the meaning of the response categories may stimulate the response. Alwin and Krosnick (1991b) suggested that the respondent may choose a middle category because it requires less effort and may provide an option in the face of uncertainty, and Alwin (1992) found that questions with three-category response scales were less reliable than two- and four-category scales, due to the ambiguities introduced.

9.2.3 Feeling Thermometers

One rather unique approach to attitude measurement using magnitude estimation in which many response categories are employed (at the other extreme from forced-choice questions) is the *feeling thermometer*. Here the purpose is not simply to measure the direction of attitude or belief but also to assess the *degree or intensity of feeling*. This approach, and related applications, takes the point of view that the respondents' feelings or affect toward a political candidate, or some other attitude object, can be registered on a response continuum with many scale points analogous to a thermometer. Here the respondent is given an opportunity to make more fine-grained distinctions using what is essentially a continuous scale, using the thermometer metaphor.

The feeling thermometer is particularly useful in the measurement of subjective variables, such as attitudes, that may be conceptualized as latent continua reflecting predispositions to respond. Although there are some social psychological theorists who define attitude continua in terms of discrete categories, as "latitudes" or "regions" on a scale (e.g., Sherif, Sherif, and Nebergall, 1965), most subjective variables can perhaps be thought of in terms of continua that reflect direction and intensity and perhaps even have a zero or neutral point.

One of the most notable exemplars of the use of this approach is the feeling thermometer used in the University of Michigan National Election Studies (NES) [Miller (1993); Weisberg and Miller (no date)]. Figure 9.1 gives an example of the approach used in the NES surveys for several of the measures examined here. The example is one in which attitudes or feelings toward social and political figures are

I'd like to get your feelings toward some of our political leaders and other people who are in the news these days. I'll read the name of the person and I'd like you to rate that person using this feeling thermometer. You may use any number from 0 to 100 for a rating. Ratings between 50 and 100 degrees mean that you feel favorable or warm toward the person. Ratings between 0 and 50 mean that you don't feel too favorable toward the person. If we come to a person whose name you don't recognize, you don't need to rate that person. Just tell me and we'll move on to the next one. If you do recognize the name, but don't feel particularly warm or cold toward the person, you would rate that person at the 50 degree mark.

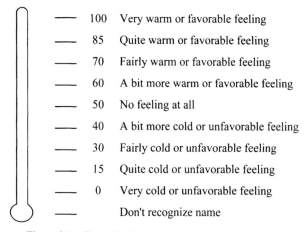

Figure 9.1.   Example of the Michigan feeling thermometer.

assessed on what is essentially a nine-category scale. Because this approach involves the image of a thermometer, survey measures such as that shown in Figure 9.1 are sometimes thought of as having 100 or more scale points, as in the case of a Centigrade thermometer. Our analysis of the NES feeling thermometer data indicates that hardly more than 3–5% of respondents give responses other than the nine numeric options shown in this figure (see Alwin, 1992, 1997).[3]

In our present sample of questions there are 47 feeling thermometer measures, assessing a range of attitudes toward political figures, candidates, groups, and parties. In light of what we said earlier for our present purposes, we consider these feeling thermometers to have nine response categories. The use of feeling thermometers often requires rather lengthy introductions, as we noted earlier, and is sometimes seen as a burden to the respondent and therefore an argument against their use.

---

[3]In our data analysis we recoded the responses that did not exactly equal the nine numeric options. This was necessary in order to compute polychoric correlations using the PRELIS program (Jöreksog and Sörbom, 1996a). This recoding had no consequences for our results inasmuch as the reliability estimates calculated in this way were no different from those calculated without the recodes.

Commenting on this aspect of the approach, Converse and Presser (1986, p. 86) noted that "despite its length, the thermometer question seems to be clear to respondents because of the familiar image of this measuring device."

It is important, however, to separate the length of the question from the length of the introduction, as we do here, because as Andrews' (1984) research suggests each may contribute independently to levels of reporting error. In the previous chapter we examined the issue of the role of the introduction and its length in the estimated reliability for questions in series and batteries. We showed that among series, those having long introductions (16 or more words) appear to have lower reliability, whereas questions in batteries having *any* introduction appear to have lower reliability. These results were obviously affected by the presence of feeling thermometers in this category. We return to this issue in a later section (see Section 9.8) when we take up the role of question length in producing errors of measurement.

9.2.4 Other Rating Scales

In addition to the question forms discussed in the foregoing, rating scales are widely used in sample surveys and social research generally (see Sudman, Bradburn, and Schwarz, 1996). Here we make a distinction between bipolar versus unipolar rating formats. We often find that some types of content, such as attitudes, are always measured using bipolar scales, whereas others, such as behavioral frequencies, are always measured using unipolar scales. Some constructs can be measured with either. For example, a scale that ranges from "completely satisfied" to "not at all satisfied" is considered to be unipolar. In contrast, a scale with endpoints of "completely satisfied" and "completely dissatisfied" is considered to be bipolar and has an explicit or implicit neutral point.

As we noted earlier, the typical *agree-disagree* Likert-type question with four or five categories can be thought of as a rating scale with either implicit or explicit numeric quantities linked to the verbal labels (strongly agree, agree, neither agree nor disagree, disagree, and strongly disagree). This question form exemplifies the traditional type of *bipolar* rating scale in that it explicitly focuses on assessing both direction and strength of attitudes. Similarly, the feeling thermometer approach described above is also a *bipolar* rating scale given that degrees of feeling are assessed around a neutral point of "no feeling at all" (see Figure 9.1). In both cases there is an implicit "zero point" corresponding to a neutral category. These two forms are *not* included in the category of "other rating forms—bipolar" in Table 9.1.

Some *bipolar scales* assess direction, but do not explicitly include a neutral category. Examples of this are the two-, four-, or six-category adaptation of the traditional *Likert scale*, where there is no inclusion of a middle category. The two-category agree-disagree scale and other types of *forced-choice* questions assessing the direction of attitudes, values, or beliefs represent prominent examples of bipolar rating scales. However, not all *forced-choice* questions reflect the measurement of a bipolar concept—some are clearly *unipolar* in character. Such scales, in contrast to *bipolar scales*, include a "zero point" at one end of the scale rather than at the neutral point.

Unipolar rating scales typically assess questions of "How many?," "How much?," or "How often?," where the zero-point is associated with categories such as "none," "not at all," and "never." In the present sample of questions we have rather large sets of unipolar (n = 143) and bipolar (n = 86) measures, which are not otherwise included as Likert agree-disagree scales, forced-choice questions, or feeling thermometers. The differences reported in Table 9.2 appear to indicate that in general unipolar response scales have somewhat higher reliabilities than bipolar rating scales. We suspect that this is related to the fact that whereas unipolar response scales are used when the researcher wants information on a single unitary concept, e.g., the frequency or amount of some quantity, in bipolar questions, several concepts are required for the respondent to handle simultaneously, namely, neutrality, direction, and intensity. This result requires further examination in the context of more closely controlling for other attributes of the questions, a topic to which we return subsequently.

9.3 NUMBER OF RESPONSE CATEGORIES[4]

Normally, with respect to "fixed form" or "closed" questions, research in survey methods considers the question and the response framework as one package. There is, however, some value in disentangling the role of the response format per se in its contribution to the likelihood of measurement error. Although the early research on question form essentially ignored the question of the number of response options used in survey measurement, there is a small literature developing around the issue of how many response categories to use in the gathering of information in survey interviews (e.g., Andrews, 1984; Alwin, 1992, 1997; Alwin and Krosnick, 1991b; Krosnick, 1999; Schuman and Presser, 1981; Scherpenzeel and Saris, 1997).

This research can be evaluated from the point of view of information theory, which posits that using more response categories improves the accuracy of measurement (Garner, 1960; Garner and Hake, 1951; Shannon and Weaver, 1949). This assumes that more categories transmit more information, which in turn affects reliability (see Alwin, 1992, 1997). Based on what is known, there appears to be a general increase in reliability with increases in the number of categories, although two- and three-category response options provide an instructive exception (see Alwin, 1992). Three-category questions are less reliable than two- and four-category questions, in part we suspect because the introduction of a middle alternative presents room for ambiguity (Schuman and Presser, 1981).

In any event, because of the logic and assumptions drawn from information theory, historically the conclusion has been drawn that higher reliability is a salutary consequence of using more response categories. In past research seven-category response scales are even more reliable than five-point scales, presumably because

---

[4]Portions of the introductory material presented in this section rely on the author's treatment of similar material (Alwin, 1992, 1997). Results presented in this section, however, are new and utilize several innovative changes in both theory and method.

they permit more accurate descriptions, and nine- and eleven-category feeling thermometers achieve even greater reliability (see Alwin and Krosnick, 1991b; Alwin,
1992, 1997; Andrews, 1984).

There is considerable room, however, for debate on this issue. On one hand, cognitive theorists would suggest that there may be some practical upper limit on the
number of response categories people can handle. Certainly, given the potential cognitive difficulties that most people have in making discriminations along a scale with
numerous categories, it seems plausible to argue that the quality of measurement will
improve up to some point, say, up to seven categories, but beyond that information
actually will be lost because the scale points tend to mean less. Indeed, one famous
article on this topic—titled "The magic number seven" (Miller, 1956)—argued on
the basis of a set of experiments using magnitude estimation of judgments of a number of physical stimuli that the optimal number of categories was seven, plus or minus
two (Miller, 1956). Despite the seeming importance of these results, it is not clear
whether Miller's conclusion can be generalized to survey measures (Alwin, 1992).

Motivational theorists, on the other hand, would argue against survey questions
with large numbers of response categories on the grounds that respondents may not
be sufficiently motivated to take the questions seriously if they are bombarded with
the difficult task of making meaningful discriminations along a continuum with a
large number of categories. It has been suggested that when faced with such complex tasks, many respondents may tend to "satisfice" rather than "optimize" (Alwin,
1991; Krosnick and Alwin, 1989; Tourangeau, 1984; Tourangeau, Rips and Rasinski,
2000). On this point, Cox (1980, p. 409) noted, for example, that as the number of
response categories increases beyond some hypothetical minimum, response burden
also increases. Cox argued that the result of more categories is an increasing number
of discrepancies between the true score and the observed score. Thus, although it
may appear that the "information carrying capacity" of a set of response categories
is increased when more categories are used, it may actually be possible that by doing
so the reliability of measurement is diminished.

A problem with assessing the reliability of measurement of questions with different numbers of response categories is that the results depend on the approach
used to estimate reliability. With few exceptions (see Alwin, 1992), past results have
been based on the use of Pearson-based correlation coefficients and have ignored the
problem addressed in Chapter 2 (above) concerning the role of number of response
categories in the calculation of reliability estimates. Following discussions in the
statistical literature (e.g., Jöreskog, 1990), we argued that when continuous latent
variables are measured with a small number of response categories, the estimated
level of association is attenuated with respect to what would be estimated with direct
measurement of an "ideal" continuous variable. Given that the degree of attenuation
due to crude categorization is directly related to the number of response categories,
this aspect of our methods of estimation must be controlled in the analysis of this
issue.

We examined reliability data for closed-form self reports (n = 369) by the number
of response categories offered to respondents, separately for facts and nonfacts. This
is the largest and most comprehensive set of survey measures that have been used

Table 9.3. Comparison of reliability estimates for closed-form questions by number of response categories and type of survey content

	Reliability Estimates[1]									Number of Questions		
	Pearson-based Estimates			Polychoric-based Estimates			Difference					
Categories	Non-Facts	Facts	Total	Non-Facts	Facts	Total	Non-Facts	Facts	Total	Non-Facts	Facts	Total
2	0.474	0.804	0.485	0.697	1.000	0.707	0.223	0.196	0.222	31	1	32
3	0.452	0.388	0.451	0.582	0.497	0.581	0.130	0.109	0.130	62	1	63
4	0.531	0.697	0.551	0.648	0.792	0.665	0.117	0.095	0.114	66	9	75
5	0.535	0.632	0.543	0.650	0.710	0.655	0.115	0.078	0.112	89	8	97
6	0.402	0.654	0.528	0.572	0.702	0.637	0.170	0.048	0.109	7	7	14
7	0.519	0.802	0.533	0.575	0.892	0.590	0.056	0.090	0.057	39	2	41
8	---	---	---	---	---	---	---	---	---	0	0	0
9	0.629	---	0.629	0.647	---	0.647	0.018	---	0.018	47	0	47
Total	0.522	0.668	0.533	0.631	0.750	0.640	0.109	0.082	0.107	341	28	369
F-ratio	10.689	1.251	8.717	3.616	2.045	3.761	26.784	1.248	26.233			
p-value	0.000	0.320	0.000	0.002	0.112	0.001	0.000	0.321	0.000			

[1]Reliability estimates are average Wiley-Wiley estimates.

to address this issue to date within the present framework. In Table 9.3 we present two sets of results—those employing Pearson-based covariance structures and those employing polychoric-based covariance structures. Regardless of the method used to assess reliability, these results suggest a significant relationship of reliability and number of response categories for nonfactual questions, but not among factual ones. Unfortunately, the number of factual questions (n = 28) is too small to cover the breadth of variation in number of response categories, so we restrict our attention here primarily to the nonfactual measures (n = 341). Notice that these comparisons cross-cut some of the question form distinctions covered above, and we have made no effort here to control for question form per se because of the confounding with number of response categories. For example, all forced-choice questions involve two-category response scales, all of our nine-category response scales come from feeling thermometers, all Likert agree-disagree questions involve four- and five-category responses, and so forth. It is therefore not possible to separate the issue of "form" from the issue of "number of categories," so we focus on the package rather than any particular element.

The results here confirm our expectations regarding the effect of method of reliability estimation, in that the differences presented in Table 9.3 between the Pearson-based estimates and the polychoric-based estimates are quite systematic. As we expected, the difference in estimated reliability between the two approaches for two-category scales is substantial—.22 in estimated reliability on average—whereas the difference in the case of nine-category scales—.02 in estimated reliability—borders on the trivial. The *differences* among these differences across numbers of response categories are statistically significant. Clearly, from the point of view of the polychoric-based estimates, reliability is underestimated by the Pearson correlation/covariance methods, and the degree of bias is systematically related to the number of response categories.

The question is whether these results change the overall conclusion that one would draw with respect to the role of the number of response categories. In some cases we find similar things, but in others we come to very different conclusions. Our different conclusions are, however, not entirely due to the use of improved methods of reliability estimation. In some cases the differences of the results from those reported in our previous investigations result from having a broader and more extensive range of measures in the present analysis, e.g., in the representation of seven-category response scales. We should note that the following discussion of findings relies exclusively on the polychoric results presented for measures of nonfactual content.

The often-found conclusion that three-category scales are less reliable than two- and four-category scales is strongly upheld, and the degree of difference in the present case is even more impressive than in previous studies. Indeed, when two-category scales are considered within the present framework, they are found to have the highest estimated reliabilities on average. They are followed in magnitude by the estimated reliabilities of four-, five-, and nine-category scales, all of which have reliabilities of roughly .65. As in past research (see Alwin, 1992), three-category response scales present a poor showing with respect to reliability of measurement. As we have argued before, this is probably due to the ambiguity introduced by the middle category.

Two strong conclusions emerge from these polychoric-based results, reinforcing some of our earlier comparable results based on many fewer measures (Alwin, 1992). First, *there seems to be little if any support for the information-theoretic view that more categories produce more reliability.* There is *no* monotonic increase in estimated reliability with increases in the number of categories in the present sample of measures. Indeed, if one were to leave aside the nine-category scales, the conclusion would be that more categories produce systematically less reliability. We would argue that little support exists for our initial hypothesis about improvements in measurement reliability linked to using more response categories. Second, *there seems to be only marginal support for the finding that questions with middle categories are systematically more or less reliable.* We noted above the replication of the finding with respect to the lower reliability of three-category scales compared to two- and four-category scales. However, this result seems to be unique to three-category scales and does not generalize to other cases. We do not find any support for the suggestion that five-category response formats are less reliable than four- and six-category scales, nor that seven-category scales are superior to all others (see Alwin and Krosnick, 1991b). One aspect of previous analyses that is upheld is the conclusion that nine-category scales are superior to seven-point scales in levels of assessed reliability, but this may be due to the particular area in which such longer scales are used and has little to do with the existence of a neutral category (see Alwin, 1992, 1997).

The possibility exists, of course, that these results are due to the confounding of category length with the topic of the questions, and this deserves further attention. Moreover, the above comparisons of estimates of reliability by number of response categories ignores the complexity of other variations in the formal characteristics of the questions. One of the most serious of these is the fact that the above analysis does not control for whether the measurement scales are assessing unipolar versus bipolar concepts. We turn now to an investigation of these issues.

9.4   UNIPOLAR VERSUS BIPOLAR SCALES

As we noted earlier in the chapter, a key distinction needs to be made among rating scales that measure "unipolar" versus "bipolar" concepts. We noted that one often finds that some types of content, such as attitudes, are always measured using bipolar scales, whereas others, such as behavioral frequencies, are always measured using unipolar scales. The key difference is that the ends of a bipolar scale are of the same intensity but opposite valence, while the ends of a unipolar scale tend to differ in amount or intensity, but not valence. Unipolar scales rarely use more than five categories, and it can be readily seen when viewed from this perspective that using three-, four-, or five-category scales to measure unipolar concepts is quite different from using similar scales intended to measure bipolar concepts. Clearly, the issue of what the "middle category" means for three- and five-category scales is quite different depending on whether the concept being measured is unipolar or bipolar.

For the present analysis each of the 426 questions in our sample was coded as having either a unipolar or bipolar response scale, although here we focus only on the closed-form measures of nonfactual content. Measures coded as unipolar include questions of frequency, count, amount, degree and duration, where one end of the scale is anchored by the absence of the quality found in abundance at the other end of the scale. This includes scales with endpoints: (1) "almost always" and "never" as well as (2) "very happy" and "not at all happy" and (3) "more than once a week" to "never." Measures using a "yes–no" scale were coded as unipolar, as the "yes" response is the affirmation of the question content, while the "no" response was interpreted as a lack of affirmation, rather than the endorsement of an alternative position. Measures in which more than one concept is presented and the underlying continuum is abstract were typically coded as bipolar. An example in this group is the question that asks if "it's better to plan your life a good way ahead" or if "life is too much a matter of luck."

Feeling thermometers were coded as bipolar, as the 50 degree point is described as feeling "neither warm nor cold toward" the subject in question. Questions that ask respondents to position themselves relative to others were coded as bipolar. For example, one question asked about the equity of the respondent's wages ("Do you feel that the money you make is much less than your fair share, somewhat less than your fair share, about your fair share, or more than your fair share?") and is coded bipolar. There were a few questions in the present sample in which the polarity is considered to be ambiguous. For example, one question asks respondents to rate their health as "Excellent," "Very good," "Good," "Fair," or "Poor." Another asks "When you think of what your family needs to get along on comfortably right now, is your husband's income high enough, a little low, or quite a bit too low?" Both of these were coded as unipolar. Finally, it should be noted that in a few cases we employ constructed variables (coded as synthesized/redundant) which were coded for polarity based on the constructed variable, not its subparts. For example, the seven-category party identification self-assessment is constructed based on both unipolar and bipolar items, but is coded as bipolar based on the resultant seven-point scale.

In Table 9.4 we present a reanalysis of the issue of the relationship of measurement reliability to the number of response options in the measurement of nonfactual material using closed-form questions. Here we have implemented the distinction drawn between unipolar and bipolar measurement and have controlled for these differences. In addition, we present the results separately for two groupings of content: (1) beliefs, attitudes, and values (BAV), and (2) self-assessments and self-perceptions (SASP). These results add considerable clarity to the above results in that they reveal that most measures falling into the BAV category are measured using bipolar scales, and most measures falling into the SASP category rely on unipolar measurement. It is therefore important to keep this fact in mind when evaluating the differences in the two types of measurement approaches. There are relatively few differences between unipolar and bipolar measures, net of number of response categories, and this cannot be attributed to the content measured.

It is also important to note that the measurement of unipolar concepts using closed-form questions rarely involves more than five categories (see Table 9.4). Evaluating the differences in reliability of measurement across categories of different lengths reveals the superiority of four- and five-category scales for unipolar concepts, a finding that supports the kind of "information theoretic" logic mentioned earlier. For bipolar measures, the two-category scale continues to show the highest levels of measurement reliability, and following this, the three- and five-category scales show an improved level of reliability relative to all others. There are many fewer substantive differences among the bipolar measures, although it is relatively clear that seven-category scales achieve the poorest results.

9.5   DON'T KNOW OPTIONS

Because of the vagueness of many nonfactual questions and response categories and the perceived pressure on the part of respondents to provide information on many such questions, even when they have little knowledge of the issues or have given little thought to what is being asked, there is concern that respondents will behave randomly, producing unreliability. In their classic study of question wording, Rugg and Cantril (1944, pp. 48–49) noted that response errors were less likely if persons "had standards of judgment resulting from stable frames of reference," as opposed to situations "where people lack reliable standards of judgment and consistent frames of reference." We discussed this issue in Chapter 2, noting the proposal by Converse (1964) that this is often the product of people responding to survey attitude questions, when in fact they have no attitudes. The argument is that the culture values "opinionated" people, and that respondents therefore assume that people with opinions are presumed to be more respected than persons without opinions. Respondents

Table 9.4. Comparison of reliability estimates for unipolar versus bipolar closed-form rating scales by number of response categories and type of survey content

Number of Response Categories	Beliefs, Values and Attitudes				Self-assessments and Self-perceptions				Total Unipolar and Bipolar				Total
	Unipolar	Bipolar	F-ratio	P-value	Unipolar	Bipolar	F-ratio	P-value	Unipolar	Bipolar	F-ratio	p-value	
2	0.738 (2)	**0.733** (17)	---	---	**0.593** (7)	**0.705** (5)	1.40	0.264	**0.626** (9)	**0.727** (22)	3.25	0.082	0.697 (31)
3	**0.504** (11)	**0.619** (26)	5.41	0.026	**0.532** (17)	**0.675** (8)	4.82	0.039	**0.521** (28)	**0.632** (34)	9.39	0.003	0.582 (62)
4	**0.673** (24)	**0.720** (5)	0.70	0.411	**0.690** (21)	**0.531** (16)	7.62	0.009	**0.681** (45)	**0.576** (21)	6.50	0.013	0.648 (66)
5	**0.707** (12)	**0.637** (56)	2.60	0.112	**0.659** (20)	0.515 (1)	---	---	**0.677** (32)	**0.635** (57)	1.94	0.167	0.650 (89)
6	---	---	---	---	**0.572** 7	---	---	---	**0.572** (7)	---	---	---	0.572 (7)
7	---	**0.530** (32)	---	---	0.686 (2)	**0.821** (5)	---	---	0.686 (2)	**0.569** (37)	---	---	0.575 (39)
8	---	---	---	---	---	---	---	---	---	---	---	---	---
9	---	**0.647** (47)	---	---	---	---	---	---	---	**0.647** (47)	---	---	0.647 (47)
Total	0.646	0.630	0.48	0.490	0.625	0.630	0.02	0.886	0.633	0.630	0.05	0.830	0.631
Total n	(49)	(183)			(74)	(35)			(123)	(218)			(341)
F-ratio	8.94	5.65			4.26	3.28			8.26	3.39			3.62
p-value	0.001	0.000			0.004	0.034			0.000	0.006			0.002

Note 1: Only **bold** values are used in the statistical tests.

Note 2: The number of questions on which reliability estimates are based is given in parentheses.

therefore assume that interviewers expect them to offer opinions and consequently produce responses that contain substantial amounts of measurement error [see Krosnick (2002) for an alternate view].

If Converse's hypothesis is true, then the explicit offering of the option to declare that they "don't know" may forestall such contributions to error. We should hasten to add, however, that even if the use of Don't Know options does not reduce such contributions to error, Converse's thesis could still be essentially correct. Still, there is some information that bears on this question, in that numerous split-ballot experiments have found that when respondents are explicitly asked if they have an opinion, the number of Don't Know responses is significantly greater than when they must volunteer a "no opinion" response (e.g., Schuman and Presser, 1981; Bishop, Oldendick and Tuchfarber, 1983). Little consistency has resulted, however, in research focusing on the quality of reporting, or reliability, of these various approaches. Andrews (1984) found that offering respondents the option to say "don't know" increased the reliability of attitude reports. Alwin and Krosnick (1991b) found the opposite for seven-point rating scales and no difference for agree-disagree questions. McClendon and Alwin (1993) and Scherpenzeel and Saris (1997) found no difference between forms. We nonetheless hypothesize, consistent with Converse's (1964) prediction, that the offering of a Don't Know alternative will improve the quality of the data, compared to those that force respondents to produce attitudes or judgments they may not have.

There are at least three possible ways in which Don't Know options are provided to respondents in surveys: (1) they may be provided in the question itself, by including a phrase, such as " . . . or don't you have an opinion about that?"; (2) they may be provided in the introduction to the series or battery in which they appear, e.g., see the example in Figure 9.1; or (3) they may be provided in a filter question preceding the question of interest, as in the following case (typical in the 1950s NES survey):

> Around election time people talk about different things that our government in Washington is doing or should be doing. Now I would like to talk to you about some of the things that our government *might* do. Of course, different things are important to different people, so we don't expect everyone to have an opinion about all of these.
>
> I would like you to look at this card as I read each question and tell me how you feel about the question. First of all, if you don't have an opinion, just tell me that. If you do have an opinion, choose one of the other answers.
>
> "The government should leave things like electric power and housing for private businessmen to handle." Do you have an opinion on this or not?

Then, only if the respondent answers "yes" to this question are they then asked to report the direction and intensity of their views on this matter. As Hippler and Schwarz (1989) have suggested, from the perspective of conversational norms, the fact that a person is asked a question may imply that he/she can answer it. Unless

Table 9.5.   Comparison of reliability estimates for nonfactual closed-form questions by number of response categories and Don't Know options

Number of Response Categories	Don't Know Options		Total	F-ratio	p-value
	Don't Know Offered	Don't Know Not Offered			
3	0.646	0.632	0.635	0.04	0.842
	(6)	(27)	(33)		
5	0.575	0.533	0.549	0.85	0.367
	(8)	(13)	(21)		
7	0.542	0.643	0.569	2.52	0.121
	(27)	(10)	(37)		
Total	0.563	0.609	0.588	1.90	0.171
Total n	(41)	(50)	(91)		
F-ratio	1.42	1.87	2.47		
p-value	0.254	0.166	0.090		

Note: The number of questions on which reliability estimates is based is given in parentheses.

the interviewer indicates the legitimacy of having "no opinion," as in the above case, a Don't Know option may not provide enough of an incentive to decline making a substantive response. They argue that a strongly worded "full filter" approach offers a stronger basis for respondents to opt out of answering the question (see Hippler and Schwarz, 1989).

In Table 9.5 we present three comparisons among measures of nonfactual content that are relevant to this hypothesis. These comparisons are within three-, five-, and seven-category bipolar rating scales in the National Election Studies (NES50s, NES70s and NES90s) where there are sufficient numbers of measures to detect the presence of an effect. Interestingly, virtually none of the questions in the Americans' Changing Lives (ACL) or the Study of American Families (SAF) used explicit Don't Know options, so it is not possible to test this hypothesis for measures from those studies.[5] The comparisons presented here provide relatively uniform support for the null hypothesis; that is, no significant differences arise in the comparisons of the two forms. Nonfactual content assessed using questions without an explicit Don't Know option are no less reliable than comparable questions that do provide such an option, a result that is consistent with the majority of studies on this topic.

---

[5]We were actually surprised by the overall *infrequency* of the use of explicit Don't Know options in these surveys. Among the 341 nonfactual questions employed in this study, only one-fourth used such an option, and these *all* occurred in the NES studies. Roughly 44% of the NES nonfactual measures employed a Don't Know option, occurring only in three-, five-, seven- and nine-category scales. None of the nine-category scales occurred without such an option.

The three- and five-category questions employing a Don't Know filter shown in Table 9.5 are all from the 1950s NES, where a great deal of concern was shown for filtering out those persons who had no opinion. They all follow the form of the question quoted above. These all represent the *strong form* of the Don't Know option. The seven-category questions shown in Table 9.5 are all from the 1970s and 1990s NES panels, and most of these follow the format wherein the Don't Know option is given as a part of the question, typically in the first question in a series or battery in which they are contained. In this case, if a respondent gave a Don't Know response to the first question, they were skipped out of the entire battery. We would argue, thus, that in virtually all instances examined here a relatively strong form of the Don't Know response option was given.

9.6 VERBAL LABELING OF RESPONSE CATEGORIES

It has often been hypothesized that the more verbal labeling that is used in the measurement of subjective variables, the greater will be the estimated reliability (Alwin and Krosnick, 1991b). Labels reduce ambiguity in translating subjective responses into the categories of the response scales. In the present data set there was extensive labeling of response options, so variation in patterns of usage are restricted in this study. Factual questions are either open-ended, with implicit verbal labels, or they are given extensive labeling of all response categories. Based on simple notions of communication and information transmission, the better labeled response categories are hypothesized to be more reliable. Despite this strong theoretical justification, there are some results that suggest that data quality is below average when all categories are labeled (see Andrews, 1984, p. 432).

The main problem with testing this hypothesis using this database is that there are very few questions employed in these surveys where there are unlabeled or partially labeled scales. The overwhelming majority (upward of 90%) of the closed-form items are represented by fully labeled scales. There do exist 26 measures in the NES series in which seven-point rating scales are employed that label the endpoints only. These can be compared with 11 fully labeled seven-point scales available in the NES studies. Our analysis of these data suggests that a significant difference in reliability exists for fully vs. partially labeled response categories (see Table 9.6). This reinforces the conclusions of Alwin and Krosnick (1991b), who found that among seven-point scales, those that were fully labeled were significantly more reliable.

Concerned that these results might be confounded with the previous results for Don't Know options in seven-point scales, we examined the labeling hypothesis among these seven-point rating scales while controlling for whether a Don't Know option was provided. These results, shown in Table 9.7, indicate there are important differences favoring labeling, an effect that is strong and significant among those questions with no explicit Don't Know option presented. Unfortunately, the number of cases in which we can simultaneously assess the effects of labeling and Don't Know options while controlling for the number of response categories is seriously limited by the available data. A more complete analysis of these issues must await further empirical investigation, but the results appear to indicate that more complete labeling enhances the reliability of measurement.

Table 9.6. Comparison of reliability estimates for seven-category nonfactual closed-form questions by the nature of verbal labeling

Nature of Verbal Labeling	
Fully labeled	0.719
	(11)
Endpoints only	0.506
	(26)
Total	0.569
Total n	(37)
F-ratio	16.01
p-value	0.000

Note: The number of questions on which reliability estimates are based is given in parentheses.

One way in which survey researchers deal with questions having large numbers of response categories is to employ visual aids or "show cards" to enhance communication between the interviewer and respondents. In our discussion of the Michigan feeling thermometers above, we included an example of the show card that was used to accompany these questions. This practice is so widespread in face-to-face interviews that it is difficult to find enough comparable questions with which to examine the differences in the estimated reliabilities of those questions that employ show cards with those that do not. In Table 9.8 we cross-tabulate the closed-form nonfactual questions in our sample according to the number of response categories and the presence or absence of the use of show cards or visual aids. These results show that such visual aids are never used with two-category scales, and always used with nine-

Table 9.7. Comparison of reliability estimates for seven-category nonfactual closed-form questions by the nature of verbal labeling and Don't Know options

Nature of Verbal Labeling	Don't Know Options		Total	F-ratio	p-value
	Don't Know Offered	Don't Know Not Offered			
Fully labeled	0.639	0.785	0.719	3.51	0.094
	(5)	(6)	(11)		
Endpoints only	0.519	0.430	0.506	1.20	0.283
	(22)	(4)	(26)		
Total	0.542	0.643	0.569	2.52	0.121
Total n	(27)	(10)	(37)		
F-ratio	2.34	42.64	16.01		
p-value	0.139	0.000	0.000		

Note: The number of questions on which reliability estimates are based is given in parentheses.

Table 9.8. Cross-tabulation of number of response categories and use of visual
aids for closed-form questions assessing nonfacts

Use of Visual Aid	Number of Response Categories							
	2		3		4		5	
No	31	(100%)	37	(60%)	29	(44%)	43	(48%)
Yes	---	---	25	(40%)	37	(56%)	46	(52%)
Total n	31	(100%)	62	(100%)	66	(100%)	89	(100%)
	6		7		9		Total n	
No	---	---	4	(10%)	---	---	144	(42%)
Yes	7	(100%)	35	(90%)	47	(100%)	197	(58%)
Total n	7	(100%)	39	(100%)	47	(100%)	341	(100%)

category scales. With a few exceptions, the relationship is more or less monotonic—three-category scales employ show cards 40% of the time, four-category scales 56% of the time, and seven-point scales 90% of the time.

The use of visual aids in survey research depends heavily on face-to-face interviewing methods, or self-administered questionnaires, and their prevalence in the surveys employed in this study must be taken into account in the interpretation of the above results. This observation should provide some caution in generalizing our results to the more common use of telephone surveys in contemporary research. Aural cues, rather than visual cues, may be much more important in telephone surveys, and what the respondents hear and when they hear it may have more of an effect on levels of measuement error in such surveys (e.g., Krosnick and Alwin, 1988). There may well be significant interactions between the mode of administering a questionnaire, i.e., telephone, face-to-face, or self-administered, and the formal properties of survey questions, and until such effects can be studied, it may be risky to draw strong conclusions from the present results.

9.7 SURVEY QUESTION LENGTH

One element of survey question writing to which many subscribe is that questions should be as short as possible, although there is still some debate on this issue. Payne's (1951, p. 136) early manual on surveys suggested that questions should rarely number more than 20 words. Thus, among the general rules given about how to formulate questions and design questionnaires, it is routinely suggested that questions be short and simple (e.g., Brislin, 1986; Fowler, 1992; Sudman and Bradburn, 1982; van der Zouwen, 1999). On the other hand, some researchers have argued, for example, that longer questions may lead to more accurate reports in some behavioral domains (see Cannell, Marquis, and Laurent, 1977; Marquis, Cannell, and Laurent,

1972; Bradburn and Sudman, 1979). Advice to researchers on *question length* has therefore been quite mixed.

In Chapter 2 we argued that one of the most basic elements of survey quality is the respondent's *comprehension* of the question, and the issue of question length is germane to this objective. We stressed the fact that if a question is unclear in its meaning, or if parts of the question can have more than one meaning, then the likelihood of measurement error will be increased. Thus, the ability to reduce measurement errors may presuppose questions and response categories that are precise in meaning regardless of how many words it takes to get that precise meaning across to the respondent. The key element here is probably not question length, per se, but question clarity (Converse and Presser, 1986, p. 12). From the point of view of communication, however, too many words may get in the way of the respondent's comprehension. On the other hand, some experts encourage *redundancy* in survey questions (Brislin, 1986). Noting the tradeoffs between adding redundancy and question length, Converse and Presser (1986, p. 12) write: "One should consider the use of redundancy now and then to introduce new topics and also to flesh out single questions, but if one larded all questions with 'filler' phrases, a questionnaire would soon be bloated with too few, too fat questions."

There is very little evidence on this point, although it is clear that there are a variety of points of view. Where there is evidence, the support seems to favor longer questions. Andrews (1984, p. 431) combined the consideration of question length and the length of introductions to questions. He found that "short introductions followed by short questions are not good . . . and neither are long introductions followed by long questions." Errors of measurement were lowest when "questions were preceded by a medium length introduction (defined as an introduction of 16 to 64 words)" followed by medium or long questions.[6] By contrast, Scherpenzeel and Saris (1997) find long questions with long introductions to be superior, but like Andrews (1984), they do not separate conceptually the two issues of battery/series introduction length and question length.

By focusing on question length we may be sidestepping the crucial issue, question clarity, although one does not normally have independent assessments of question clarity. We focus instead on the criterion of reliability, which can be measured, as illustrated here. Further, we can contribute to this discussion by suggesting that a distinction be made between the length of the introduction to a question and the length of the question itself. It is argued that either type of question length is fundamentally related to the overall burden felt by the respondent, but lengthy introductions may be viewed by the researcher as a helpful aid to answering the questions. Respondents

---

[6]We are somewhat confused by Andrews' (1984) and Scherpenzeel and Saris' (1997) analyses because in the typical survey questions by themselves *do not* have introductions. Series of questions, or batteries, or entire sections, on the other hand, do typically have introductions (see Chapter 6). As can be seen in the above discussion, we see the introduction to series and/or batteries of questions to be a separate one conceptually from that of question length, and we therefore distinguish between question length and the length of introductions to organizational units larger than the question.

Table 9.9. Question length by question content and question context

| | Question Context | | | | | |
Question Content	Alone	Series	Battery	Total	F-ratio[1]	p-value
Respondent self-report and proxy facts	**17.86**	**24.73**	45.50	23.43	3.50	0.065
	(21)	(56)	(2)	(79)		
Respondent self-report nonfacts	47.67	**28.10**	**11.54**	18.25	231.89	0.000
	(9)	(121)	(217)	(347)		
Total	26.80	27.03	11.85	19.21		
Total n	(30)	(177)	(219)	(426)		
F-ratio[2]	14.91	3.25	----			
p-value	0.001	0.073	----			

[1]Test within facts excludes two battery fact triads; test within nonfacts excludes the nine stand-alone items.
[2]Test within batteries was not carried out.
Note: The number of questions on which reliability estimates are based is given in parentheses.

and interviewers, on the other hand, may find them time-consuming and distracting. Converse and Presser (1986, pp. 12–13) suggest that "the best strategy is doubtless to use short questions when possible and slow interviewer delivery—always—so that respondents have time to think." At the same time, they concede "in other cases, long questions or introductions may be necessary to communicate the nature of the task" (p. 12).

We examined several attributes of question length, including both the number of words in the "prequestion text" and the number of words in the actual question. In our previous discussion of the reliability of questions in series and batteries, we considered the role of the presence and length of subunit introductions on estimates of measurement reliability, finding that it had modest effects. Here we consider the number of words appearing in the text for a given question following any prequestion text that might be included as an introduction. Thus, our measure of "question length" refers only to the question itself and does not include a count of the words in the prequestion text.

As our results in Table 9.9 indicate, not surprisingly, question length is associated with several other attributes of question content and context. Nonfactual questions included in a stand-alone or series are generally a few words longer. Questions in batteries are much shorter, but this is due to the fact that often the actual question stimulus is just a few words, given the existence of a lengthy introduction to the battery that explains how the respondent should use the rating scale. The results in Table 9.9 reinforce the observation made earlier that the length of a question should not be considered independently of the length of the unit (series or battery) introduction. In fact, as noted above, the situation is even more complicated in the case of some batteries. What we have considered "questions" in batteries, namely, that question text that follows the "introduction," are often not questions at all, but rather stimulus words or phrases. This accounts for the fact that on average questions in batteries are as a group much shorter than questions offered in other contexts—and not only this, as we discuss below; in many instances the battery introduction includes part of the question.

However, we hasten to point out that in the case of batteries the concept of question length, apart from the length of the introduction, is somewhat ambiguous. It will therefore be valuable for our subsequent discussion and analysis of the effects of question length if we introduce further clarification of the types of question we have considered to be part of questionnaire batteries: (1) those cases in which none of the actual question appears in the introduction and the postintroduction question text consists of a self-contained question, (2) those cases in which the basic form of the question to be answered in each question contained in the battery is presented in the introduction and the postintroduction text refers to that question; and (3) those cases in which the question to be answered is contained in the introduction and the post-introduction text contains only a word or a phrase to be addressed with respect to that question. Table 9.10 provides examples of these three types of batteries from the NES surveys. Note that questions in series are most closely comparable to batteries

Table 9.10. Examples of different types of questions appearing in batteries

Type 1—Cases in which none of the actual question appears in the introduction and the post-introduction question text is comprised of a self-contained question.

Example from the 1970s National Election Study:

Now I'm going to ask you some questions that you can answer by telling me the response from this card that best describes how you feel.

    a. Over the years, how much attention do you feel the government pays to what the people think when it decides what to do? A good deal, some or not much?

    b. How much do you feel that political parties help to make the government pay attention to what the people think?

    c. And how much do you feel that having elections makes the government pay attention to what people think?

    d. How much attention do you think most Congressmen pay to the people who elect them when they decide what to do in Congress?

Example from the 1990s National Election Study:

We would like to find out about some of the things people do to help a party or a candidate win an election.

    a. During the campaign, did you talk to any people and try to show them why they should vote for or against one of the parties or candidates?

    b. Did you wear a campaign button, put a campaign sticker on your car, or place a sign in your window or in front of your house?

    c. Did you go to any political meetings, rallies, speeches, dinners, or things like that in support of a particular candidate?

    d. Did you do any (other) work for one of the parties or candidates?

Table 9.10 *Continued.*   **Examples of different types of questions appearing in batteries**

Type 2—Cases in which the basic form of the question to be answered or the scale to be used in each question contained in the battery is presented in the introduction and the post-introduction question text refers to that question.

Example from the 1970s National Election Study:

(Interviewer: Hand respondent show card) Some people feel that the government in Washington should see to it that every person has a job and a good standard of living. Suppose that these people are at one end of this scale—at point number 1. Others think the government should just let each person get ahead on his own. Suppose that these people are at the other end—at point number 7. And, of course, some other people have opinions somewhere in between.

   a. Where would you place yourself on this scale, or haven't you thought much about this?
   b. (Where would you place) the Democratic Party (on this scale)?
   c. (Where would you place) the Republican Party (on this scale)?

Example from the 1990s National Election Study:

(Please turn to page 9 of the booklet) Here are three statements about taxes and the budget deficit. Please tell me whether you agree or disagree with each:

   a. We need to reduce the federal budget deficit, even if that means ordinary people will have to pay more in taxes. Do you agree strongly, agree somewhat, disagree somewhat, or disagree strongly?
   b. We need to reduce the federal budget deficit, even if that means spending less money for health and education. (Do you agree strongly, agree somewhat, disagree somewhat, or disagree strongly?)
   c. We need to spend more money on health and education, even if that means ordinary people will have to pay more taxes.

---

of the first type in that in neither case is there any part of the actual question contained in the introduction.

In order to assess the effects of question length on the reliability of measurement, we estimated the parameters of several multivariate linear regression models in which question length and other relevant variables were included as predictors. For purposes of this analysis we have included only those questions from batteries that are comparable to series with respect to the absence of any part of the question in the introduction. We also excluded 74 questions from batteries that involved a "short stimulus" (e.g., "labor unions"). We present results for 27 stand-alone questions, 172 questions in series, and 138 questions in batteries.[7]

---

[7]These results also exclude 15 cases (three stand-alone questions, five from series, and seven from batteries) that had standardized residuals greater than 2.0. These exclusions do not substantially alter the results.

Table 9.10 *Continued.*   **Examples of different types of questions appearing in batteries**

Type 3—Cases in which the question to be answered is contained in the introduction and the post-introduction question text contains only a word or a phrase to be addressed with respect to that question.

Example from the 1970s National Election Study:

Some people think that certain groups have too much influence in American life and politics, while other people feel that certain groups don't have as much influence as they deserve. On this care are three statements about how much influence a group might have. For each group I read to you, just tell me the number of the statement that best says how you feel. The first group is labor unions.

    a. Labor unions
    b. Poor people
    c. Big business
    d. Blacks
    e. Liberals
    f. Young people
    g. Women
    h. Republicans
    i. People on welfare
    j. Old people
    k. Democrats

Example from the 1990s National Election Study:

(Looking at page 8 of the booklet) Think about Bill Clinton. In your opinion, does "moral" describe Bill Clinton extremely well, quite well, not too well, or not well at all?

    a. Moral?
    b. Provides strong leadership?
    c. Really cares about people like you?
    d. Knowledgeable?
    e. Gets things done?

    The results in Table 9.11 show that for stand-alone questions and questions in series there is consistently a negative relationship between question length and the level of estimated measurement reliability. The combined results for standalone questions and questions in series are presented in Figure 9.2. These results show the gradual linear decline in reliability associated with questions of greater length. Note that for the results presented in the table we have scaled question length in 10-word units in order to produce a somewhat more meaningful metric with which to discuss the effects of question length. These results indicate that for each 10-word increase in question length, the associated reduction in estimated reliability is between .034 and .044 (see models 1 and 2).

Table 9.11.   Regression of reliability estimates on length of question, position in series/batteries, and length of introduction

Questions in Series

Model

Predictors	1 Stand-alone Questions	2 All Questions	3 All Questions	4 First Questions	5 Second+ Questions	6 Second+ Questions
Intercept	0.956 ***	0.799 ***	0.796 ***	0.857 ***	0.786 ***	0.797 ***
Question length	-0.045 ***	-0.033 ***	-0.036 ***	-0.039 **	-0.033 **	-0.030 **
First question in s/b			0.051			
Introduction length						-0.012 **
R^2	0.381	0.098	0.115	0.181	0.069	0.127
Total n	30	177	177	42	135	135

Questions in Batteries

Model

Predictors	7 All Questions	8 All Questions	9 First Questions	10 Second+ Questions	11 Second+ Questions
Intercept	0.614 ***	0.615 ***	0.637 ***	0.600 ***	0.720 ***
Question length	0.005	0.004	0.002	0.013	-0.007
First question in s/b		0.010			
Introduction length					-0.025 ***
R^2	0.007	0.008	0.001	0.006	0.116
Total n	145	145	40	105	105

Key: $*p \leq .05$, $**p \leq .01$, $***p \leq .001$

Because of the problem mentioned above—that question length in batteries is ambiguous in many cases—we can only assess the relationship between question length and estimated measurement reliability in the case of batteries of the first type mentioned above (see Table 9.10) in which none of the actual question appears in the introduction and the postintroduction question text consists of a self-contained question. The results for batteries in Table 9.11 indicate that there is no relationship between reliability and question length for questions in batteries. This exception to the findings noted above does pose an interesting puzzle with respect to the possible interaction between question context, question length, and reliability of measurement.

Given the likelihood of a relationship between reliability and the order of presentation of questions in a series or battery, we included a dummy variable indicating whether the case was the first question in the series/battery (see models 3 and 8). The results indicate that there is a significant advantage for first questions in the case of series but not in the case of batteries. We also examined the relationship between reliability and question length separately for the first questions in the series/battery

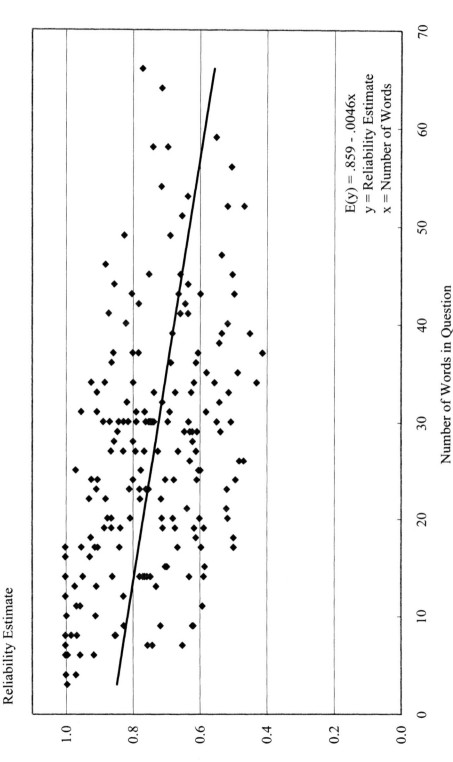

Reliability Estimate

Number of Words in Question

$E(y) = .859 - .0046x$
y = Reliability Estimate
x = Number of Words

Figure 9.2. Regression of reliability estimates on question length for stand-alone and series questions.

and questions appearing second or later (see models 4 and 5 for series and models 9 and 10 for batteries). The results of these regressions reveal a great deal of uniformity across the categories defined by position—a consistently negative effect of question length for questions in series and no relationship for questions in batteries.

In the case of questions in series, the relationship between length and reliability is slightly stronger for questions placed first than for questions placed later. Recall that for questions placed first in a series or battery, the question text includes the introduction to the question, and this stronger effect is, thus, likely due to the effects of length of introduction. In order to capture the potential effect of the length of the introduction on the reliability of questions placed later, we included a measure of introduction length in the regressions for questions placed second or later (models 6 and 11). For both questions in series and batteries there is a significant effect of length of introduction, reducing reliability slightly.

The bulk of these results present a relatively convincing case for the conclusion that, at least with respect to some types of survey questions, there are declining levels of reliability associated with greater numbers of words in questions and further support for the typical advice given to survey researchers that their questions should be as short as possible. The results reinforce the results of previous studies that verbiage in survey questions—either in the question text or in the introduction to the question—has negative consequences for the quality of measurement.

9.8 CONCLUSIONS

In the previous chapter we speculated about whether the *context effects* discovered there might result from the type of question form employed in questions in stand-alone versus grouped (e.g., series versus batteries) contexts. We suggested the possibility that the differences observed in reliability of measurement between questions in different contexts might be due to the formal properties of questions appearing in one or the other type of questionnaire unit, e.g., type of rating scale used, the length of the question, the number of response categories, the labeling of response categories, or the explicit provision of a Don't Know option. The reliability of measurement might have little to do with question context (stand-alone, series, or batteries) *per se*, and we suggested that if we can account for the observed differences in reliability by reference to the formal attributes of questions rather than the context in which they are embedded, then the focus of our attention can be simplified, by focusing instead on formal characteristics of the questions themselves.

This chapter has addressed the question of whether salient aspects of survey questions might be predictive of differences in measurement reliability, and we find there are several clear relationships that may help us understand some of the earlier patterns reported. The most obvious issue involves the fact that stand-alone questions are more often open-ended, and I find substantial differences in reliability favoring the open-ended format within both categories of facts and nonfacts. There are too few cases involved in these analyses to warrant a strong generalization, but this does suggest that researchers might more profitably exploit the uses of the open-ended approach, especially when measuring factual content. In Chapter 7 we noted that the use of "vague quantifiers" in questions seeking information on behavior frequencies seemed to produce lower levels of estimated reliability for some factual questions, and that an open-ended approach to measurement, along with the use of online computer-based coding of responses might improve the quality of the data.

Among the "fixed-form" or "closed" questions, research in survey methods tends to consider the question and the response framework as one package, and I argued that there is some value in disentangling the distinct roles of the question and response format with respect to their contributions to measurement reliability. There are potentially a number of aspects of "questions" per se, as separate from response formats, that are relevant to reliability, but we focused on one dimension here—that of question (and introduction) length. I indicated in the above discussion that although the literature on questionnaire development is relatively consistent on the advice to researchers that they keep their questions short and simple, there has heretofore been very little evidence for this in terms of the quality of the data. The analysis of question length presented above strongly reinforces this desiderata at least with respect to stand-alone questions and questions in series—questions with more words in the text of the question and in the text of the introduction to the question have consistently lower levels of reliability.

This chapter also investigated the effects of several formal aspects of the response framework provided by the survey question. Although there is much theory and research supporting the idea that Don't Know options should improve the quality of the information gained from survey questions, our findings indicate that they do not provide improvements in reliability. On the other hand, we find that verbal labeling of response options does contribute to higher levels of reliability, although this issue was difficult to assess given that most of the questions in our database labeled all options.

With respect to the number of response categories, we found fewer differences than expected. Our analysis noted that previous investigations of this issue were flawed in two important respects. First, they have failed to acknowledge the complexity of the characteristics of the concepts measured—specifically whether they are assessing unipolar versus bipolar concepts. Second, they have failed to take into account the linkage between the number of categories and the methods for assessing reliability. Past analyses have ignored the fact that standard approaches to assessing statistical relationships (on which reliability estimates are based) are themselves affected by the number of categories in the scale. Taking these two issues into account, the results here are mixed. Although there is support for the conclusion that questions with greater numbers of response categories can produce higher levels of reliability, e.g., nine-category scales have higher levels of reliability than seven-point scales, there are other findings that suggest it is not the number of categories per se. For example, consistent with past literature, I find that three-category scales have among the lowest levels of estimated reliability, but I find no support for the suggestion that seven-category scales are superior to all others, nor that five-category scales are superior to four-category scales.

From the point of view of assessing survey measurement quality, there is probably no such thing as a perfect question. The most we can ask for are principles that help in reducing the amount of error. The advice one would glean from the present analysis is, thus, not all that different from what one would learn from experienced question writers, but it may help their cause to know that some of our findings support their hunches. Probably the two most basic elements of question format for survey quality are that the investigator be clear about what information they are seeking and to communicate this in a manner that enhances the respondent's comprehension of the question. If a question is unclear in its meaning, or if parts of the question are ambiguous or can have more than one meaning, then the likelihood of measurement error will be increased. As we noted in the previous discussion, by focusing on the formal attributes of questions we may be sidestepping the crucial issue of question clarity, but until we devise some objective means for assessing the *comprehensibility* of survey questions we must settle for the type of information summarized here.

CHAPTER TEN

Attributes of Respondents

... what does it mean to say that an instrument is perfectly reliable *vis-à-vis* one class of people and totally unreliable *vis-à-vis* another ... ? The property of reliability is certainly not inherent *solely* in the instrument of measurement, contrary to what is commonly supposed.

Philip E. Converse, "The nature of belief systems in the mass public" (1964, p. 244)

Responses to survey questions are affected by a number of factors that produce errors of measurement. As we discussed at length earlier, it is generally agreed that key sources of measurement errors are linked to aspects of survey questions, the cognitive processes of information processing and retrieval, the motivational context of the setting that produces the information, and the response framework in which the information is then transmitted (see Alwin, 1989, 1991b; Alwin and Krosnick, 1991b; Bradburn and Danis, 1984; Cannell, Miller and Oksenberg, 1981; Hippler, Schwarz and Sudman, 1987; Knäuper et al., 1997; Schaeffer, 1991b; Schaeffer and Presser, 2003; Schwarz and Sudman, 1994; Strack and Martin, 1987; Tourangeau, 1984, 1987; Tourangeau and Rasinski, 1988; Tourangeau, Rips and Raskinski, 2000). In that earlier development I argued that with respect to assessing survey measurement errors, there are essentially six critical elements of the response process that directly impinge on the reliability of survey measurement: (1) content validity, (2) comprehension, (3) accessibility, (4) retrieval, (5) motivation, and (6) communication (see Chapter 2).

Although traditionally reliability is thought of as being primarily a characteristic of the measuring instrument and the conditions of measurement, it is also inherently conditional on the attributes of the population to which the survey instrument is applied and from which the data come. The argument I develop in this chapter is that, although it is no doubt partially correct to think of reliability as inherently a characteristic of the measuring instrument, it is also the case that *reliability can only be defined with respect to a particular population and its attributes*. In order to illustrate the importance of this claim, in this chapter I investigate the relationship of respondent characteristics—specifically age and education—to the reliability of measurement. These variables in no way exhaust the possible range of factors that contribute to differences in reliability; we focus on these in part because they are easily measured in the data sets we use for this study and in part because a substantial amount has been written about their contributions to measurement error. One approach I use to address these issues is to partition the data used in previous chapters into groups defined by these factors and present reliability estimates for these groups. In addition, later in the chapter I introduce a statistical framework that provides another approach to the problem, employing more rigorous tests of hypotheses regarding subpopulation differences in components of variance that contribute to the estimation of reliability.

10.1   RELIABILITY AS A POPULATION PARAMETER

Throughout the foregoing chapters we have defined the reliability of measurement within the framework of an approach that assumes a finite population of persons (S) for whom the classical assumptions of the true-score model apply. Thus, *reliability* is defined as a population parameter, namely, the proportion of the observed variance that is accounted for by true-score variance, expressed as the squared correlation between Y_g and T_g in a specific population, namely (see Chapter 3):

$$\rho_g^2 = COR[Y_g, T_g]^2 = VAR\ [T_g]\ /\ VAR[Y_g] = (VAR[Y_g] - VAR[E_g])\ /\ VAR[Y_g])$$

As a generic concept, then, *reliability* refers to the relative proportion of random error versus true variance in the measurement of Y_g, in a population whose properties VAR $[T_g]$, VAR$[E_g]$, and VAR$[Y_g]$ depend on the attributes of that population. To the extent that the relative proportion of error variance in VAR$[Y_g]$ differs as a result of the attributes of a given population, we may expect reliability to vary systematically with respect to those attributes. I stress these matters in order to reinforce the fact that the level of reliability is influenced not only by the properties of the survey questions that we use for measuring constructs of interest and the conditions of survey measurement; as a population parameter expressed by the ratio of true-score to observed score variance, it is also influenced by the characteristics of the population to which the measures are applied.

The theory on which the development of the concept of reliability is based, as I reviewed in detail in Chapter 3, allows us to partition the population into H subpopulations for which the model holds. Given a population of persons for whom the above model holds, we can write the model for the gth measure of Y, for the hth subpopulation as follows: $^{(h)}Y_g = {}^{(h)}T_g + {}^{(h)}E_g$, where the properties of the model for the population as a whole apply to the subpopulations as well. In this model the expectation of $^{(h)}E_g$ is zero, from which it follows that $^{(h)}E[Y_g] = {}^{(h)}E[T_g]$. Also under this model the covariance of the true and error scores is zero, i.e., $^{(h)}COV[T_g, E_g] = 0$, from which it follows that the variance of the observed score equals the sum of the variance of the true score and the variance of the error score, i.e., $^{(h)}VAR[Y_g] = {}^{(h)}VAR[T_g] + {}^{(h)}VAR[E_g]$. For the present purposes we can represent subpopulation reliability using the following simple notation

$$^{(h)}\rho_g^2 = {}^{(h)}VAR\ [T_g]\ /\ {}^{(h)}VAR[Y_g] = {}^{(h)}(VAR[Y_g] - {}^{(h)}VAR[E_g])\ /\ {}^{(h)}VAR[Y_g]$$

where these quantities are simply the result of partitioning the population into H subpopulations and defining the subpopulation variances accordingly. It is a straightforward extension of this formulation that can lead to hypotheses about reliability differences across subpopulations.

10.2 RESPONDENT ATTRIBUTES AND MEASUREMENT ERROR

A critical question that follows from the assumption that reliability varies across subpopulations, then, is what are the critical attributes of respondents that define subgroups across which estimates of reliability may be expected to vary? In this section we focus on this issue, although we should again note that even though theories of measurement are formulated at the individual level, the empirical study of reliability *cannot* focus on the individual as the unit of analysis. In a study of reliability such as this, the unit of analysis is by necessity the single survey question, applied under a particular set of measurement constraints and *specific to a particular population*. In keeping with this requirement, in previous chapters our analyses were conducted at *the level of the survey question* applied to a sample of a broadly defined population, e.g., a national sample of households. Here we shift our focus to specific subgroups of that population, specifically defined in terms of subpopulations characterized by attributes of respondents that produce greater or lesser amounts of measurement error.

The factors we have discussed in previous chapters that contribute to measurement error play a role regardless of the characteristics of the subject population, but they are often theorized to be especially relevant to the response process for some subgroups of the population. Age, for example, may have a bearing on the response process, via its interaction with the six critical elements we discussed in Chapter 2. Similarly, several studies report that educational attainment is positively related to memory performance and reliability of measurement, and here again these may be related to the survey response process via these factors (Alwin, 1989, 1999; Alwin and Krosnick, 1991a; Arbuckle et al., 1986; Perlmutter, 1978; Herzog and Rodgers, 1989). Given the relevance of cognitive skills to survey measurement, one must also consider the extent to which other age-related phenomena are confounded with differences among age groups. For example, differences in cohort-related experiences, such as level of schooling, may be linked to factors that contribute to the response process. If so, the unique role of aging may be confounded with cohort differences in levels of schooling, to the extent these are inferred from comparisons of measurement errors across age groups.

10.2.1 Strategies of Analysis

This chapter investigates the relationship of levels of reporting reliability to attributes of respondents. The chapter also addresses how the potential relationship of age group differences to estimated reliability intersects with relationships of schooling levels to reliability, and how the differing cohort experiences relevant to cognitive development, particularly the differing schooling levels of respondents of different ages, contribute to the understanding of the link between age differences and errors of measurement. I do so by examining adjustments to reliability estimates based on the relationships between age, schooling, and reliability. This is important because there are many aspects of cohort experiences that are tied to schooling—both the amount of schooling and its quality—and these experiences are relevant to performance on survey measures. If cohort differences in schooling account for the lower reliability of older people in some content domains, then processes of aging would not represent a complete explanation.

There are a number of different ways in which researchers have approached the relationship of respondent attributes and levels of measurement error. For example, one can compare age groups in the experimental effects of varying question forms (e.g., Knäuper et al. 1997). This approach, however, focuses on mean levels or biases between experimental categories and does not address the question of the contributions of measurement error to response variance. By contrast, the approach employed here formulates the problem of measurement error as a component of response variance, using information from variances and covariances in a within-subjects measurement design, specifically a repeated-measures design, which permits a focus on the reliability of measurement, as described in previous chapters.

As we have argued in previous chapters, there are two generally accepted design strategies for estimating the reliability of single survey questions: (1) the use of similar measures in the same interview or (2) the use of replicate measures in reinterview designs (see Alwin, 1989). The application of either design strategy is problematic, and in some cases the estimation procedures used require assumptions that are not appropriate. Estimation of reliability from information collected within the same interview is especially difficult, owing to the virtual impossibility of replicating questions. Researchers often employ similar although not identical questions, and then examine correlation or covariance properties of the data collected. It is risky to use such information to estimate question-specific reliabilites, since questions that are different contain specific components of variance, orthogonal to the quantity measured in common; and difficulties in being able to separate reliable components of specific variance from random error variance present a major obstacle to this estimation approach. Such designs require relatively strong assumptions regarding the components of variation in the measures, notably that measures are univocal and have properties of congeneric measures (see Alwin and Jackson, 1979; Jöreskog, 1971b). In Chapter 4 we illustrated the MTMM method as an approach that permits the estimation of reliability, and we employ this method here as well, disaggregating the sample used in that analysis by age and education.

A second approach to the estimation of the reliability of survey data uses a reinterview, or panel design. Such designs also have problems for the purpose of estimating reliability. For example, the test-retest approach using a single reinterview must assume that there is no change in the underlying quantity being measured (Lord and Novick, 1968; Siegel and Hodge, 1968). This is a problem in many situations, since with two waves of a panel study, the assumption of perfect correlational stability is unrealistic, and without this assumption little purchase can be made on the question of reliability in designs involving two waves. Because of the problematic nature of the assumption of perfect stability, several efforts have been made to analyze panel surveys involving three waves, where reliability can be estimated under certain assumptions about the properties of the error distributions (see Heise, 1969; Wiley and Wiley, 1970). These are the models used to produce the results presented in Chapters 5–9, and we employ this approach here as well. This model is discussed in detail in Chapter 5 (see also Alwin, 1989; Saris and Andrews, 1991).

Recall that this strategy employs a class of just-identified quasi-simplex (or autoregressive) models that specify two structural equations for a set of three over-time measures of a given variable. As we noted in Chapter 5, the limitation of the test-retest approach using a single reinterview is that it must assume that there is no change in the underlying quantity being measured. To address the issue of taking individual-level change into account, this class of autoregressive or quasi-simplex models specifies two structural equations for a set of P over-time measures of a given variable Y_t (where $t = 1, 2, \ldots, P$) as follows:

$$Y_t = T_t + E_t$$
$$T_t = \beta_{t,t-1} T_{t-1} + Z_t$$

The first equation represents a set of measurement assumptions indicating that (1) over-time measures are assumed to be τ-equivalent, except for true-score change, and (2) measurement error is random. The second equation specifies the causal processes involved in change of the latent variable over time. Here it is assumed that Z_t is a random disturbance representing true-score change over time. This model assumes a lag-1 or Markovian process in which the distribution of the true variables at time t is dependent only on the distribution at time $t - 1$ and not directly dependent on distributions of the variable at earlier times. If these assumptions do not hold, then simplex models may be inappropriate. However, our previous analyses indicate that, as a first-stage analytic approach, this model is extremely valuable.

In this chapter we conduct several sets of analyses of the relationship of subpopulations defined by age and education to reliability. We examine age and education differences in reliability using partitions of the samples analyzed in previous chapters. First, we partition the NES, ACL, and SAF samples (described in Chapter 6) examined previously into groups defined by age and education and estimate the quasi-simplex models separately within these subpopulations. Second, we partition the 1978 Quality of Life sample into groups defined by age and education analyzing these data using the MTMM true-score model applied in Chapter 4. Third, we illustrate a somewhat novel set of results from our recent study of aging and reliability, presenting a set of analyses from an ongoing study using data from the Health and Retirement Study (HRS), an innovative study of a national sample of older Americans, wherein we have substantial numbers of cases for groups defined by birth year. These analyses are formulated within a statistical framework that allows us to test hypotheses about age differences in reliability in a somewhat more rigorous way.

10.3 AGE AND RELIABILITY OF MEASUREMENT

Because of the cognitive demands of some survey questions (see Sudman, Bradburn and Schwarz, 1996), and given well-documented laboratory and survey results regarding declines associated with aging in some domains of cognitive functioning (e.g., Herzog and Rodgers, 1989; Park, Smith, Lautenschlager, Earles, Frieske, Zwahr, and Gaines, 1996; Salthouse, 1991, 1996; Schaie, 1996, 2005; Alwin, McCammon, Wray and Rodgers, 2007), it has been suggested that there may be identifiable differences in errors of measurement associated with the age of the respondent in actual surveys.

Aging may affect respondents' behavior with respect to several of the key factors involved in the survey response process. Specifically, with regard to comprehension of the question, it is well known that cognitive capacities for recognition of meaning and understanding concepts appear to decline in old age (see Salthouse, 1991; Schaie, 1996), and this may be the basis for declines in the reliability of survey reporting in old age (Andrews and Herzog, 1986). Similarly, given differences in memory faculties, accessibility to requested information may decline with age. However, with regard to subjective content, the obverse may be true. An example of this is Converse's (1964) famous example of nonattitudes—respondents who do not have opinions or attitudes may feel pressure to respond to survey questions because they assume interviewers want them to answer, and because of cultural norms, they believe that opinionated people are held in higher esteem than the ignorant and uninformed. One interpretation of this is that respondents wish to conform to these expectations and project positive self-images and, consequently, they frequently concoct attitude reports, making essentially random choices from among the offered response alternatives. Critics of this viewpoint (e.g., Achen, 1975) argue that measurement errors are not the fault of respondents' random reports, but are due primarily to the vagueness of the questions. Whatever the case, older respondents may be more subject to these processes, since they may be less informed of current events and less likely to have attitudes about them. There is some support for this in studies showing that older respondents are more likely to answer "don't know" to survey questions (Gergen and Back, 1966; Ferber, 1966; Francis and Bush, 1975; Rodgers and Herzog, 1992). There is reason to expect, however, that this pattern is more prevalent for attitudes and expectations, and less so for factual content.

Retrieval of requested information is often viewed as one of the most serious elements of the response process, given the importance of both memory and cognitive processing in formulating responses to many survey questions. Age differences in memory have been widely documented, although research is often characterized by weak research designs based on limited samples. An exception to this is the study by Herzog and Rodgers (1989), which reports a clear, age-related decline in memory performance in a cross-sectional sample of the Detroit, Michigan metropolitan area. Obviously, when memory fails or when retrieved information is ambiguous or imprecise, there is a tendency to create measurement errors in approximating the "true" value. To the extent that there are *communication* difficulties associated with age, it is likely to be linked to the response framework offered the respondent in mapping retrieved information to the response categories provided. Older respondents may not be as facile as younger respondents with more difficult and complicated response schemes. Similarly, *motivational* differences among respondents could produce differences in measurement error, although it is not altogether clear that motivation per se varies by age. That is, older respondents may not be any more likely to "satisfice" than "optimize" in producing answers to survey questions (see Alwin, 1991b). Despite the reasonableness of these expectations based on theories of cognitive aging, however, survey researchers studying older populations have generally not been able to demonstrate that there are serious age-related errors of measurement in routine applications of the survey interview. In one important study using comparisons of survey reports to administrative records, census counts, and maps, Rodgers and Herzog (1987a, p. 387) found few age differences in the accuracy in self-reports of factual material (e.g., voting behavior, value of housing, and characteristics of neighbors). They concluded that evidence to date "does not indicate that [problems of measurement error] are consistently more serious for older respondents than for any other age group." Similarly, research on the measurement of subjective variables has failed to show large differences in reporting errors by age, although it is sometimes the case that the responses of older people (e.g., over the age of 60) are the least reliable (Alwin, 1989; Alwin and Krosnick, 1991b; Rodgers et al., 1992).

The possibility that age is linked to the quality of measurement in surveys raises a number of additional questions. For example, it is important to ask whether the effects of age are monotonic, or whether the decremental effects of aging occur only after a certain age, say, age 60. Also, we need to know whether the possible measurement errors associated with aging are reflected in the measurement of all phenomena, or whether they are linked specifically to the nature of the content being measured. In the analysis presented later in this chapter, I examine these issues for factual material and nonfactual material separately.

In assessing the relationship of respondent age and reliability, it is important to understand the two-level aspect of the analysis. As noted above, even though we conceptualize errors of measurement for individuals, these errors are unobservable, and we cannot analyze reliability as if it were a property of individuals. Because reliability can be defined only at the level of population or subpopulation, we must examine variations in reliability over subpopulations rather than over individuals. In other words, it is not possible to predict reliability of reporting for an individual, and when looking at effects of respondent characteristics, it is not a simple matter of writing an equation for an individual in a multilevel model that predicts his or her reporting reliability from his/her attributes and question attributes. Thus, in the first part of the analysis the individual is the unit of analysis in the sense that we estimate a set of model parameters using data gathered from individuals. However, when we focus on the relationship of respondent age to reliability, we focus on subgroup differences in levels of reliability where the survey question is the unit of analysis.

In the NES and ACL data we employ four different age groups—(1) 18–33, (2) 34–49, (3) 50–65, and (4) 66 and older—and compare the four age groups across the age range, as defined above, within each separate study. We simply analyze reliability variation by age across questions as a simple ANOVA (analysis of variance) problem within question content type (e.g., facts versus subjective variables). This analysis

Table 10.1. Reliability estimates for self-
reports by age group for questions in four panel
study questionnaires[1]

	Reliability Estimates		
Age Group	Facts	Nonfacts	Total
18 - 33	0.760	0.623	0.649
34 - 49	0.789	0.643	0.671
50 - 65	0.795	0.640	0.669
66 +	0.792	0.571	0.613
Total	0.784	0.619	0.651
Total n	44	189	233
N of cases	176	756	932
F-ratio	0.37	6.01	4.37
p-value	0.777	0.001	0.005

[1]ACL, NES 50s, NES 70s, and NES 90s.

of age and reliability will make use of the question/subpopulation database described in Chapter 6. Here we enter reliability estimates for each of several subpopulations defined by age, as well as the "pooled" estimate. In this analysis age is the treatment factor whose main effect is to be estimated. This analysis will be performed *within* each of the two major content categories of questions—factual and nonfactual questions—in order to see if the age effects observed are generalizable across types of question.

Table 10.1 displays reliability estimates by age category and the content of the questions for 233 measures from the three NES panels and the ACL panel (the SAF samples are omitted here because of their restricted range of age). These estimates are based on Pearson correlations using the three-wave quasi-simplex model described in Chapter 5. In a subsequent analysis (see Table 10.5), I disaggregate these results for the four studies involved.

As I have indicated previously, it is really not appropriate to apply statistical tests to these differences in reliability, given that the sampling assumptions have not been met. That is, survey questions have not been sampled from a universe of such questions, and the items included are by no means independent of one another. Nonetheless, this offers a rough gauge of the extent of differences by age. The present set of results suggests that, consistent with prior research, there is very little variation by age in reliability of measurement of factual content (see Rodgers and Herzog, 1987a). On the other hand, there does appear to be some relationship between reliability and age in the measurement of nonfactual content. The level of estimated reliability is lowest for the oldest age group, but both younger and older respondents tend to have significantly lower levels of reporting reliability than do other groups.

Probably the most serious challenge to obtaining reasonable estimates of age differences in reporting reliability in survey data is the confounding of age with cohort factors. If cohort experiences were not important for the development of cognitive functioning, there would be no reason for concern. However, there are clear

Table 10.2. Reliability estimates for self-reports by years of schooling for questions in six panel study questionnaires

	Reliability Estimates		
Years of Schooling	Facts	Nonfacts	Total
Less than high school	0.732	0.540	0.570
High school graduate	0.759	0.629	0.650
Some college	0.782	0.631	0.655
College graduate or more	0.825	0.673	0.697
Total	0.774	0.618	0.643
Total n	49	258	307
N of cases	196	1,032	1,228
b-coefficient	0.015	0.020	0.019
β-coefficient	0.183	0.234	0.216
F-ratio	6.76	59.52	60.26
p-value	0.010	0.000	0.000

differences among age groups in aspects of experience that are relevant to the survey response. Specifically, since age groups differ systematically in their amount of schooling attained, cohort factors may contribute to the spuriousness of the empirical relationship between age and measurement errors.

10.4 SCHOOLING AND RELIABILITY OF MEASUREMENT

Previous research has demonstrated that levels of schooling are potentially related to levels of reporting reliability (e.g., Alwin, 1989; Alwin and Krosnick, 1991b), and therefore one of the central goals of this chapter is to test the hypothesis that, if there are differences in measurement reliability by age, one possible explanation for this is the differences in the educational levels of younger versus older respondents. Table 10.2 presents data on reliability estimates by levels of schooling within content categories of self-reports. Overall, these results suggest there are reliability differences in populations defined with respect to levels of schooling. Shown here is the relationship between education and reliability, which is basically monotonic—that is, reliability increases as a function of higher levels of schooling. The relationship is nearly linear, with a correlation (given here as the ß-coefficient) that ranges around .22.

The regression coefficients given in Table 10.2—the b-coefficients—indicate that a one-year increase in education produces roughly a .02 change in reliability for the average survey question. Thus, persons falling in the category "college graduate or more" report information in surveys at a level of reliability that is .10 higher than that for high school graduates. This evidence signals the possibility that cohort differences in education could play a role in patterns of age differences in reliability estimates, a possibility that we explore further subsequently.

Table 10.3. Reliability estimates for self-reports by years of schooling and panel study for questions in six panel study questionnaires

	Panel Study											
	ACL		NES 50s		NES 70s		NES 90s		SAF Children		SAF Mothers	
Years of Schooling	Facts	Non-facts	Facts	Non-facts	Facts	Non-facts	Facts	Non-facts	Facts	Non-facts	Facts	Non-facts
Less than high school	0.690	0.565	0.914	0.634	0.815	0.556	0.721	0.502	---[1]	0.443	0.572	0.593
High school graduate	0.696	0.617	0.928	0.589	0.824	0.614	0.744	0.672	---	0.599	0.744	0.644
Some college	0.716	0.609	0.923	0.597	0.842	0.628	0.804	0.654	---	0.597	0.754	0.670
College graduate or more	0.765	0.612	0.938	0.664	0.875	0.664	0.869	0.706	---	0.670	0.777	0.711
Total	0.717	0.601	0.926	0.621	0.839	0.616	0.784	0.633	---	0.577	0.712	0.654
Total n	21	42	6	15	8	71	9	61	---	33	5	36
N of cases	84	168	24	60	32	284	36	244	---	132	20	144
b-coefficient	0.012	0.007	0.003	0.005	0.010	0.017	0.025	0.030	---	0.034	0.031	0.019
β-coefficient	0.133	0.100	0.102	0.056	0.196	0.227	0.326	0.307	---	0.348	0.479	0.212
F-ratio	1.48	1.67	0.23	0.18	1.20	15.32	4.03	25.25	---	17.93	5.35	6.71
p-value	0.228	0.198	0.635	0.671	0.282	0.000	0.053	0.000	---	0.000	0.033	0.011

[1]There are no measures of facts available in this sample.

Table 10.3 performs a similar analysis disaggregated by data set. The results in Table 10.3 demonstrate that, although there is some variation in results, the above conclusions pertain to most of the data sets that we rely on here. With few exceptions, the least educated group has the lowest level of reporting reliability, and the most educated groups have the highest levels of reliability, for both facts and nonfacts. Generally speaking, there is a monotonic relationship between amount of schooling and level of reliability, although the differences are not always statistically significant given the groups of questions considered here.

10.5 CONTROLLING FOR SCHOOLING DIFFERENCES

One difficulty with the attribution of differences in cognitive performance to aging is the fact that in such cross-sectional studies, age is confounded with unique cohort experiences. If cohort experiences were not important for the development of cognitive functioning, there would be no reason for concern, but there are clear differences among cohorts in aspects of experience that are relevant for cognitive scores, e.g., the nature and amount of schooling (Alwin, 1991a). Thus, on logical grounds, before entertaining explanations of age differences in measurement errors involving the possibilities of the impact of aging and processes associated with it, one should probably try to rule out the role of different cohort experiences. More than 30 years ago, Matilda Riley (1973) warned that comparing different age groups could lead to an "aging" or "life cycle" fallacy. Cohort experiences are temporally prior to the overall experience of aging, so it is logical and necessary to rule out cohort differences among age groups that may be creating a spurious association between age and reliability of measurement. This, of course, requires the specification of theoretically relevant cohort-related factors that are also related to measurement error. For example, are aging interpretations of age differences robust with respect to controls for variables that reflect cohort experiences linked to cognitive development and that are associated with errors of measurement? Specifically, several studies report that educational attainment is positively related to memory performance (Arbuckle et al., 1986; Perlmutter, 1978; Herzog and Rodgers, 1989). Since cohorts differ systematically in their amount of schooling attained, it is natural to wonder whether cohort factors may be contributing spuriously to the empirical relationship between age and errors of measurement.

10.5.1 Age Differences in Schooling

In order to reinforce the need for controls for education in assessing age differences in reliability, Table 10.4 presents data on schooling differences among the cohorts (or age groups) studied in the 1970s NES panels. Age is strongly related to level of education due to the influence of cohort factors. The standard deviation of years of schooling in these data is about 3.2, so the oldest and the youngest differ by more

Table 10.4. Mean years of schooling by age group in six panel studies[1]

Panel Study

Age Group	ACL Years of Schooling	N	NES 50s Years of Schooling	N	NES 70s Years of Schooling	N	NES 90s Years of Schooling	N	SAF	Years of Schooling	N
18 - 33	13.3	499	11.6	329	13.0	425	13.5	331	Children	13.8	955
34 - 49	12.9	707	11.0	435	12.4	379	14.0	327	Mothers	12.6	966
50 - 65	11.4	689	9.9	253	11.0	300	12.8	178			
66 +	10.7	503	8.4	98	9.2	160	11.6	162			
Total	12.1	2,398	10.7	1,115	11.9	1,264	13.2	998		13.2	1,921

[1]Means are unweighted.

than a standard deviation, on average about 4 years. Given the link between schooling and measurement reliability reported above (see Tables 10.2 and 10.3), differences in the amount of schooling across age groups could explain any tendency in these data for older respondents to have lower reliability.

10.5.2 Schooling-Adjusted Age Differences in Reliability

In this part of this study we are concerned with the intersection of the above sets of findings. Specifically, we ask whether age group variation in measurement reliability can be explained by (1) differences in levels of reliability with respect to schooling and (2) differences among age groups in levels of schooling. In other words, are age differences in schooling experiences partly or totally responsible for measurement errors apparently linked to age? If so, one needs to be cautious in drawing inferences about the causal role of processes of cognitive aging in producing measurement errors. This is, however, more than a statistical issue of making sure that adequate controls have been undertaken to rule out alternative hypotheses. The issue I raise is more fundamental—observations of differences among groups of different ages need not be evidence of anything having to do with processes of aging.

To put the issue somewhat more narrowly, to what extent are differences in response effects or measurement errors heretofore identified as linked to processes of aging really due to cohort experiences in exposure to schooling? Do age differences in reliability of survey reporting remain when we control for cohort differences in schooling? Schooling is one of the clear-cut differences among birth cohorts in contemporary society, with vast differences between those born at the beginning of the century and those born more recently. Schooling contributes to the reduction of survey errors, since more educated respondents, regardless of age, systematically produce fewer errors in survey responses (Alwin and Krosnick, 1991b). In order to control for schooling differences, I apply covariance adjustments of reliability estimates for age groups, adjusted for compositional differences in schooling of the age groups. This exercise essentially equates the groups statistically in their exposure to schooling. Because schooling is strongly related to both age and reliability, some substantial changes are recorded.

In Table 10.5, I present the relationship between age and reliability within each of the content categories considered—facts and nonfacts—for each of the four studies where we have variation across the full age range, from age 18 to 66 and above. This table presents both the unadjusted mean reliabilities by age (column 1) as well as the adjusted mean reliabilities (column 2). Considering the unadjusted reliability estimates first, note that these results confirm the hypothesis based on earlier research that there are few age differences in the reliability of reporting factual information—the older respondents are no more or less likely to produce greater errors of measurement—but in the case of nonfacts (i.e., beliefs, attitudes and, self-perceptions) the oldest age groups have systematically lower reporting reliabilities in three of the four studies. These unadjusted results by age are, however, not always

Table 10.5. Reliablity estimates for self-reports by age group adjusted for years of schooling in four panel study questionnaires

	Panel Study							
	ACL				NES 50s			
Age Group	Facts		Nonfacts		Facts		Nonfacts	
	(1)[1]	(2)	(1)	(2)	(1)	(2)	(1)	(2)
18 - 33	0.702	0.687	0.531	0.523	0.908	0.903	0.617	0.610
34 - 49	0.706	0.696	0.627	0.622	0.925	0.922	0.622	0.618
50 - 65	0.748	0.756	0.634	0.638	0.900	0.901	0.626	0.627
66 +	0.736	0.753	0.573	0.582	0.903	0.909	0.656	0.664
Total	0.723	0.723	0.591	0.591	0.909	0.909	0.630	0.630
Total n	21	21	42	42	6	6	15	15
N of cases	84	84	168	168	24	24	60	60
F-ratio	0.26	0.69	4.58	5.12	0.13	0.10	0.09	0.17
p-value	0.851	0.558	0.004	0.002	0.940	0.961	0.964	0.914

	Panel Study							
	NES 70s				NES 90s			
Age Group	Facts		Nonfacts		Facts		Nonfacts	
	(1)	(2)	(1)	(2)	(1)	(2)	(1)	(2)
18 - 33	0.805	0.789	0.629	0.602	0.756	0.744	0.682	0.667
34 - 49	0.846	0.836	0.636	0.619	0.841	0.815	0.667	0.636
50 - 65	0.831	0.835	0.633	0.640	0.803	0.808	0.657	0.662
66 +	0.884	0.906	0.517	0.554	0.770	0.804	0.612	0.653
Total	0.841	0.841	0.604	0.604	0.793	0.793	0.655	0.655
Total n	8	8	71	71	9	9	61	61
N of cases	32	32	284	284	36	36	244	244
F-ratio	0.79	1.69	7.35	2.90	0.41	0.31	1.34	0.29
p-value	0.512	0.193	0.000	0.035	0.746	0.820	0.261	0.829

[1]Column (1) contains unadjusted reliability estimates; column (2) contains estimates of reliability adjusted for cohort differences in years of schooling.

statistically significant, and it is not always the case that the oldest age group has the lowest reliability. These age patterns in the reliability of nonfactual questions are present for the ACL, NES70s and NES90s data sets, but in the ACL the youngest age group has distinctively lower reliability as well.

To the extent that there is an age-related pattern in the unadjusted reliabilities for nonfactual content, the question remains as to the source of these patterns, and whether they may be due to differences in the schooling experiences of different cohort groups. Column 2 in Table 10.5 presents the adjusted differences in mean reliability, once educational differences among cohorts are taken into account. The adjusted reliability for the gth item in the hth subgroup is as follows:

$$^{(h)}\rho_g^{2*} = {}^{(h)}\rho_g^2 - b_{\rho x} \, [\mu_{.h} - \mu_{..}]$$

where $^{(h)}\rho_g^2$ is the unadjusted reliability for the hth age group, $b_{\rho x}$ is the unstandardized regression coefficient in the regression of reliability on years of schooling (see Table 10.3), $\mu_{.h}$ is the mean level of schooling for the hth age subgroup, and $\mu_{..}$ is the grand mean of schooling for the population (see Table 10.4). All estimates here are based on survey specific data and carried out within content categories (see Table 10.5), and thus, the regression coefficients used in these calculations varied as a function of the content considered.

These adjusted reliability results show that when we take into account the age-education and education-reliability relationships, the differences in nonfact reliability by age groups change in some cases. In the NES90s data, the adjustments essentially remove whatever differences were observed initially. In the NES70s data, the adjustments reduce the differences, but the lower average reliability in the oldest age group is still present. In the ACL data, the adjustments have no appreciable effect, and the lower reliability of the oldest age group persists, as does the lower reliability in the youngest age group. In the NES50s data, there is no change introduced by the adjustments, but there the oldest age group has the highest average reliability. The adjustment removed the distinctiveness of the oldest age group in only one case—the NES90s data—indicating that once education differences among the age groups are removed, reporting reliability in the oldest age categories is no lower than in the younger groups. Still, in two of the four cases (the ACL and NES70s data) after adjustments for education, the oldest age groups have distinctively low reliability.

10.6   GENERATIONAL DIFFERENCES IN RELIABILITY

We can also examine the issue of age differences in reporting reliability further by comparing the reliability estimates for SAF mothers and children. These samples were excluded from the above analyses (in Section 10.5) because of the restricted age ranges represented in the two samples (see Chapter 6). Table 10.6 presents a comparison of average estimates of reliability for the SAF mothers and SAF children samples using two alternative pools of nonfactual questions. There were two few factual questions in the SAF children sample to implement a reasonable comparison for this content category, so we limited our analysis to the nonfactual questions. Set A includes all available questions in the two samples; set B includes only those questions that are identical for the two samples—24 in each group.

As above, column 1 presents the unadjusted reliability estimates and column 2 presents the reliability estimates adjusted for differences in schooling and the relationship of schooling to reliability. These results indicate that the mothers' survey reports are slightly more reliable than those of their children, roughly .66 versus .60—differences that are marginally significant, given the number of measures involved. This result is compatible with the findings reported above that the youngest respondents produced survey data that were marginally less reliable than middle-aged respondents.

Table 10.6.   Reliability estimates for nonfactual questions
by generation group in the Study of American Families

Reliability Estimates

Group	Set A		Set B	
	(1)[1]	(2)[2]	(1)	(2)
Children	0.607	0.591	0.608	0.592
Mother	0.639	0.655	0.650	0.665
Total	0.623	0.623	0.629	0.629
Total n	89		48	
N of cases	89		48	
F-ratio	1.15	4.35	1.50	4.50
p-value	0.286	0.040	0.227	0.039

[1]Column (1) contains unadjusted reliability estimates.
[2]Column (2) contains estimates of reliability adjusted for cohort
differences in years of schooling.

10.7   MTMM RESULTS BY AGE AND EDUCATION

Here we reexamine the MTMM results we presented in Chapter 4 for measures of life satisfaction measured using 7- and 11-category scales. We examine variation in estimates of reliability, and true-score validity and invalidity estimates by categories of age and education. Recall the design of the study [see Alwin (1997) and Chapter 4] involved the comparison of measures using 7- and 11-category rating scales in the measurement of life satisfaction. The study employed the 1978 national survey conducted by the Survey Research Center of the University of Michigan in which 17 domains of life satisfaction were each measured using these two types of rating scales. The 1978 Quality of Life (QoL) survey was a probability sample (n = 3,692) of persons 18 years of age and older living in households (excluding those on military reservations) within the coterminous United States. The sample design and response rates, as well as the measurement design, are discussed in detail in Chapter 4. In brief, the use of multiple measures of several domains of life satisfaction permits the analysis of both "trait" and "method" components of variation in each measure, given that each domain was assessed using multiple response formats.

Here we employ 16 domains of satisfaction—community, neighborhood, dwelling, United States, education, health, ways to spend spare time, friendships, family, income, standard of living, savings and investments, life, self, housework, and marriage (for those married). Satisfaction with one's job was omitted from these analyses because of the comparisons by age and the fact that older respondents (those 66 years of age and older) who were retired had no job. These measures were assessed using both 7-point and 11-point response scales. Three of these domains—

place of residence, standard of living, and life as a whole—were rated using three separate scales. The three methods used were (1) a 7-point "satisfied-dissatisfied" scale, (2) a 7-point "delighted-terrible" scale, and (3) an 11-point "feeling thermometer." The order of presentation of measurement approaches was, unfortunately, the same across topics and across respondents. Although our analysis considered all three methods of measurement, we examine only those measures assessed using the 7-point "satisfied-dissatisfied" scale and the 11-point "feeling thermometer."

As above, the data were analyzed using the structural equation models, implemented using the AMOS software (Arbuckle and Wothke, 1999). In the analysis presented here we employ Pearson correlations and handle missing data using full-information maximum-likelihood (FIML) methods. This approach, described in detail by Wothke (2000) and implemented in AMOS (Arbuckle and Wothke, 1999), yields model parameter estimates that are both consistent and efficient if the incomplete data are *missing at random* (MAR) (Wothke, 2000). Given our previous analyses of these data, we do not expect our conclusions to be affected by these choices. Preliminary results using polychoric correlational methods for ordinal-polytomous variables, as discussed in Chapter 3, suggested that patterns by age and education were similar to those reported here.

The covariance matrix among the variables was analyzed using the CTST representation of the MTMM model given in Chapter 3. In these models the trait factors are allowed to correlate, whereas the method factors are uncorrelated with the trait factors and with each other. The overall goal of this approach is to partition the Quality of Life sample into categories of age and education and estimate the model separately within each group. In each case the sample response variances of the measures are decomposed into three parts: (1) reliable trait variance, (2) reliable method variance, and (3) unreliable variance (Alwin, 1974; Groves, 1989; Saris and Andrews, 1991). For each measure three coefficients are presented for each trait-method combination: (1) the estimated reliability for each item, (2) the standardized factor pattern coefficient linking the trait factor to the true score in question, and (3) the standardized factor pattern coefficient linking the method factor involved to the true score in question. The definitions we use follow those given by Saris and Andrews (1991, pp. 581–582), in which the coefficients of *validity* and *invalidity* are scaled such that the square of the two loadings sum to 1.0. In the tables these are denoted as h^2, b_{T*}, and b_M, respectively (see notation used in Figure 4.4 discussed above), a slight variant of the notation used by Saris and Andrews (1991).

Our strategy in this part of the analysis is to (1) examine the estimates of reliability, validity, and invalidity by age; (2) examine these estimates by education; and (3) adjust the age differences in reliability, validity, and invalidity by differences in education, as we have in the previous sections. Table 10.7 presents reliability, validity, and invalidity estimates by age group averaged across the 16 measures of life satisfaction for two methods of measurement—the 7-point "satisfied-dissatisfied" scale and the 11-point "feeling thermometer." As above, we present a test statistic from a simple ANOVA in which the observations are the 16 measures within each of the four age groups. Between group differences among the treatments, i.e., age groups, signify differences in reliability by age.

Table 10.7.   Reliability, validity and invalidity estimates by age group for 16 measures of life satisfaction: 1978 Quality of Life Survey based on full-information maximum-likelihood (N = 3,687)

		Reliability (h^2)		Validity (b_{T*})		Invalidity (b_M)	
Age	N	7-pt.	11-pt.	7-pt.	11-pt.	7-pt.	11-pt.
18-33	1,339	0.605	0.836	0.970	0.946	0.183	0.316
34-49	870	0.637	0.818	0.964	0.964	0.234	0.249
50-65	881	0.590	0.794	0.929	0.978	0.358	0.147
66+	597	0.545	0.762	0.886	0.962	0.448	0.217
Total	3,687	0.594	0.802	0.937	0.963	0.306	0.232
F-ratio		2.86	1.44	12.78	2.44	14.83	5.04
p-value		0.044	0.239	0.000	0.073	0.000	0.004
r^2		0.076	0.066	0.362	0.040	0.419	0.107
η^2		0.125	0.067	0.390	0.109	0.426	0.201
n		64	64	64	64	64	64

Table 10.7 indicates, as we observed above, that the oldest group of respondents has the lowest levels of reliability. In this case, this is equally true whether one considers the 7-point or the 11-point scale measurements of quality of life, although the latter differences are not judged to be significant. Using this approach, levels of reliability can be attributed to one or both of two sources—the reliable measurement of substantive traits and/or the reliable measurement method factors. As can be seen in Table 10.7, the valid portion of the measures tends to be lower for older respondents in the case of the seven-point rating scales, but there is hardly any difference in the validity estimates for the feeling thermometers. The results for the validity estimates, thus, parallel those for the reliability estimates.

Interestingly, however, the MTMM approach may reveal a reality more complex than that available to the traditional CTST approach. It could be that older respondents (or any group of respondents, for the matter) may be more influenced by method factors, which would increase the level of reliability attributed to them. Thus, if older respondents are differentially affected by method factors, this will tend to increase reliability, rather than simply reduce it through the contribution of greater random measurement error. It appears that this is the case in the instance of seven-point scale measurement of life quality. There seems to be a monotonic increase in the role of method variance (i.e., invalidity) for the 7-point scales in this case; however, older respondents are not significantly more or less likely to be affected by the method factors in the case of the 11-point scales in these data.

As above, we are concerned that any differences in reliability estimates attributed to age differences take account of the differences in levels of schooling of the age groups and the differences in reliability estimates by level of schooling. In Table 10.8, we present estimates of reliability, validity and invalidity for four education categories. These results reveal an expected pattern, namely, monotonic increases

Table 10.8.   **Reliability, validity and invalidity estimates by years of schooling for 16 measures of life satisfaction: 1978 Quality of Life Survey based on full-information maximum-likelihood (N = 3,660)**

		Reliability (h^2)		Validity (b_{T*})		Invalidity (b_M)	
Education	N	7-pt.	11-pt.	7-pt.	11-pt.	7-pt.	11-pt.
0-11	1,153	0.574	0.761	0.906	0.971	0.407	0.183
12	1,096	0.608	0.823	0.953	0.960	0.287	0.260
13-15	875	0.614	0.828	0.962	0.959	0.248	0.275
16+	536	0.648	0.869	0.965	0.971	0.237	0.216
Mean	3,660	0.611	0.820	0.947	0.966	0.295	0.234
F-ratio		1.40	3.46	9.67	0.66	8.42	2.44
p-value		0.250	0.022	0.000	0.578	0.000	0.073
r^2		0.065	0.143	0.268	0.000	0.258	0.016
η^2		0.066	0.147	0.326	0.032	0.296	0.109
n		64	64	64	64	64	64

in levels of reliability and validity, and monotonic decreases in levels of invalidity across the education categories. Although the patterns are not always judged significant, in general the group with the highest level of schooling have the highest levels of reliability and validity and the lowest levels of estimated method variance. In general the patterns are more pronounced for the seven-point "satisfied-dissatisfied" measurement scales.

Following through with the adjustments of the age-group differences for differences in mean schooling levels and differential levels of reliability, we find, as shown in Table 10.9, no age differences in reporting reliability, once the schooling factors are taken into account. In the case of the seven-point scales (but not the feeling thermometers), however, we continue to see the pattern observed above, namely, lower average validity and higher average invalidity among the oldest respondents. Therefore, it appears that while there are few, if any, differences in reliability among the different age groups, these null findings are masking an underlying pattern linked to age that is not attributable to age differences in education. Older respondents provide ratings of life quality that are slightly less true variance and slightly more method variance. We find this pattern, however, only for the seven-point measurement scales, and such results will need further replication before one can safely attribute the patterns of measurement error to differences in age.

10.8   STATISTICAL ESTIMATION OF COMPONENTS OF VARIATION

In our investigation of subpopulation differences in estimates of reliability, I introduce another approach to the examination of respondent attributes and reliability of measurement that relies on examples from our recent National Institute on Aging

Table 10.9.   Reliability, validity and invalidity estimates by age group, adjusted for years of schooling for 16 measures of life satisfaction: 1978 Quality of Life Survey based on full-information maximum-likelihood (N = 3,660)

		Reliability (h^2)		Validity (b_{T^*})		Invalidity (b_M)	
Age	N	7-pt.	11-pt.	7-pt.	11-pt.	7-pt.	11-pt.
18-33	1,153	0.596	0.823	0.962	0.947	0.204	0.311
34-49	1,096	0.633	0.811	0.960	0.964	0.245	0.247
50-65	875	0.596	0.803	0.934	0.978	0.343	0.151
66+	536	0.563	0.787	0.901	0.962	0.406	0.226
Mean	3,660	0.597	0.806	0.939	0.963	0.300	0.234
F-ratio		1.59	0.32	7.17	2.42	8.67	4.50
p-value		0.201	0.810	0.000	0.075	0.000	0.007
r^2		0.030	0.016	0.240	0.039	0.297	0.085
η^2		0.074	0.016	0.264	0.108	0.302	0.184
n		64	64	64	64	64	64

research project focusing on aging and reliability of measurement (Alwin, McCammon and Rodgers, 2006). This research examines the reliability of measurement in large samples of individuals aged 51–91 from the HRS. Although our larger project evaluates all measures in the HRS, we report the results for a limited number of measures here in order to illustrate a more rigorous approach to the assessment of reliability by age and education that we hope will find application in future research in this area. For each of these measures, we construct a multiple-group model in which we compare several versions of the model in which the measurement parameters are constrained across age groups. We use state-of-the-art likelihood-ratio statistics (see Muthén and Muthén, 2001–2004) to evaluate the fit of these models relative to those that do not place such constraints on the model. This will provide question-specific results regarding whether a common measurement error structure fits all age groups equally well. If the results lead us to reject the null hypothesis, we will conclude there are important differences in error structures that would be reflected in different reliabilities of measurement by age.

As noted, we employ data from the HRS, a biennial survey of middle-aged and older Americans that began in 1992, which I briefly describe here. The original HRS was a national panel study of typically preretirement men and women aged 51–61 assessed in 1992 (n = 9,824) and reinterviewed every 2 years thereafter through to 2004. In 1993, a second national panel study—the Study of Asset and Health Dynamics among the Oldest Old (AHEAD)—was added, which interviewed adults aged 70 and above in 1993 (n = 7,443) and reinterviewed them in 1995, 1998, 2000, 2002, and 2004. The HRS and AHEAD were designed as parallel studies of current employment (retirement) and job history, family and social supports, health and function as well as economic status (Juster and Suzman 1995; Soldo, Hurd, Rodgers and Wallace, 1997). In the HRS and AHEAD studies, spouses were interviewed.[1] Both studies included oversamples of African-Americans and Hispanic Americans. After two data collection follow-ups of HRS (1994, 1996), and one of AHEAD (1995), the studies were fully integrated in 1998, and respondents in both panels were reinterviewed at that time and since.

In order to more completely represent all birth cohorts from the early 1900s to midcentury, two additional cohort samples were added to the HRS in 1998, samples representing persons born during the Great Depression and persons born during World War II. Since 1998, the HRS sample was comprised of four subsamples: (1) the original HRS subsample consisting of people (and their spouses or partners at the time of the initial interview or any follow-up interview) who were born 1931 through 1941 and were household residents of the coterminous United States in spring 1992; (2) the AHEAD subsample consisting of people (and their spouses or partners at the time of the initial interview or any follow-up interview) born in 1923 or earlier, who were household residents of the coterminous United States in spring 1992, and who were still household residents at the time of their first interview in 1993 or 1994; (3) a supplemental sample of "war babies," consisting of people (and their spouses or partners at the time of the initial interview or any follow-up interview) who were born 1942 through 1947 (N = 2360), who were household residents of the coterminous United States in the spring of 1992, and were still household residents at the time of their first interview in 1998, and reinterviewed in 2000, 2002, and 2004; and (4) a supplemental sample of "children of the depression" (CODA) consisting of people who were born in 1924–1930 (n = 2189), who were household residents of the coterminous United States when first interviewed in 1998, and who, at that time, did not have a spouse or partner who was born before 1924 or between 1931 and 1947. The CODA sample explicitly excludes households that were eligible to be in AHEAD, HRS, or the "war babies" subsamples. We employ data for these samples from the 1998 through 2002 waves.

We employ two sets of self-report measures from the HRS interviews to investigate age differences in reporting reliability. The first domain involves measures of physical health and functioning, specifically the types of measures that are typical of survey measures of health:

- Rating of current health: Would you say your health is excellent, very good, good, fair, or poor?
- Rating of current health relative to past health (three categories): Compared to your health when we talked with you (last interview), would you say that your health is better now, about the same, or worse?
- Rating of current health relative to past health (five categories—same question as above with follow-up): Compared with your health when we talked with you

---

[1]We pool the respondent and spousal data for purposes of this analysis, and although a general concern about the nonindependence of cases from the same household is admittedly justified, we have consulted with senior HRS design specialists and data analysts about whether we need to be concerned about our inferences, and in their view we do not. It is their judgment that including both spouses in our analyses will not bias the results appreciably.

(last interview), would you say that your health is better now, about the same, or worse? *If better*, is it much better or somewhat better? *If worse*, is it much worse or somewhat worse?

- Rating of vision: Is your eyesight excellent, very good, good, fair, or poor (using glasses or corrective lenses as usual)?
- Rating of distant vision: How good is your eyesight for seeing things at a distance, like recognizing a friend across the street? Is it excellent, very good, good, fair, or poor (using glasses or corrective lenses as usual)?
- Rating of near vision: How good is your eyesight for seeing things up close, like reading ordinary newspaper print? Is it excellent, very good, good, fair, or poor (using glasses or corrective lenses as usual)?
- Rating of hearing: Is your hearing excellent, very good, good, fair, or poor (using hearing aid as usual)?
- Trouble with pain: Are you often troubled by pain? Yes/No
- Weight: About how much do you weigh? (Expressed in pounds)

The second domain of interest is what the economists affiliated with the HRS refer to as *expectations*. This set of measures consists of a battery of questions requesting respondents to rate the likelihood that an "event" will happen in the future. We have selected a subset of these expectations questions for our present purposes—this series of questions is introduced as follows:

> Next I have some questions about how likely you think various events might be. When I ask a question I'd like for you to give me a number from 0 to 100, where "0" means that you think there is absolutely no chance, and "100" means that you think the event is absolutely sure to happen.
>
> For example, no one can ever be sure about tomorrow's weather, but if you think that rain is very unlikely tomorrow, you might say that there is a 10% chance of rain. If you think there is a very good chance that it will rain tomorrow, you might say that there is an 80% chance of rain.

Following a practice example, the HRS interview requests responses to a battery of such questions on a range of topics pertaining to the respondent's life chances. Because not all questions were asked of everyone in the sample, we include an analysis of only a subset of these measures here. When a given question was not asked of everyone, we provide a description of the subsample of persons who were asked each specific question, in the following description of questions:

- What do you think are the chances that your income will keep up with inflation for the next 5 years?
- Including property and other valuables that you might own, what are the chances that you [and your (husband/wife/partner)] will leave an inheritance totaling $10,000 or more?
- What are the chances that you [and your (husband/wife/partner)] will leave an inheritance totaling $100,000 or more? (Asked only of those who gave some chance of leaving an inheritance in response to the previous question.)

- Sometimes people are permanently laid off from jobs that they want to keep. On the same scale from 0 to 100, where 0 equals absolutely no chance and 100 equals absolutely certain, what are the chances that you will lose your job during the next year? (Asked only of persons reporting that they are currently working, but are not self-employed.)
- Suppose that you were to lose your job this month. What do you think are the chances that you could find an equally good job in the same line of work within the next few months? (Asked only of persons reporting that they are currently working, but are not self-employed.)
- (On this same 0–100 scale), what are the chances that you will be working for pay at some time in the future? (Asked only of those persons who are not currently working.)
- Thinking about work in general and not just your present job, what do you think the chances are that you will be working full-time after you reach age 62? (The question was asked only of respondents who were in the 1940–1947 birth cohorts.)
- What about the chances that you will be working full-time after you reach age 65? (The question was asked only of respondents who were in the 1940–1947 birth cohorts who reported a nonzero chance of working past age 62—see the preceding question.)
- What about the chances that your health will limit your work activity during the next 10 years? [Respondents reporting (in 1998, 2000, or 2002) that they are currently not working were not asked this question.]
- Now using the same scale as before, where "0" is absolutely no chance and "100" means that it is absolutely certain, please tell me what you think are the chances that you will move in the next two years? (Asked only of respondents who were 65 years or older.)
- (What is the percent chance) that you will live to be 75 or more? (Asked only of respondents from the 1937–1947 birth cohorts, i.e., aged 51–61 in the 1998 survey.)
- (Using a number from 0 to 100) What are the chances that you [and your (husband/wife/partner)] will give financial help totaling $5000 or more to grown children, relatives, or friends over the next 10 years?
- (What is the percent chance) that you will move to a nursing home in the next five years? (Asked only of those persons under 65 years of age at the time of the interview.)

Using the classification we developed in Chapter 6, these expectation measures are *self-perceptions*, the measures of self-rated health and functioning are primarily *self-evaluations*, and the estimate of weight is a *fact*.

Table 10.10 gives the total sample estimates of the reliability of measurement for the physical health and functioning measures using both the Pearson- and polychoric-based correlation approaches (see Chapter 3). Note that in the case of the estimate of weight, since it is a continuous measure, there is no polychoric-based estimate. In the following analyses we follow the same rule employed in previous chapters; that is, we base our analyses on polychoric techniques where it is appropriate—when

Table 10.10.   Total sample reliability estimates for selected
measures of health: Health and Retirement Study

Question Content	Sample Size	Pearson Estimate	Polychoric Estimate
Health			
Rating of current health	15,852	0.727	0.780
Rating of health relative to past (3 categories)	12,122	0.326	0.417
Rating of health relative to past (5 categories)	12,104	0.332	0.394
Sensory Function			
Rating of vision	15,710	0.572	0.630
Rating of distant vision	15,693	0.542	0.602
Rating of near vision	15,684	0.519	0.574
Rating of hearing	15,834	0.675	0.734
Pain			
Trouble with pain	15,828	0.557	0.787
Weight			
Estimate of weight	15,428	0.971	---

the number of response categories is less than 16 (see Chapter 6). Consistent with previous results, the measure of nonfactual content in the HRS survey involves less-than-perfect reliability. The estimated reliability of the typical self-rating of personal experiences falls in the .60–.80 range. Only the estimate of weight (About how much do you weigh?) has a reliability near the theoretical limit of unity. The remaining measures have much lower levels of reliability, which is typical of self-reports of nonfactual content. Indeed, for the two measures involving the self-rating of health relative to past health [Is your health better or worse? (If better/worse) is it much better/worse or somewhat better/worse?] the reliability level is relatively low. As we argued in previous chapters, this level of reliability is among the lowest obtained in current practice, and it raises questions about the utility of such measures.

Table 10.11 gives the estimates of measurement reliability for the ratings of expectations about selected topics for the total sample and two subsamples defined by level of schooling. These estimates are based on Pearson correlations, because given the 0–100 scale employed in these measures, we can readily assume that they are continuous measures. Inspecting these results, we can conclude that, consistent with all our previous results, these subjective measures involve substantially less than perfect reliability. The estimated reliability of the typical self-rating of personal expectations is .56, not altogether different from other survey measures of self-perceptions (see Chapter 7). The range extends from a reliability of .34 for the measurement of the likelihood of losing one's job to a reliability of .70 for the measurement of the likelihood of working after age 62. We return subsequently to an examination of the differences among the two subpopulations defined by level of schooling displayed in this table.

Table 10.11.   Reliability estimates of ratings of expectations about selected topics for the total sample and two categories of schooling: Health and Retirement Study

	Total Sample		High School or Less		More than High School	
Expectations	Sample Size	Reliability Estimate	Sample Size	Reliability Estimate	Sample Size	Reliability Estimate
Income and inflation	16,904	0.524	10,360	0.451	6,542	0.581
Leave heirs $10K	17,148	0.663	10,588	0.648	6,557	0.588
Leave heirs $100K	12,543	0.687	6,837	0.657	5,704	0.679
Lose your job	3,503	0.337	1,850	0.334	1,652	0.338
Find new job	3,504	0.606	1,851	0.547	1,652	0.664
Work after 62	2,594	0.703	1,227	0.660	1,366	0.736
Work after 65	1,747	0.630	786	0.606	961	0.644
Health limits work	4,721	0.364	2,353	0.257	2,367	0.474
Move next two years	8,388	0.479	5,507	0.461	2,880	0.494
Live to 75 or more	6,716	0.621	3,796	0.574	2,918	0.670
Give financial help	17,175	0.610	10,616	0.580	6,556	0.581
Go to nursing home	7,840	0.451	5,111	0.417	2,729	0.516
Average		0.556		0.516		0.580

10.8.1   Age and Reliability of Measurement in the HRS

We examined the relationship of age to reliability in all of these measures, with the general expectation that reliability would decline monotonically with age. Our initial approach to this examined the linear regression of the reliability estimates on age, as shown in Tables 10.12 and 10.13. In virtually all cases, there was no significant linear relationship with age, and thus, very little encouragement for the basic hypothesis. Only for the measures of self-rated health and two of the expectations ratings—expectations about income and inflation and expectations about giving financial help to children, relatives, or friends—is there any substantial variance in reliability accounted for by its linear relationship to age. In the case of the health ratings, only the rating of current health shows a relationship in the expected direction. Although in a negative direction, the relationship of reliability and age for the two ratings of past health are of nontrivial magnitude.

Figures 10.1 and 10.2 present a graphic depiction of the linear regression of reliability on age for the self-rated health measure and the measure of expectations about personal income and inflation. These patterns give visual clarity to the nature of the regression results presented above for these two variables.

10.8.2   Education and Reliability of Measurement in the HRS

We also expected, on the basis of the foregoing analyses, that there would be a systematic relationship between level of schooling and reliability. Indeed, our estimates

Table 10.12.   Linear regression of reliability estimates on age for selected
measures of health: Health and Retirement Study

Question Content	Intercept	Slope	Standard Error	p-value	R^2
Health					
Rating of current health	0.722	0.002	0.001	0.010	0.158
Rating of health relative to past (3 categories)	0.610	-0.008	0.003	0.026	0.127
Rating of health relative to past (5 categories)	0.539	-0.006	0.003	0.042	0.104
Sensory Function					
Rating of vision	0.618	0.000	0.001	0.857	0.001
Rating of distant vision	0.577	0.001	0.001	0.648	0.006
Rating of near vision	0.598	-0.001	0.001	0.451	0.015
Rating of hearing	0.694	0.001	0.001	0.280	0.031
Pain					
Trouble with pain	0.744	0.002	0.001	0.210	0.040
Weight					
Estimate of weight	0.966	0.000	0.000	0.848	0.001

Table 10.13.   Linear regression of reliability estimates on age for ratings of expectations: Health
and Retirement Study

Expectations	Cohorts	Ages	Intercept	Slope	Standard Error	p-value	R^2
Income and inflation	1910-1947	51-88	0.571	-0.003	0.001	0.023	0.135
Leave heirs $10K	1910-1947	51-88	0.671	-0.001	0.001	0.371	0.022
Leave heirs $100K	1910-1947	51-88	0.699	-0.001	0.001	0.275	0.033
Lose your job	1929-1947	51-69	0.331	0.001	0.008	0.923	0.001
Find new job	1929-1947	51-69	0.652	-0.007	0.005	0.205	0.093
Work after 62	1940-1947	51-58	0.704	0.003	0.021	0.877	0.004
Work after 65	1940-1947	51-58	0.602	0.005	0.012	0.675	0.031
Health limits work	1926-1947	51-72	0.395	-0.001	0.006	0.805	0.003
Move next two years	1910-1933	65-88	0.485	-0.001	0.003	0.789	0.003
Live to 75 or more	1937-1947	51-61	0.616	0.002	0.005	0.766	0.010
Give financial help	1910-1947	51-88	0.657	-0.004	0.002	0.024	0.134
Go to nursing home	1910-1933	65-88	0.416	0.002	0.004	0.567	0.015

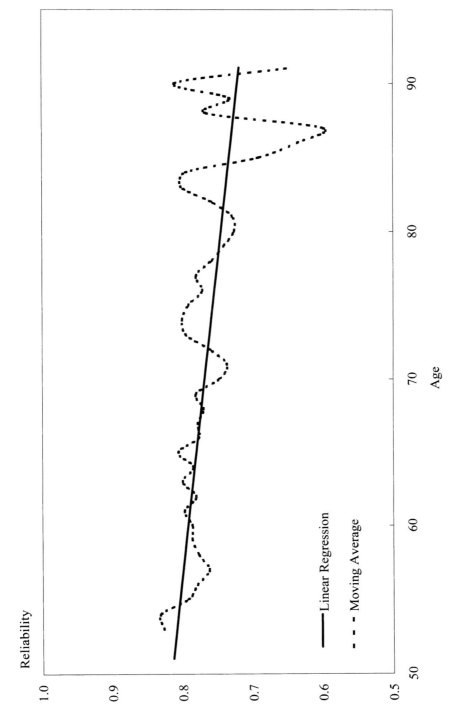

Figure 10.1.   Reliability estimates of a rating of self-reported health by age: Health and Retirement Study.

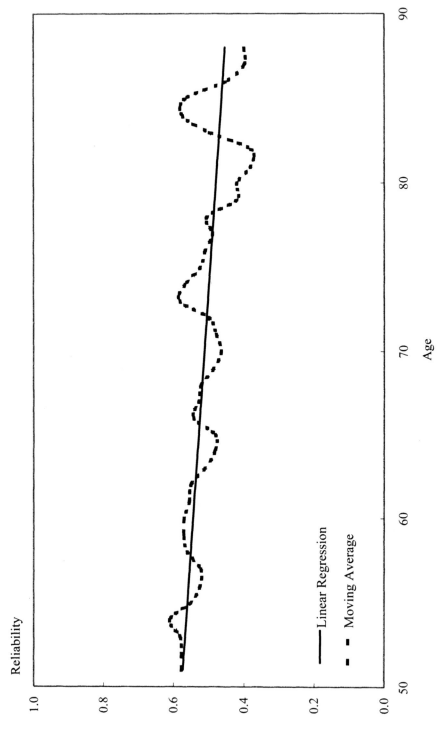

Figure 10.2. Reliability estimates of a rating of expectations about income and inflation by age: Health and Retirement Study.

Table 10.14.   Reliability estimates of selected measures of health for two categories of school-ing: Health and Retirement Study

	High School or Less			More than High School		
Question Content	Sample Size	Pearson Estimate	Polychoric Estimate	Sample Size	Pearson Estimate	Polychoric Estimate
Health						
Rating of current health	9,832	0.713	0.765	5,979	0.711	0.773
Rating of health relative to past (3 categories)	7,773	0.325	0.415	4,329	0.330	0.423
Rating of health relative to past (5 categories)	7,761	0.337	0.395	4,323	0.319	0.390
Sensory Function						
Rating of vision	9,733	0.547	0.603	5,936	0.562	0.624
Rating of distant vision	9,718	0.512	0.568	5,935	0.546	0.612
Rating of near vision	9,707	0.497	0.550	5,936	0.493	0.550
Rating of hearing	9,817	0.655	0.712	5,976	0.689	0.753
Pain						
Trouble with pain	9,810	0.562	0.787	5,977	0.538	0.779
Weight						
Estimate of weight	9,553	0.969	---	5,835	0.973	---

of the reliability of both facts and nonfacts from the six panel studies employed throughout this study (see Table 10.2) produced major differences across groups with respect to levels of schooling. This pattern is not uniformly borne out in the HRS data. In Table 10.11 (see above), we presented reliability estimates for the HRS expectations measures for two categories of schooling—respondents with a high school education or less and respondents with more than a high school education. Although there are systematic differences in the aggregate across these measures, the differences are not large. Similarly, as can be seen in Table 10.14, where we display both Pearson and polychoric-based reliability estimates for the nine HRS physical health and functioning measures, hardly any differences exist among two education groups—high school or less and more than high school. There are a few cases—rating of hearing and vision—where the expected patterns are present, but generally speaking, the hypothesis of education subpopulation differences is not sup-ported insofar as reliability is concerned. There are also some exceptions to this pat-tern among the expectations measures, however, where the differences by schooling level in reliability is substantial, e.g., expectations about income and inflation. The lack of an association with level of schooling may not be the case for the components of variance involved in the reliability estimates, and we therefore pursue this issue of the relationship between reliability, age and education further below.

10.8.3   Age and Components of Variance

Recall from the discussion presented at the beginning of this chapter that reliability for the hth subpopulation is defined as the ratio of two variance estimates, as follows:

$$^{(h)}\rho_g^2 = {}^{(h)}\text{VAR} [T_g] / {}^{(h)}\text{VAR}[Y_g] = ({}^{(h)}\text{VAR}[Y_g] - {}^{(h)}\text{VAR}[E_g]) / {}^{(h)}\text{VAR}[Y_g]$$

or alternatively

$$^{(h)}\rho_g^2 = {}^{(h)}\text{VAR} [T_g] / {}^{(h)}\text{VAR}[Y_g] = {}^{(h)}\text{VAR}[T_g] / ({}^{(h)}\text{VAR}[T_g] + {}^{(h)}\text{VAR} [E_g]).$$

We see from this expression that the key population components estimated from the reliability models are the variance of the errors, i.e., $\text{VAR}[E_g]$ and the variance of the true scores, i.e., $\text{VAR}[T_g]$. We therefore recovered these estimates from our models and examined them separately by age and education. Tables 10.15 and 10.16 present these estimated components of variance, along with the reliability estimates for the self-reported health measure and the measure of expectations about income and inflation by age group in the HRS. Moving averages of the components of variance estimates, along with the best-fitting linear regression lines based on these tables are given in Figures 10.3 and 10.4 for these two measures.

The estimates presented here come from an autoregressive (or simplex) model discussed above estimated using the M*plus* structural equation statistical package, treating the self-rated health variable as ordinal and the expectations variable as continuous. In the case of the self-rated health variable, the estimator used is WLSMV—weighted least squares parameter estimates using a diagonal weight matrix with standard errors and mean- and variance-adjusted chi-square (χ^2) test statistic that uses a full weight matrix (Muthén and Muthén, 2001–2004). These methods were developed based on the paper by Satorra and Bentler (2001).

These displays reveal an interesting pattern of results—a monotonically increasing level of error variance in these measures with age and a declining level of true variance—which combine to produce the declining estimate of reliability for these measures depicted in Figures 10.1 and 10.2 above. Thus, with increasing age the amount of error variance increases, contributing to greater observed variance in the denominator of the reliability estimator, while at the same time the numerator—the true variance of the measure—appears to be shrinking with age. The systematic nature of these trends can be verified through the inspection of Figures 10.3 and 10.4. In the case of self-rated health, the numbers do not fit this pattern beyond the age of 80, where the true variance in self-rated health begins to increase. Overall, at least for these two measures, these trends appear to account for the age-related decrements to reliability identified earlier.

Table 10.15. Estimates of reliability, error variance, and true-score variance of a rating of self-reported health by age for the total sample and two categories of schooling: Health and Retirement Study

Age in 1998	Sample Size	Total			High School or Less			More than High School		
		Reliability	Error Variance	True-score Variance	Reliability	Error Variance	True-score Variance	Reliability	Error Variance	True-score Variance
51	525	0.825	0.159	0.748	0.822	0.202	0.937	0.809	0.088	0.375
52	522	0.871	0.119	0.803	0.890	0.112	0.907	0.816	0.095	0.420
53	448	0.777	0.190	0.664	0.849	0.173	0.969	0.698	0.147	0.340
54	498	0.838	0.135	0.696	0.818	0.185	0.832	0.845	0.078	0.428
55	562	0.764	0.205	0.663	0.773	0.261	0.887	0.735	0.146	0.405
56	606	0.745	0.208	0.608	0.739	0.276	0.780	0.721	0.134	0.345
57	853	0.773	0.186	0.632	0.750	0.266	0.798	0.764	0.106	0.343
58	851	0.803	0.166	0.673	0.778	0.223	0.779	0.778	0.099	0.348
59	815	0.783	0.171	0.618	0.793	0.214	0.819	0.716	0.127	0.319
60	800	0.771	0.171	0.575	0.783	0.214	0.769	0.689	0.130	0.287
61	798	0.838	0.143	0.737	0.822	0.199	0.918	0.865	0.075	0.483
62	730	0.729	0.173	0.466	0.718	0.239	0.607	0.701	0.103	0.242
63	719	0.830	0.127	0.621	0.806	0.197	0.817	0.875	0.052	0.362
64	712	0.796	0.154	0.600	0.762	0.237	0.760	0.834	0.072	0.360
65	623	0.790	0.148	0.558	0.806	0.178	0.739	0.742	0.114	0.329
66	653	0.745	0.168	0.490	0.735	0.243	0.675	0.743	0.095	0.274
67	628	0.802	0.155	0.628	0.798	0.208	0.824	0.767	0.102	0.336
68	552	0.764	0.155	0.503	0.738	0.239	0.672	0.761	0.086	0.273
69	487	0.773	0.184	0.629	0.766	0.249	0.816	0.755	0.118	0.364
70	536	0.708	0.192	0.464	0.624	0.316	0.525	0.781	0.080	0.285
71	504	0.732	0.159	0.433	0.735	0.203	0.562	0.688	0.129	0.285
72	465	0.853	0.110	0.638	0.879	0.126	0.916	0.824	0.072	0.336
73	467	0.804	0.146	0.596	0.756	0.214	0.662	0.907	0.053	0.521
74	430	0.740	0.190	0.541	0.685	0.296	0.643	0.863	0.069	0.433
75	394	0.833	0.119	0.594	0.821	0.162	0.744	0.827	0.060	0.287
76	443	0.747	0.187	0.553	0.762	0.214	0.684	0.723	0.159	0.416
77	417	0.766	0.187	0.612	0.781	0.209	0.746	0.682	0.161	0.345
78	365	0.766	0.217	0.709	0.752	0.267	0.808	0.767	0.137	0.451
79	321	0.683	0.244	0.526	0.690	0.315	0.701	0.592	0.176	0.256
80	321	0.726	0.202	0.535	0.691	0.247	0.551	0.778	0.108	0.381
81	250	0.791	0.205	0.777	0.767	0.270	0.887	0.762	0.125	0.401
82	200	0.771	0.223	0.750	0.652	0.373	0.698	1.000	0.000	0.935
83	187	0.847	0.113	0.625	0.911	0.073	0.748	0.529	0.296	0.333
84	156	0.764	0.218	0.706	0.839	0.174	0.911	0.668	0.250	0.502
85	182	0.468	0.449	0.395	0.510	0.535	0.557	0.407	0.427	0.293
86	134	0.713	0.249	0.619	0.760	0.201	0.636	0.707	0.252	0.607
87	120	0.674	0.291	0.602	0.595	0.420	0.617	---[1]	---	---
88	78	1.000	0.000	1.334	1.000	0.000	1.572	---	---	---
89	77	0.552	0.527	0.649	0.494	0.587	0.574	0.624	0.404	0.672
90	48	0.915	0.059	0.638	1.000	0.000	1.178	0.479	0.521	0.479
91	130	0.607	0.393	0.607	1.000	0.000	1.000	---	---	---
Total	18,607	0.768	0.190	0.630	0.772	0.227	0.786	0.743	0.143	0.391

[1]Estimate not obtained due to insufficient sample size. Cases combined with adjacent age group.

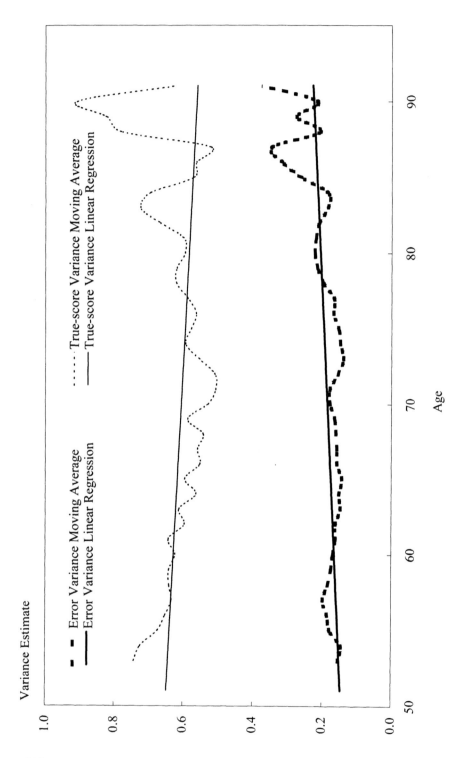

Figure 10.3. Estimates of true and error variance of a rating of self-reported health by age: Health and Retirement Study.

Table 10.16. Estimates of reliability, error variance, and true-score variance of a rating of expectations about income and inflation by age for the total sample and two categories of schooling: Health and Retirement Study

Age in 1998	Sample Size	Total			High School or Less			More than High School		
		Reliability	Error Variance	True-score Variance	Reliability	Error Variance	True-score Variance	Reliability	Error Variance	True-score Variance
51	496	0.555	465.887	582.002	0.339	718.930	369.351	0.772	220.065	747.194
52	497	0.490	567.143	544.700	0.478	616.546	564.974	0.451	513.757	422.667
53	417	0.686	318.479	697.311	0.550	469.568	574.585	0.798	170.435	674.406
54	466	0.567	465.261	609.726	0.583	444.443	621.659	0.538	449.762	523.668
55	532	0.580	445.712	614.304	0.334	640.389	321.221	0.834	178.794	900.326
56	556	0.509	513.981	532.915	0.517	461.673	494.877	0.484	570.510	535.681
57	790	0.484	523.785	492.209	0.487	501.847	476.838	0.435	546.618	420.428
58	796	0.575	435.651	590.381	0.466	538.347	470.341	0.641	333.230	595.343
59	749	0.623	390.312	645.673	0.552	448.555	553.034	0.691	319.667	716.388
60	727	0.512	525.591	551.711	0.413	626.049	440.892	0.633	356.288	614.764
61	729	0.572	454.816	607.640	0.506	501.494	514.630	0.661	347.831	677.322
62	661	0.579	451.571	622.300	0.686	351.256	767.327	0.412	542.048	380.368
63	656	0.498	505.988	502.951	0.430	584.511	441.291	0.561	401.624	513.309
64	658	0.451	553.963	454.641	0.389	595.525	379.182	0.507	486.796	501.209
65	568	0.498	509.228	504.281	0.384	564.433	352.345	0.600	398.569	598.715
66	599	0.498	547.667	543.360	0.408	607.623	419.071	0.604	459.158	700.138
67	585	0.633	408.966	705.823	0.499	571.844	569.449	0.770	233.638	780.503
68	506	0.452	616.347	507.654	0.464	550.988	476.643	0.339	713.031	365.553
69	439	0.478	618.123	566.416	0.331	701.284	346.871	0.664	404.305	799.251
70	486	0.526	548.018	608.623	0.428	638.519	477.353	0.554	443.589	550.909
71	468	0.390	640.669	409.452	0.336	704.448	356.913	0.479	519.328	478.221
72	426	0.521	516.174	561.094	0.396	570.954	374.875	0.651	429.633	803.036
73	417	0.589	443.132	634.757	0.578	466.259	637.613	0.585	398.615	560.826
74	396	0.637	385.926	677.021	0.481	544.414	504.998	0.861	136.549	846.029
75	357	0.471	596.489	530.855	0.356	732.479	405.319	0.549	397.157	483.056
76	401	0.462	627.960	539.795	0.514	569.286	602.462	0.409	627.885	433.884
77	374	0.595	464.983	683.883	0.444	604.292	482.095	0.809	229.103	973.141
78	333	0.410	627.625	436.331	0.449	600.100	489.311	0.369	605.043	354.191
79	284	0.509	540.785	560.666	0.513	529.078	558.035	0.443	578.200	459.512
80	282	0.340	772.908	398.350	0.200	880.162	220.615	0.543	526.953	627.162
81	218	0.409	695.154	480.826	0.448	662.247	536.802	0.388	668.532	423.235
82	160	0.390	669.641	428.176	0.218	728.357	202.986	0.578	498.242	682.870
83	164	0.340	713.682	368.114	0.235	798.583	245.181	0.562	421.271	541.550
84	130	0.761	249.494	793.990	0.677	315.306	660.104	0.886	137.070	1,060.071
85	148	0.621	445.122	729.873	0.394	770.053	500.659	---[1]	---	---
86	112	0.317	837.063	389.376	0.502	565.474	569.879	0.208	895.747	235.398
87	86	0.419	716.809	517.425	0.279	810.061	313.496	0.481	474.868	440.703
88+	235	0.460	725.343	618.809	0.536	605.281	700.358	0.293	878.858	363.887
Total	16,904	0.524	518.082	569.666	0.451	582.692	477.774	0.581	430.042	596.220

[1] Estimate not obtained due to model failure.

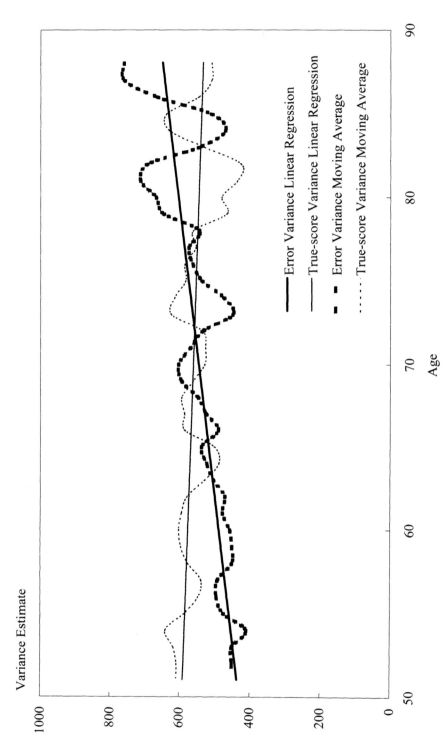

Figure 10.4. Estimates of true and error variance of a rating of expectations about income and inflation by age: Health and Retirement Study.

10.8.4 Education and Components of Variance

We conjectured above that, contrary to our reports of differences in reliability estimates across subsamples defined by level of schooling earlier in the chapter (for the NES and ACL panels), our analysis of the HRS measures did not appear to provide much support for the hypothesis of education subpopulation differences in reliability. This conclusion is supported for the measure of self-rated health; however, as noted earlier, it is not true for the estimates of reliability for the measure of expectations about income and inflation. These strikingly different sets of results can be seen in Figures 10.5 and 10.6, which present the estimates of reliability for the two levels of schooling groups, for self-rated health and expectations about income and inflation, respectively. Figure 10.5 shows very few differences in reliability by schooling level, and Figure 10.6 shows more differences, consistent with Table 10.11, where the education group difference in reliability is .45 for the low education subsample and .58 for the high education subsample. In the case of self-rated health, the difference was .71 versus .76 by education.

Tables 10.15 and 10.16 also present components of observed variance by age for the two education groups—high school or less (low education) and more than high school (high education). As can be seen in these tables, the case base for the high education group in some of the early-born cohorts was relatively sparse. We tested whether the estimates were equal across the two education groups shown in Table 10.15 for the measure of self-rating health using a matched-pairs test (see Blalock, 1972, p. 233) in order to get a rough idea of how different the estimates were across groups. This test essentially employs a direct pair-by-pair comparison across age groups, examining the significance of the difference between the two estimates under the null hypothesis that the mean of the pair-by-pair differences is zero. The t statistics for the differences are as follows: reliability (n = 38, mean = .022, t = .941, p = .353), error variance (n = 38, mean = .091, t = 4.042, p = .000), and true-score variance (n = 38, mean = .373, t = 12.446, p = .000). These results indicate that, while there are few education group differences in the reliability estimates, as we surmised from the results presented in Tables 10.11 and 10.14, there are important differences by education groups in the true-score variances, and error variances for the two schooling groups. These patterns are shown in Figures 10.7 and 10.8 for the estimates of true variance and error variance in the measure of self-reported health, respectively.

The story is much the same for the measure of expectations about income and inflation, although different in the case of schooling differences in reliability. The application of the direct pair-by-pair comparison across age groups for the estimates given in Table 10.16 revealed a significant difference between the two estimates in all three cases under the null hypothesis that the mean of the pair-by-pair differences is zero. The t statistics for the differences are as follows: reliability (n = 37, mean = .125, t = .3.777, p = .001); error variance (n = 37, mean = 143.455, t = 4.275, p = .000); and true-score variance (n = 37, mean = 115.998, t = 2.840, p = .007). These results indicate, again, that for this measure of expectations about income and inflation there are important differences across the groups defined by level of schooling, not only in reliability, but also in the true-score and error variances. The patterns of estimated true variance and error variance for the measure of expectations about income and inflation are shown in Figures 10.9 and 10.10, respectively.

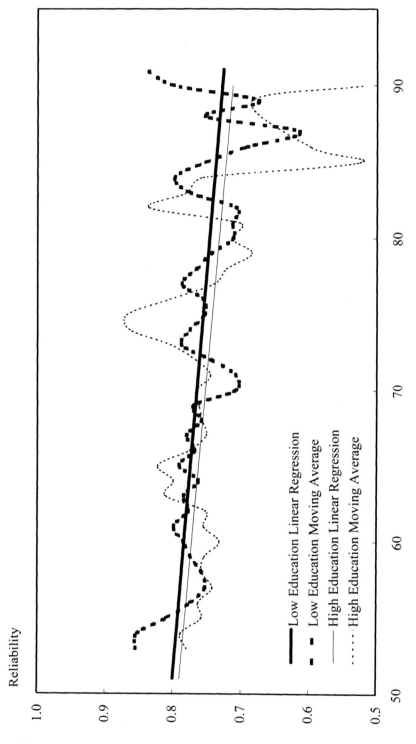

Figure 10.5. Reliability estimates of a rating of self-reported health by age for two categories of schooling: Health and Retirement Study.

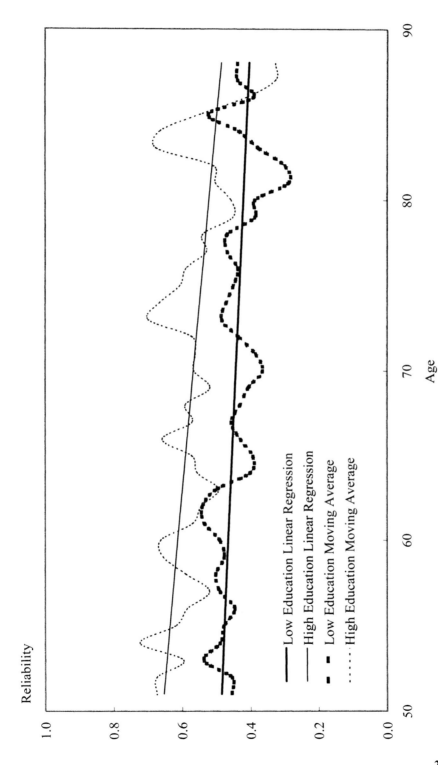

Figure 10.6. Reliability estimates of a rating of expectations about income and inflation by age for two categories of schooling: Health and Retirement Study.

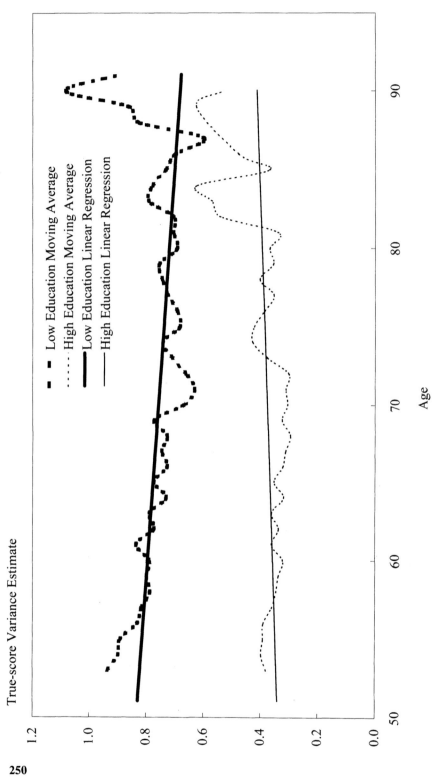

Figure 10.7. Estimates of true variance of a rating of self-reported health by age for two categories of schooling: Health and Retirement Study.

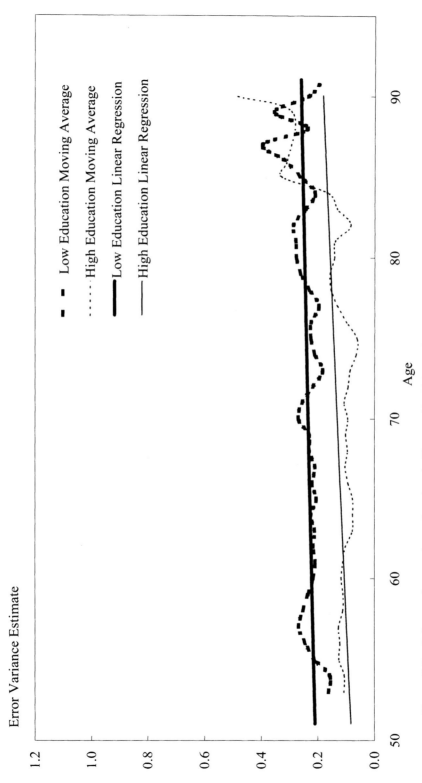

Figure 10.8. Estimates of error variance of a rating of self-reported health by age for two categories of schooling: Health and Retirement Study.

True-score Variance Estimate

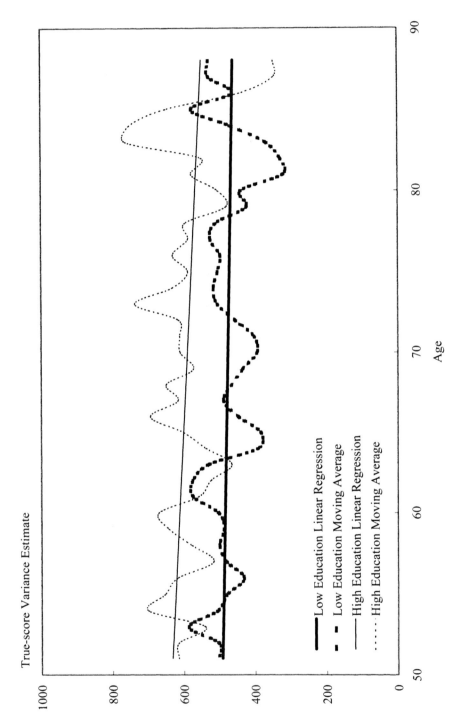

Figure 10.9. Estimates of true variance of a rating of expectations about income and inflation by age and years of schooling: Health and Retirement Study.

Error Variance Estimate

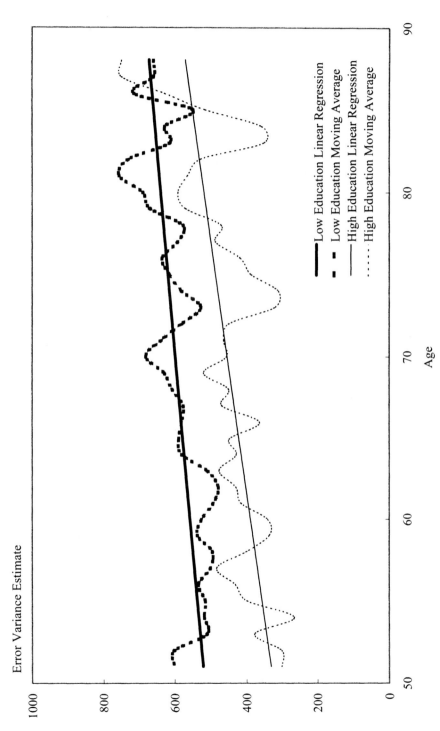

— Low Education Linear Regression
■ ■ Low Education Moving Average
—— High Education Linear Regression
······ High Education Moving Average

Age

Figure 10.10. Estimates of error variance of a rating of expectations about income and inflation by age and two categories of schooling: Health and Retirement Study.

10.9   TESTS OF HYPOTHESES ABOUT GROUP DIFFERENCES[2]

Although we have spent considerable time in this chapter dealing with descriptive differences in reliability and components of variance estimates across groups defined by age and education, the question of group differences in estimates of variance can be addressed somewhat more formally for all the HRS variables considered here. There are standard methods within the SEM framework for examining differences between the components of variance across groups. The most basic and conclusive of these tests is the hypothesis of the *equality of covariance structures* (see Jöreskog, 1971b, p. 419):

$$H_0 : \Sigma_1 = \Sigma_2 = \cdots = \Sigma_H$$

(where, as above, H is the number of subpopulations—not to be confused with H_0, the standard notation for "null hypothesis"). If the examination of the hypothesis of a common covariance structure across all subpopulations yields a conclusion that the covariance structures are equivalent, then it follows that the parameters of models generating these covariances may be assumed to be equivalent. There may be a large number of such models, but whatever model one chooses, there is a common set of parameter estimates that will fit the data across groups. The approach in this case is to fit a model with .5 P (P + 1) parameters across the H subpopulations (recall that P is the number of occasions of measurement). The test statistic for the null hypothesis in this case is a function of departures in the fit of the common covariance structure to the subpopulations. The test statistic for the null hypothesis in this case has degrees of freedom equal to .5(H − 1) P (P + 1) and is distributed as χ^2. If the value of this test statistic is large and judged to be statistically significant, the hypothesis of the equality of covariance structures may be rejected—the conclusion being that some components of true-score and measurement error variance estimated from the observed covariance structure are meaningfully different across subpopulations. It does not, however, necessarily follow from this that there are subpopulation differences in reliability. However, if the hypothesis cannot be rejected, the common covariance structure can be said to adequately fit the data in the H subpopulations, a conclusion that is consistent with the idea that reliability of measurement is the same across groups.

Recall from the above that reliability (for the gth measure in the hth subpopulation) is defined as a ratio of variances, that is, $^{(h)}VAR [T_g] / ^{(h)}VAR[Y_g] = {}^{(h)}(VAR[Y_g] - ^{(h)}VAR[E_g]) / ^{(h)}VAR[Y_g]$. And, returning to the development of the quasi-simplex model in Chapter 5, where we developed the rationale for the longitudinally-based estimates of reliability, we can see that the quantity $^{(h)}VAR[E_g])$—the measurement error variance for measure g in subpopulation h—is defined as[3]

---

[2]The author acknowledges the important contributions of Ryan McCammon to the development of the approach employed in this part of the chapter.

[3]In the equation that follows, we have dropped the notation designating that these quantities refer to the gth measure, and have included numerical subscripts here that refer to occasions of measurement.

$$^{(h)}VAR(E_t) = {}^{(h)}VAR(Y_2) - [{}^{(h)}COV(Y_2,Y_3) / [{}^{(h)}COV(Y_1,Y_3) / {}^{(h)}COV(Y_1,Y_2)]]$$

Thus, if we cannot reject the hypothesis of equality of covariance structures, then we can easily draw the conclusion of equal measurement error variances, and from this the conclusion of equal reliability.

In the case where we must reject the hypothesis of equality of covariance structures, there are several alternative interpretations. One focuses on the equality/inequality of measurement error variances across subpopulations (in contrast to equal reliabilities). Note from the above formulation that the measurement error variance for the three-wave Wiley-Wiley model used here, i.e., $VAR(E_t)$, is defined entirely in terms of the observed variance at time 2, i.e., $VAR(Y_2)$, and the three covariance terms, $COV(Y_2,Y_3)$, $COV(Y_1,Y_3)$, and $COV(Y_1,Y_2)$. From this, one can reason that there is an additional model that can be tested that follows from the logic of the above development. This is a model in which the measurement error variances are constrained equal across groups. Thus, if the conclusion is reached that a common covariance structure cannot be fit to the H subpopulations, it is possible to examine the hypothesis of equal error variances.

A second alternative interpretation in the case where we must reject the hypothesis of equality of covariance structures is that while the measurement error variances are not equal across groups, the ratios of true variances to observed variances (i.e., reliabilities) are equal. Because reliability is a unique quantity, expressing a ratio of variances, and because it is not possible within presently available SEM software to implement constraints involving equalities among ratios of quantities across populations, we must pursue the examination of this hypothesis using the Heise version of the simplex model in which we produce estimates of reliability using standardized variables (see below).

In the analysis presented here, we present the results for three models: (1) the equal covariance or fully constrained model, (2) the unequal error variance model, and (3) the equal reliability model. We also present a comparison of the first two models as a way of testing for unequal measurement error variances. We examined these cross-group constraints using, as we have throughout this book, a six-parameter just-identified quasi-simplex model (refer to Figure 5.3 in Chapter 5 and the discussion surrounding it), in which the parameters involved were as follows: $VAR(Z_1)$, β_{21}, $VAR(E_t)$, β_{32}, $VAR(Z_2)$, and $VAR(Z_3)$. Recall that the key assumption of the model is that the measurement error variances are constrained equal across waves, that is, $VAR(E_1) = VAR(E_2) = VAR(E_3)$. Hence, we refer to the estimated measurement error variance as $VAR(E_t)$.

The equality of covariances model can be implemented within this framework by setting these six parameters equal across groups. In the particular case of 41 groups defined by age in the HRS (e.g., in the case of respondent's weight), there are $.5\ P\ (P + 1) \times H = 246$ unique variances and covariances across the 41 groups. The model uses 6 of the available pieces of information, leaving 240 degrees of freedom, or $df = .5(H - 1)\ P\ (P + 1)$. In the standard variance/covariance case, the same result would be achieved if we actually had a model, as implied by the above discussion, in which the actual variances and covariances are constrained across groups (see Alwin and Jackson, 1979, p. 100).[4]

Within the framework of the quasi-simplex model employed here, the "unequal error variance" case, the parameters $VAR(Z_1)$, β_{21}, $VAR(Z_2)$, β_{32}, and $VAR(Z_3)$ are all constrained equal across groups, whereas a unique $^{(h)}VAR(E_t)$ is estimated for each subpopulation. This can be thought of as constraining the true-score variances equal across subpopulations, since these parameters combine to produce the true-score variances as follows (see Chapter 5):

$$VAR(T_1) = VAR(Z_1)$$
$$VAR(T_2) = \beta^2_{21} VAR(T_1) + VAR(Z_2)$$
$$VAR(T_3) = \beta^2_{32} VAR(T_2) + VAR(Z_3)$$
$$= \beta^2_{32} [\beta^2_{21} VAR(T_1) + VAR(Z_2)] + VAR(Z_3)$$

This model is thus implemented by allowing the $^{(h)}VAR(E_t)$ to be unconstrained across groups, leaving the remaining five parameters constrained equal across groups. The model therefore estimates the five parameters—$VAR(Z_1)$, β_{21}, $VAR(Z_2)$, β_{32}, and $VAR(Z_3)$—constrained to be the same across the 41 groups, plus $^{(h)}VAR(E_t)$, where $H = 41$. The model therefore estimates $Q = 5 + 41 = 46$ parameters, and the model therefore has 200 degrees of freedom.

Given the logic of the argument developed above, we may think of model 1 as nested within model 2, and we present information on the comparison of these two models. The fit statistic for this comparison in the standard variance/covariance case is simply the difference between the fit of the fully constrained, or "equality of covariance" model, and the fit of the unequal error variance model. The degrees of freedom associated with the comparison is simply the difference between the degrees of freedom associated with the two models, that is, $5(H - 1) P (P + 1) - Q$, i.e., $240 - 200 = 40$.

In the case of the "equal reliability" model a single $VAR(E_t)$ is estimated, constrained equal across subpopulations, but all of the other parameters—$VAR(Z_1)$, β_{21}, $VAR(Z_2)$, β_{32}, and $VAR(Z_3)$ are unconstrained across groups. These constraints are implemented on the correlation matrix instead of the covariance matrix. The logic behind this model is that we estimate a model in which the error variances were constrained equal across groups, but the remaining parameters were allowed to vary, and given the standardized metric of the variables the error variance is a proportion of variance set equal across groups, hence equal reliabilities. This model estimates 1 error variance, $VAR(E_t)$, plus $^{(h)}VAR(T_1)$, $^{(h)}\beta_{21}$, $^{(h)}VAR(Z_2)$, $^{(h)}\beta_{32}$, and $^{(h)}VAR(Z_3)$ in the standardized case, where again $H = 41$. Such a model therefore estimates $1 + (5 \times 41) = 206$ parameters, leaving 40 degrees of freedom.

In Tables 10.17 and 10.18 we present the results that test for age-related differences in the variance components of reliability models within the framework of

---

[4]In the foregoing discussion, the degrees of freedom were calculated under the assumption that there were 41 groups. This is not the case for all variables in our subsequent analyses, and not all such models will have 240 degrees of freedom.

structural equation modeling for the physical health measures and expectation measures, respectively. As described above, we here present the results for four tests: (1) the equal covariance or fully constrained model (model 1), (2) the unequal error variance model (model 2), (3) the comparison of models 1 and 2, and (4) the equal reliability model. As noted above, given the logic of the argument developed above, we may think of model 1 as nested within model 2, and the tables therefore present information on the comparison of these two models.

For each of these models we present the sample size, the χ^2 value associated with the minimum of the likelihood function for a given fitted model, the degrees of freedom of the model, the ratio of χ^2 / df, and the probability of a type I error (using the χ^2 probability distribution, given the degrees of freedom). In all cases the alternative hypothesis is the fully saturated model (i.e., $\chi^2 = 0$ with no degrees of freedom); that is, the statistical test in each case may be thought of as a difference between two models, the fitted model and the fully saturated model. A caveat needs to be added at this point regarding sample size. It is generally known that fit statistics in SEM models are partially a function of sample size. Thus, with very large samples (as in the present case) it is quite often the case that the statistical tests have far too much power against trivial alternative hypotheses. With large sample sizes, one may be forced to reject quite reasonable null hypotheses, simply because the sample size is large.

The logic of the above development is straightforward in the case of continuous measures of continuous latent variables—these are variables for which standard Pearson-based correlations and covariances are defined (see Chapter 3). There are, however, some complications that arise in the case of categorical measures of continuous latent variables. As noted above, for ordinal variables we employ estimates from an autoregressive (or quasi-simplex) model using the M*plus* structural equation statistical package, treating the variables as ordinal. The estimator used is WLSMV— weighted least square parameter estimates using a diagonal weight matrix with standard errors and mean- and variance- adjusted chi-square (χ^2) test statistic that uses a full-weight matrix (Muthén and Muthén, 2004). The goodness-of-fit information provided from this approach does not conform to the standard variance/covariance approach which is involved when there are continuous measures of continuous latent variables, as the fit statistics and degrees of freedom do not "add up" for nested models. The basis for the calculations involved are discussed by Muthén and Muthén (2001–2004, App. 4, pp. 19–20). Essentially, there is no good intuitive interpretation of the fit statistics and the degrees of freedom produced by the M*plus* statistical package. The only available interpretation is that the χ^2 test statistic and its degrees of freedom are data dependent because they draw on the estimates, their derivatives, and the asymptotic covariance matrix of the sample statistics with the aim of choosing the degrees of freedom that give a trustworthy χ^2-based p-value. What this means is that for the ordinal variable case (which involves 8 of the 9 variables analyzed in this section), the equality of covariances test cannot be neatly partitioned into two additive components, as above. One can obtain goodness-of-fit statistics and degrees of freedom, and a p-value, but these are data dependent and are not (directly) comparable to anything else (see Muthén and Muthén, 2001–2004, App. 4, pp. 19–20). It

Table 10.17. Comparison of models involving cross-group constraints on variance components in reliability models for selected measures of health: Health and Retirement Study

	Sample Size	Goodness of Fit Information															
		Model 1 Fully Constrained				Model 2 Unequal Error Variances				Model 1 vs. Model 2				Model 3 Equal Reliability			
Question Content		χ^2	df	χ^2/df	p-value	χ^2	df	χ^2/df	p-value	χ^2	df	χ^2/df	p-value	χ^2	df	χ^2/df	p-value
Health																	
Rating of current health	18,607	616.9	250	2.5	0.000	424.8	246	1.7	0.000	324.5	39	8.3	0.000	102.4	40	2.6	0.000
Rating of health relative to past (3 categories)	18,248	441.7	208	2.1	0.000	335.7	171	2.0	0.000	119.3	40	3.0	0.000	174.2	40	4.4	0.000
Rating of health relative to past (5 categories)	18,248	585.5	307	1.9	0.000	---[1]	---	---	---	---	---	---	---	117.5	36	3.3	0.000
Sensory Function																	
Rating of vision	18,583	648.5	307	2.1	0.000	484.2	276	1.8	0.000	238.3	40	6.0	0.000	120.8	40	3.0	0.000
Rating of distant vision	18,579	443.8	310	1.4	0.000	356.4	276	1.3	0.001	109.5	39	2.8	0.000	58.0	39	1.5	0.026
Rating of near vision	18,577	495.3	320	1.5	0.000	382.0	285	1.3	0.000	143.9	40	3.6	0.000	69.9	40	1.7	0.002
Rating of hearing	18,607	419.1	264	1.6	0.000	311.0	251	1.2	0.006	180.1	40	4.5	0.000	91.8	40	2.3	0.000
Pain																	
Trouble with pain	18,608	104.3	99	1.1	0.338	71.4	77	0.9	0.658	46.5	40	1.2	0.221	281.3	40	7.0	0.000
Weight																	
Estimate of weight	18,518	1,392.2	240	5.8	0.000	841.2	200	4.2	0.000	551.0	40	13.8	0.000	215.9	40	5.4	0.000

[1]Estimate not obtained due to model failure.

would be nice to have survey data that are continuous in all cases, but the exigencies of categorical measures of latent continuous variables require us to adopt these more rigorous approaches, even though they are more difficult to interpret.[5]

With these caveats and understandings in mind, we can provide an interpretation of the results in Table 10.17. In all cases, save one—the measure of "trouble with pain"—taking the test statistics at face value, we must reject the null hypotheses of equal reliability and equal measurement error variances. Evaluating the goodness of fit of these models strictly in terms of statistical criteria, we are in all other cases required to reject the hypothesis of no difference in errors of measurement by age group—whether we consider the equality of covariances model (model 1), the unequal error variances model (model 2), or the equal reliability model (model 3). Furthermore, when we examine the test of the inequality of error variances (i.e., the comparison of the fully constrained model and the unequal error variances model—model 1 versus model 2), we are forced to conclude that there are differences among the age groups, even when we constrain measurement errors to be equal across groups. In other words, according to these statistical tests, everything varies by age group!

These results are reinforced by the consideration of the results in Table 10.18 for 12 measures of expectations about selected topics. Here, again, in the majority of cases we must reject the null hypotheses of equal reliability and equal error variances. There are a few instances where we fail to reject equal reliabilities—expectations about working after age 65 and the limitations on work due to health factors—but in virtually all cases we must reject the equal error variances hypothesis. Our purchase on the latter test is based on the comparison of fit statistics for model 1 versus model 2, and these results produce a resounding conclusion that we must reject the null hypothesis. For only one of these measures—expectations about finding a new job—is there little support for differences in variance components across age groups.

We noted earlier that these tests are probably too powerful against even trivial alternatives, so we must perhaps put substantial weight on our descriptive look at the data, at least as much as these statistical tests. There is every reason to perform these formal tests, and there is a great deal to be learned from implementing them. At the same time, the sample sizes involved in most cases here are extremely large (more than 18,500 cases), and as a consequence the odds are tipped in favor of rejecting just about any null hypothesis. There are some aspects of these results that are highly instructive. First, note that for virtually all variables considered in this analysis, there is a better fit per degree of freedom (using χ^2 / df as an indicator of relative fit) for the "unequal error variances" model than the other models. The χ^2 / df ratio is substantially under 2.0 (and in many cases under 1.5) for the "unequal error variances" model, whereas the same χ^2 / df ratio is almost 4 times that in the case comparison of model 1 versus model 2, suggesting that the unequal error variances model is relatively better. These relative fit ratios are consistently lower for the "unequal error variance" model compared to the fully constrained model as

---

[5]We consider the case of categorical measures of latent categories in the next chapter.

Table 10.18. Comparison of models involving cross-group constraints on variance components in reliability models for ratings of expectations about selected topics: Health and Retirement Study

Expectations	Sample Size	Model 1 Fully Constrained				Model 2 Unequal Error Variances				Model 1 vs. Model 2				Model 3 Equal Reliability			
		χ^2	df	χ^2/df	p-value	χ^2	df	χ^2/df	p-value	χ^2	df	χ^2/df	p-value	χ^2	df	χ^2/df	p-value
Income and inflation	16,904	299.6	222	1.3	0.000	174.0	185	0.9	0.708	125.6	37	3.4	0.000	55.3	37	1.5	0.027
Leave heirs $10K	17,148	479.5	222	2.2	0.000	204.7	185	1.1	0.153	274.8	37	7.4	0.000	77.0	37	2.1	0.000
Leave heirs $100K	12,543	253.6	222	1.1	0.071	161.9	185	0.9	0.889	91.7	37	2.5	0.000	86.6	37	2.3	0.000
Lose your job	3,503	260.0	108	2.4	0.000	171.5	90	1.9	0.000	88.5	18	4.9	0.000	28.0	18	1.6	0.062
Find new job	3,504	97.7	108	0.9	0.751	73.5	90	0.8	0.898	24.3	18	1.3	0.146	32.4	18	1.8	0.020
Work after 62	2,594	68.9	42	1.6	0.006	46.4	35	1.3	0.094	22.4	7	3.2	0.002	24.4	7	3.5	0.001
Work after 65	1,747	46.2	42	1.1	0.304	23.9	35	0.7	0.922	22.3	7	3.2	0.002	6.1	7	0.9	0.526
Health limits work	4,721	183.7	126	1.5	0.001	100.5	105	1.0	0.607	83.2	21	4.0	0.000	25.0	21	1.2	0.247
Move next two years	8,388	286.1	138	2.1	0.000	208.1	115	1.8	0.000	78.0	23	3.4	0.000	37.2	23	1.6	0.031
Live to 75 or more	6,716	92.4	60	1.5	0.005	57.7	50	1.2	0.213	34.7	10	3.5	0.000	19.0	10	1.9	0.040
Give financial help	17,175	367.6	222	1.7	0.000	267.2	185	1.4	0.000	100.3	37	2.7	0.000	141.4	37	3.8	0.000
Go to nursing home	7,840	791.8	138	5.7	0.000	386.0	115	3.4	0.000	405.8	23	17.6	0.000	37.0	23	1.6	0.033

well—the model that assumes equal error variances and equal true-score variances. In other words, relatively speaking, it appears that we can fit the data better if we allow the error variances to be unconstrained across age groups, compared to any alternative. It is clear from the descriptive presentation above that it also appears important to consider the true-score variances to be different with respect to age as well, but to constrain them equal probably relinquishes a lesser loss of fit than is the case in constraining measurement errors equal.

10.10 CONCLUSIONS

I began this chapter with the observation that because it is defined as the ratio of two population variances, reliability can only be defined with respect to a particular population and its attributes. Reliability, by definition, is a function of the true variance and the observed variance (i.e., the latter being the sum of true and measurement error variances)—two population quantities. I argued that while traditionally reliability is thought of primarily as a property of the measures, consideration of the theory on which the concept is based reveals that it is inherently an attribute of the population of interest, and the goal of research is to estimate these population values from sample data. These observations motivate the investigation of differences in reliability across subpopulations where we expect differences in the relevant quantities reflecting true-score and error variances. I suggested that age and level of schooling are two variables defining subpopulations of interest that serve as a starting point for the investigation of differences in patterns of reliability.

In the case of schooling, we find mixed support for the idea that survey reports vary with respect to reliability for respondents with differing levels of schooling. Our analysis of the six panel data sets used as the main basis for this study—the three NES panels, the ACL panel, and the two SAF panels—produce results that are consistent with the overall finding in the literature that less educated respondents produce a greater amount of response error (e.g., Sudman and Bradburn, 1974; Alwin and Krosnick, 1985). The analysis of data on health measures from the Health and Retirement Study, however, shows no such differences. Clearly, future research should take up this issue—perhaps the factors contributing to less reliability in some instances have as much to do with the measures involved as they do with the level of respondents' schooling.

In the case of age, although there are strong theoretical reasons to expect that cognitive decrements known to accompany aging may impair reliable reporting of information in survey research, studies comparing rates of measurement error across age groups in actual surveys have generally not been able to demonstrate that aging is a major problem in routine use of surveys (e.g., Rodgers and Herzog, 1987a). Despite these findings, measurement error continues to be an issue of considerable interest to students of aging. The conclusions of the present research, however, seem to support the view that aging is relevant to the study of measurement error, in that there seems to be some support for the hypothesis that aging contributes to greater errors of measurement in older populations, even once differences in processes related to schooling are taken into account. In the analysis presented here, among factual measures there are virtually no detectable differences in reliability by age. Among non-factual measures, particularly measures of attitudes, beliefs, and self-descriptions, however, there is a clear pattern in the unadjusted reliabilities that those in the oldest age group may produce more measurement errors—in three of four studies. Reliability is generally highest during midlife—ages 34–65—with those either younger or older tending (although not always) to have lower levels of reliability. When we take into account age-education and education-reliability relationships, we can reduce or remove the distinctiveness of the older age pattern of lower reliability, but in two studies once age differences in schooling experiences are taken into account, lower levels of reliability in the oldest age groups persist.

In addition, our analyses of the Health and Retirement Study data show some meaningful differences in components of reliability by age—not in all cases—but, in some instances there are systematic age differences in components of true-score and measurement error variance, viz., measures of self-rated health. These results call for further examination and replication before they can be generalized, but they do suggest that there are potentially increases in measurement error with age. Our approach to these data has focused on decomposing reliability into components of true-score variance and measurement error variance. This approach has been very informative in that we have been able to show that where reliability may not differ (e.g., across education groups), its components do. Specifically, we found evidence to support the idea that there were few if any differences in reliability by education categories for the HRS health measures, but there were substantial differences in true-score variance and measurement error variance, conclusions that provide a more interesting set of possible explanations for age patterns in measurement error. In light of these findings, I would especially urge researchers studying age-related differences in measurement error to take schooling differences and other cohort experiences into account before concluding that cognitive decrements due to aging create problems for survey measurement. In the meantime, given the possibilities of measurement error differences among younger and older respondents in some domains, it is still important to examine patterns of measurement error variance, where possible. Finally, given the impact of levels of measurement error on the robustness of statistical analysis in social and behavioral research (see the development in Chapter 12), if there are differences in measurement error across age groupings, it is important to take these differences into account in the analysis of data.

CHAPTER ELEVEN[1]

Reliability Estimation for
Categorical Latent Variables

> . . . the notion of reliability only makes obvious sense when the latent classes can be
> put into a one-to-one correspondence with the manifest classes.
> Lee M. Wiggins, Panel analysis (1973, p. 26)

The foregoing discussions and analyses have applied methods for assessing reliability in cases where the measures are thought to assess continuous latent variables. In the case where the latent variable of interest is *not* continuous, but rather categorical (i.e., a set of discrete categories), other methods of reliability estimation must be employed. In Chapter 6 an array of approaches to reliability estimation was displayed for several types of models using longitudinal data; see Table 6.4. That table cross-classified the nature of the observed and latent variables, which I associated with particular estimation strategies considered to be the most appropriate for the particular cases involved. If the observed variable is categorical in the sense that the response categories cannot be ordered along a continuum (e.g., sex, region, or race), then there is little question that the underlying latent variable is composed of categories. The latent variable in such cases is composed of latent categories or "latent classes." On the other hand, if the observed variable is seemingly categorical, but can be ordered along an ordinal or interval continuum, then an argument can be made for using continuous latent variable approaches (see Chapters 3 and 6 above) and polychoric correlation methods may be appropriate (see Jöreskog, 1994). Even in this case, some analysts would prefer a categorical variable approach to estimation and analysis rather than make these assumptions. In the previous chapters we discussed the continuous latent variable approaches at length, including both ordinal and continuous/interval observed variables, but indicated along the way that there were more appropriate approaches to estimating reliability for measures of latent variables that are truly categorical, where the traditional CTST assumptions cannot

---

[1]I acknowledge the collaboration of Ryan McCammon and Jacob Felson in the development of this chapter.

be met. The estimation of measurement reliability for categorical latent variables, or "latent classes," is the topic of this chapter.

There are several intuitively based approaches to defining reliability in the case of categorical variables, but many of these are inadequate. One might think, for example, that the percentage of cases that agree in response disposition for a set of individuals across two occasions of measurement might be a sensible approach to reliability estimation. Such approaches have been used in response to the obvious inappropriateness of applying methods of reliability estimation for continuous variables and/or the perceived lack of available reliability models for categorical data. The approach that computes the percent agreement among two or more measures of the same latent variable is problematic because the extent of agreement clearly depends on the number of categories involved (see Alwin, 1992, p. 105, fn. 19). It is more productive to approach the problem within the framework of a model that specifies observed variables as reflections of latent variables and a measurement design that provides a means of estimating the relationship between observed and latent variables.

11.1 BACKGROUND AND RATIONALE

Early in this project, we made a calculated decision to restrict the main focus of the study to measures of continuous latent variables, for which the measures were at least ordinal in character. We rationalized this decision in part on the grounds that linear correlational models are not generally appropriate for categorical data and that it was inappropriate to employ methods derived from the CTST model to reliability where the latent variables are categorical. In addition, we felt the literature on statistical estimation of measurement reliability for categorical latent variables was not as fully developed as was the case for continuous latent variables. However, in the meantime, several advances have been made, primarily in the availability of software that can estimate latent class models; and while our analyses in previous chapters have focused exclusively on continuous latent variables, it is important to review methods of reliability estimation for categorical latent variables that are compatible with the approach taken here.

It is somewhat ironic that methods of reliability estimation using longitudinal data are based on concepts (if not procedures) that were originally conceptualized in terms of categorical data. Coleman's (1964, 1968) now-classic work on the measurement of change, which dealt with the separation of change from unreliability of measurement, employed an example involving categorical data (see Coleman, 1968, pp. 472–73). The conceptual apparatus advocated in Coleman's work was further developed for the continuous latent variable case by Heise (1969), Jöreskog (1970), and Wiley and Wiley (1970), whose work is the mainstay of the longitudinal approach advocated in previous chapters (see Chapter 5). This model has seen considerable application, given that the statistical identification issues were readily dealt with using structural equation modeling strategies. Wiggins (1973) developed this model for the categorical case, but despite the existence of latent class models for many years (Lazarsfeld and Henry, 1968), work on the estimation of the parameters of latent change models occurred relatively recently (see van de Pol and de Leeuw, 1986; Bassi, Hagenaars, Croon, and Vermunt, 2000; Langeheine and van de Pol, 2002). Although efforts to develop estimates of composite reliability (analogous to coefficient α) for use with latent class models have emerged (see Bartholomew and Schuessler, 1991), the problem of estimating reliability for individual survey measures of latent classes has been relatively neglected.

Several streams of recent developments have the potential to deal with estimating reliability of individual survey questions where the latent variable being measured is a set of latent classes: (1) a static latent class approach involving multiple indicators (Clogg and Manning, 1996; Flaherty, 2002); (2) a latent transition Markov model for single variables (Wiggins, 1973; van de Pol and de Leeuw, 1986; Langeheine and van de Pol, 2002); and (3) a latent transition Markov model for multiple indicators (Collins, 2001; Collins and Flaherty, 2002; Collins, Fidler, and Wugalter, 1996; Flaherty, 2002). In the following sections, we discuss the first two of these, both of which provide a straightforward approach to assessing the reliability of single survey questions, in a manner consistent with the theoretical underpinnings of this book.

11.2   THE LATENT CLASS MODEL FOR MULTIPLE INDICATORS

The first of the models considered here, formulated by Clogg and Manning (1996), is a straightforward application of latent class analysis to *multiple indicators* of a common latent variable (Clogg, 1995; Clogg and Goodman, 1984; Lazarsfeld and Henry, 1968; Goodman, 2002). What is particularly valuable about their contribution is the discussion of various ways in which reliability can be operationalized within standard latent class models, including their attempt to produce information on question-specific reliabilities for the measurement of latent classes. The latter fits well with the present agenda for generating estimates of reliability for individual survey questions; however, the question of the compatibility of these approaches with traditional CTST conceptions of reliability is one that has not been thoroughly examined. Although we do not undertake such an examination here, we explore their use in the estimation of question-specific reliabilities.

As we shall see, the latent class model proposed by Clogg and Manning (1996) is analogous to a single factor model in the case of continuous variables, where multiple indicators of a single latent variable are used to estimate the reliability of measurement (see Chapter 4, section 4.1). Thus, from the perspective of this book, a potential problem with the Clogg-Manning application of their model to the estimation of reliability is that it fails to make the distinction between "multiple-measures models" and "multiple-indicators models" (as discussed in Chapter 4). One might argue therefore that the Clogg-Manning model does not necessarily provide an estimate of reliability when the measures are not replicates of one another. However, as we illustrate subsequently using an example, where one has longitudinal reports of non-ordered variables (latent classes), and where the assumption can be made that the latent classes do not change over time, it is possible to apply the Clogg-Manning approach to produce estimates of reliability for categorical data. I return to this issue below and illustrate how the Clogg-Manning approach is also useful when membership in the latent classes corresponding to the categories of the variables of interest does not change over time.

11.2.1   The Model

First, I discuss the general form of the latent class model presented by Clogg and Manning (1996) to draw inferences about reliability of measurement for categorical variables. The technical aspects of this model can be found in Clogg (1995).[2] The present discussion, however, relies primarily on the nontechnical presentation in Clogg and Manning (1996), and I borrow heavily from their discussion. As noted above, we believe [see quote from Wiggins (1973) above] that it is important for purposes of reliability estimation that there is an isophormism between the number of latent classes and the number of categories of the observed measures. Consider the case, for example, where there are three dichotomous measures of the same underlying latent variable, X, that has two classes—classes that are identical to the categories of the observed measures. Following the Clogg-Manning notation, we denote these three measures of X as A, B, and C, and refer to their levels using subscripts i, j, and k, respectively. Similarly, we denote the number of latent classes by T (in this case T = 2), the levels of which are indexed by t. One need not restrict the model to dichotomous measures, as will be seen in the following application; however, the use of more than two latent classes introduces the complexity of whether or not the categories are ordered. A schematic diagram of the relationships between the observed and latent variable distributions in the latent class model is given in Figure 11.1.

The latent class model in this case assumes the following relationship:

$$\pi_{ABC}(ijk) = \sum_{t=1}^{T} \pi_{ABCX}(ijkt)$$

where $\pi_{ABC}{}^{(ijk)}$ denotes the probability of cell (i,j,k) in the cross-classification of the observed variables, and $\pi_{ABCX}{}^{(ijkt)}$ the probability of cell (i,j,k,t) in the unobserved contingency table cross-classifying the observed and latent variables. This latter probability embodies the relationship of the observed variables with the latent unobserved variable, i.e., the T latent classes.

Essential to the formulation of a notion of reliability consistent with that used in the case of continuous latent variables is the definition of a set of conditional probabilities relating the observed measures to the latent variables. These probabilities are defined based on the nontrivial assumption of local independence, an assumption common to all model-based assessments of reliability (see Clogg and Manning, 1996, p. 172). These relationships are defined under the latent class model as

---

[2]In this chapter we adopt Clogg and Manning's (1996) notation, and do not necessarily attempt to make it compatible with the notation found in other chapters of the book.

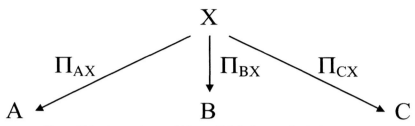

Figure 11.1. Latent class model for multiple indicators of a single construct.

$$\pi_{ABCX}(ijkt) = \pi_X(t)\pi_{A|X=t}{}^{(i)}\pi_{B|X=t}{}^{(j)}\pi_{C|X=t}{}^{(k)}$$

where $\pi_X(t)$ (the probability that $X = t$) denotes the latent class proportions, and $\pi_{A|X=t}{}^{(i)}$, $\pi_{B|X=t}{}^{(j)}$, and $\pi_{C|X=t}{}^{(k)}$ are the conditional probabilities that particular levels of the observed measures are observed in the data, given the latent classes. Specifically, $\pi_{A|X=t}{}^{(i)}=\Pr(A=i\,|\,X=t)$, $\pi_{B|X=t}{}^{(j)}=\Pr(B=j\,|\,X=t)$, and $\pi_{C|X=t}{}^{(k)}=\Pr(C=k\,|\,X=t)$. The latent class proportions are given as $\pi_X(t)$, with $\sum_{t=1}^{T}\pi_X(t) = 1$ and $\pi_X(t) > 0$ for each t.

Recall that in the continuous case, reliability was defined as a function of the correlation between the latent variable and the measures (see Chapter 3). Consistent with this notion, Clogg and Manning (1996, p. 172) point out that the conditional probability parameters given above can be used to define the reliability of measurement for the specific measures. The relationships between the latent classes and manifest response categories are given by the matrices Π_{AX}, Π_{BX}, and Π_{CX}. The following gives the $\pi_{A|X}$ elements in Π_{AX} in the Clogg-Manning notation:

		X	
		t = 1	t = 2
A	i = 1	$\pi_{A|X=t}(i)$	$\pi_{A|X=t}(i)$
	i = 2	$\pi_{A|X=t}(i)$	$\pi_{A|X=t}(i)$

These conditional probabilities express the elements of what one might consider an analogue to the *index of reliability* discussed in Chapter 3. If, for example, the level-specific probabilities $\pi_{A|X=1}{}^{(1)} = \pi_{A|X=2}{}^{(2)} = 1.0$, then measure A can be said to be a perfectly reliable measure of X. Values less than unity reflect some degree of unreliability. Thus, the degree of reliability for the specific measure in this case is a composite of two (in the general case T) separate level parameters, namely, $\pi_{A|X=1}{}^{(1)}$ and $\pi_{A|X=2}{}^{(2)}$, the conditional probability that measure A takes on the value of 1, given the latent variable takes on the value of 1, *and* the conditional probability that measure A takes on the value of 2, given the latent variable takes on the value of 2. We return to a consideration of the interpretation of this kind of measure-specific reliability using the latent class model following a presentation of an example.

At a general level, Clogg and Manning's (1996, pp. 172–173) approach to defining reliability in terms of conditional probabilities appears to have a great deal in common with classical reliability theory. Indeed, their "reliability as predictability" approach can be used to define a correlation type measure of item level reliability, for measure A (it goes without saying that the same can be done for measures B and C), as

$$Q_{AX} = (\theta_{AX} - 1)/(\theta_{AX} + 1)$$

where θ_{AX} is the partial (or conditional) odds ratio between X and A, defined as

$$\Theta_{AX} = \pi_{A|X=1}(1)\pi_{A|X=2}(2)/[\pi_{A|X=1}(2)\pi_{A|X=2}(1)]$$

Thus, the Clogg-Manning approach to item reliability defines reliability in terms of odds ratios linking the latent and manifest classes and specifies a coefficient that is a Yule's Q transform of the odds ratio. This is a workable analog to the classical true score definition of reliability (see Lord and Novick, 1968, p. 61), where reliability is defined as a *proportion of variance*, i.e., an R^2, specifically the proportion of the observed variance accounted for by the true score. It is important to note in this regard that Yule's Q, like R^2, is a PRE (i.e., proportional reduction in error) measure of association (Costner, 1965). Finally, Clogg and Manning (1996, pp. 173–174) also present an approach to defining an index of reliability for the set of measures involved in the latent class model, which they refer to as "set reliability." Naturally, the latter index is of less interest to us here, given the focus on item-level reliabilities rather than the reliability of composite scores.

11.2.2 An Example

We draw upon three dichotomous measures of social trust from the 1972–1974–1976 National Election Study (NES70), described in Chapter 6 to illustrate the latent class model.[3] These measures are as follows:

> Generally speaking, would you say that most people can be trusted or that you can't be too careful in dealing with people? [TRUST]
> Would you say that most of the time people try to be helpful, or that they are mostly just looking out for themselves? [HELP]
> Do you think most people would try to take advantage of you if they got a chance, or would they try to be fair? [FAIR]

Table 11.1 presents the cross-classification of the three indicators of the two latent categories of social trust in the 1972 wave of the 1970s NES panel survey. Specifically,

---

[3]These questions can be found in the Appendix under "beliefs about economy and society (part C.2)," variable numbers 7080, 7081, and 7082.

Table 11.1.   Cross-classification of measurement of two latent categories and probabilities of latent class membership from a latent class model of three indicators of social trust in the 1972 National Election Study (N = 1,215)

Response Pattern A B C	Observed Frequency	Probability of Latent Class Membership Pr(X=1)	Probability of Latent Class Membership Pr(X=2)	Predicted Latent Class X	Probability of Latent Class Pr(X)	Expected Frequencies X=1	Expected Frequencies X=2
111	462	0.998	0.002	1	0.998	461.1	0.9
112	23	0.858	0.142	1	0.858	19.7	3.3
121	91	0.907	0.093	1	0.907	82.5	8.5
122	36	0.098	0.902	2	0.902	3.5	32.5
211	109	0.917	0.083	1	0.917	100.0	9.0
212	39	0.110	0.890	2	0.890	4.3	34.7
221	108	0.166	0.834	2	0.834	17.9	90.1
222	347	0.002	0.998	2	0.998	0.7	346.3
Total	1,215						1,166.9

N of cases	1,215
N of invariant cases	809
Proportion invariant	0.666
N of observations	3,645
N of congruent observations	3,239
Proportion of congruent observations	0.889
Set reliability	0.960

Question text:
TRUST: Generally speaking, would you say that most people can be trusted or that you can't be too careful in dealing with people?
HELP: Would you say that most of the time people try to be helpful, or that they are mostly just looking out for themselves?
FAIR: Do you think most people would try to take advantage of you if they got a chance, or would they try to be fair?

Latent class and observed variables:
A =TRUST; B = HELP; C = FAIR
1 = People can be trusted, try to be helpful, try to be fair.
2 = You can't be too careful, people look out for themselves, take advantage.

this table presents the 2^3 table that cross-classifies three dichotomous indicators and their relationship to the latent classes of the model, i.e., the probability of latent class membership. We have implemented the Clogg-Manning model using M*plus* (Muthén and Muthén, 2001–2004) and present results here comparable to those reported by Clogg and Manning (1996, pp. 175–176).

This table also gives the *predicted latent class* and the *expected frequencies* under the model of mutual independence (see Clogg and Manning, 1996, pp. 174–175). Recall that if the latent class variable (X) could be observed, then the relationships could be arranged in a four-way contingency table cross-classifying A, B, C, and X. In this model π_{ABCX}(ijkt) denotes the probability of cell (i,j,k,t) in this "*indirectly observed* contingency table" (Clogg and Manning, 1996, p. 171). The expected frequencies in this table are simply a function of the observed frequencies for the response patterns multiplied by the probabilities of latent class membership.

We noted above that there are several intuitively based (or non-model-based) approaches to thinking about reliability in the case of categorical variables, which may be tempting to look at. We can, for example, calculate the proportion of cases that produce consistent scores across all indicators, i.e., those "invariant" cells. In Table 11.1, there are 809 cases (462 + 347) representing .67 of all cases (or 809 ÷ 1,215) that had the same value across the three indicators. One can see that the level of consistency reflected in this measure is dependent on both the number of categories and the number of indicators. If only the first two indicators (A and B) were used to calculate the invariance, the index of invariance would equal .77. Furthermore, this computation does not take the latent class model into account. Another way to define a descriptive assessment of reliability is to assess the proportion of observations in which the response is "congruent" (or in agreement) with the predicted latent class. For the data in Table 11.1, the proportion of congruent observations is .89 (or 3,239 ÷ 3,645). Because it is based on a model for measurement (i.e., the latent class model), this is probably a more sensitive measure of consistency than the "proportion invariant" measure given above.

Although these quantities may have some appeal, the approach suggested by Clogg and Manning (1996, pp. 172–173) involving conditional probabilities has a great deal more in common with classical reliability theory. Table 11.2 presents the probabilities of levels of the response variables conditional on latent class membership, along with the estimates of reliability as defined above. Specifically, these results present the following quantities: $\pi_{A|X=t}^{(i)}$, $\pi_{B|X=t}^{(j)}$, and $\pi_{C|X=t}^{(k)}$, where A, B, and C denote the three indicators, and the indices i, j and k take on the values of 1 or 2. (Note that in Table 11.2 we use the notation A_1, B_1, and C_1 to denote the fact that these measures come from wave 1 of the NES panel.) The conditional probability that a response of "1" is reported given the latent class is "1" is .882 for indicator A, .848 for indicator B, and .959 for indicator C. Similarly, the conditional probability that a response of "2" is reported given the latent class is "2" is .914 for indicator A, .909 for indicator B, and .793 for indicator C.

In modern software applications of these models, standard errors are available, and it is therefore possible to test specific hypotheses about population parameters. Specifically, these permit one to address the question of whether a particular parameter estimate is 1.0 or 0.0—the case of perfect reliability. The response probabilities presented in Table 11.2 are accompanied by the 95% confidence intervals (inside the parentheses), and in no case are the values of 1.0 and 0.0 within these intervals. It is also possible in theory to examine hypotheses about the response probabilities in alternative models where these parameter values are constrained, and for nested models comparisons of model fit examined, e.g., where the matrices Π_{AX}, Π_{BX}, and Π_{CX} are set equal, or where they are all set equal to an identity matrix.

Table 11.2. Estimated parameter values and reliability estimates for the measurement of two latent categories and probabilities of latent class membership from a latent class model of three indicators of social trust in the 1972 National Election Study (N = 1,215)

Response	Probability of Item Response Conditional on Class Membership		$\hat{\Theta}$	$\hat{Q}$	Probability of Latent Class Conditional on Item Response	
	Pr(item|X=1)	Pr(item|X=2)		Item Reliability	Pr(X=1|item)	Pr(X=2|item)
A_1=1	0.822 (0.787, 0.857)	0.086 (0.055, 0.117)	49.1	0.960	0.927	0.074
A_1=2	0.178 (0.143, 0.213)	0.914 (0.883, 0.945)			0.204	0.796
X	0.568	0.432				
Response	**Pr(item|X=1)**	**Pr(item|X=2)**	$\hat{\Theta}$	$\hat{Q}$	**Pr(X=1|item)**	**Pr(X=2|item)**
B_1=1	0.848 (0.815, 0.881)	0.091 (0.060, 0.122)	55.7	0.965	0.924	0.075
B_1=2	0.152 (0.119, 0.185)	0.909 (0.878, 0.940)			0.180	0.820
X	0.568	0.432				
Response	**Pr(item|X=1)**	**Pr(item|X=2)**	$\hat{\Theta}$	$\hat{Q}$	**Pr(X=1|item)**	**Pr(X=2|item)**
C_1=1	0.959 (0.939, 0.979)	0.207 (0.164, 0.250)	89.6	0.978	0.859	0.141
C_1=2	0.041 (0.021, 0.061)	0.793 (0.750, 0.836)			0.064	0.936
X	0.568	0.432				

Notes:

$\hat{\Theta}$ = Odds ratio expressing likelihood of item response congruent with latent class membership.

$\hat{Q}$ = Yule's Q transform of $\hat{\Theta}$. This expresses the odds ratio in a correlational metric, and is an index of item reliability.

Latent class and observed variables:

    A = TRUST; B = HELP; C = FAIR

    1 = People can be trusted, try to be helpful, try to be fair.

    2 = You can't be too careful, people look out for themselves, take advantage.

As reported in Table 11.2, these item-specific level probabilities combine to contribute to item-specific reliabilities of .960, .965, and .978, respectively, for indicators A, B, and C, as defined by Clogg and Manning's (1996) Yule Q estimate described above. These are high levels of reliability, suggesting there are hardly any errors of classification in reports of social trust using these indicators. From this example, most readers will conclude that these are very high levels of reliability, registering a much brighter picture regarding reliability of measurement than what we obtained on the basis of classical true-score theory. In this regard, I urge caution in the interpretation of these models as if they were comparable to the model-based parameter estimates derived from CTST models. A great deal more needs to be learned about these categorical latent variable models before the definitions of reliability put forward by Clogg and Manning (1996) can be compared with analogous definitions that we have derived for continuous variables.

11.2.3 Interpretation

There are two important features of the latent class model as presented by Clogg and Manning's (1996) paper. The first, and most important, is that X (the underlying set of latent classes) is modeled nonparametrically, so that reliability is assessed without having to make the assumption that X is continuous, or even quantitative (Clogg and Manning 1996, p. 171). The second is that X is assumed to have a specific number of latent categories or latent classes. In their model, the number of latent classes can be an unknown parameter of the model, but as they present the latent class model and as we use it here, there is a one-to-one correspondence between the categories of observed and latent classes. In other words, we assume that the true underlying variable is discrete with the same number of categories as expressed in the observed variable (see Clogg and Manning, 1996, p. 172). This second feature of the latent class model (as we use it) follows the principle introduced into this literature by Wiggins (1973, p. 26)—cited at the beginning of the chapter—that "the notion of reliability only makes obvious sense when the latent classes can be put into a one-to-one correspondence with the manifest classes." In terms of estimating the reliability of single survey questions, this feature is an essential component of the models developed in this chapter.

I want to again emphasize that, as Clogg and Manning (1996) developed it, the latent class model is analogous to a common factor model for *multiple indicators* of a single latent variable. In the terminology introduced earlier in this book (see Chapter 4, Section 4.1), such an application fails to adequately distinguish between "multiple measures" and "multiple indicators." Following this line of argument, I argue that the latent class model, as developed by Clogg and Manning (1996), provides a biased estimate of reliability to the extent there is reliable specific variance in the indicators. Unless the response variables are exact replicates of one another, the model cannot produce estimates that can be interpreted in terms of reliability. Therefore in the following application (see Section 11.3) of their model, rather than using "multiple indicators"—as employed in the Clogg and Manning (1996, pp. 174–178) examples—we employ the model using "multiple measures," where the measures come from longitudinal reports of nonordered variables (latent classes) and where the assumption can be made that the latent classes do not change over time. As we demonstrate later on (Section 11.4), where membership in the latent classes change over time, one must seek alternatives to the Clogg-Manning approach.

11.3 THE LATENT CLASS MODEL FOR MULTIPLE MEASURES

The second model for categorical latent variables we consider here is a special case of the latent class model for multiple indicators discussed above in which the three variables are *multiple measures*, that is, replicate measures. As we argued in Chapter 4, in the CTST tradition multiple measures are required for estimating reliability, and when the design of measurement includes multiple indicators instead, the

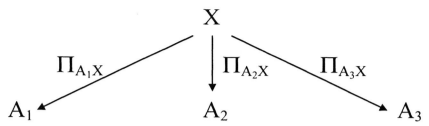

Figure 11.2.   Latent class model for multiple measures of a single construct.

application of models for estimating reliability will confuse measurement error and specific variance.

Compare the schematic diagram presented in Figure 11.2 with that presented above for the latent class model. The earlier model, I argued, is analogous to a common factor model for multiple indicators, and therefore does not necessarily fit well with the notion of reliability developed in this book. Although the models shown in Figures 11.1 and 11.2 have the same formal structure, they differ in the survey design—the model in Figure 11.1 involves three distinct indicators observed at the same point in time, whereas the model in Figure 11.2 involves three *replicate measures* of the same latent variable observed on three successive occasions. Given the equivalence of their formal structure, one can apply the Clogg-Manning estimation strategy to the case where three replicate measures were available in a longitudinal design where the latent classes could, by definition, not change. Hence, the depiction in Figure 11.2 of A_1, A_2, and A_3 denotes replicate measures of the variable A at three points in time, rather than three distinct indicators A, B, and C. Any lack of consistency among the measures reflects, it could readily be argued, unreliability rather than true change in the underlying variables. The matrices of conditional probabilities linking the observed and latent classes can be designated as Π_{A_1X}, Π_{A_2X}, and Π_{A_3X}. In the general case the model places no constraints on these parameters.

11.3.1   An Example

The example we use to illustrate the usefulness of the latent class model to the estimation of reliability for multiple measures involves data from the 1956–1958–1960 National Election Panel Study (NES50) for a question about voting preferences in the 1956 election (see Table 11.3 for the details of the questions used in these three surveys). In this case, we have three trichotomous measures of an underlying latent variable, A_1, A_2, and A_3, where the subscripts refer to the three points in time. Again, the levels of those variables are denoted by subscripts i, j, and k, respectively. In this case, level 1 refers to "did not vote," level 2 to "voted for Adlai Stevenson," and level 3 to "voted for Eisenhower."

Table 11.3 presents the cross-classification of the three-wave measurement of the three latent categories of electoral behavior in the 1950s NES panel survey.

Table 11.3. Cross-classification of three-wave measurement of three latent categories of electoral behavior in the 1950s NES Panel Study

Response Pattern	Observed Frequency	Probability of Latent Class Membership			Predicted Latent Class	Expected Frequencies		
		Pr(X=1)	Pr(X=2)	Pr(X=3)		X=1	X=2	X=3
111	101	1.000	0.000	0.000	1	101.000	0.000	0.000
112	6	0.975	0.025	0.000	1	5.850	0.150	0.000
113	20	0.995	0.001	0.005	1	19.900	0.020	0.100
121	1	1.000	0.000	0.000	1	1.000	0.000	0.000
122	17	0.004	0.996	0.001	2	0.068	16.932	0.017
123	5	0.085	0.714	0.200	2	0.425	3.570	1.000
131	1	1.000	0.000	0.000	1	1.000	0.000	0.000
133	22	0.009	0.004	0.988	3	0.198	0.088	21.736
211	1	1.000	0.000	0.000	1	1.000	0.000	0.000
212	2	0.033	0.966	0.001	2	0.066	1.932	0.002
213	1	0.436	0.401	0.162	1	0.436	0.401	0.162
222	239	0.000	1.000	0.000	2	0.000	239.000	0.000
223	25	0.000	0.981	0.019	2	0.000	24.525	0.475
232	11	0.000	0.954	0.046	2	0.000	10.494	0.506
233	21	0.000	0.051	0.949	3	0.000	1.071	19.929
311	1	1.000	0.000	0.000	1	1.000	0.000	0.000
313	2	0.092	0.008	0.901	3	0.184	0.016	1.802
322	18	0.000	0.986	0.014	2	0.000	17.748	0.252
323	12	0.000	0.155	0.845	3	0.000	1.860	10.140
332	11	0.000	0.069	0.931	3	0.000	0.759	10.241
333	416	0.000	0.000	1.000	3	0.000	0.000	416.000
Total	933						925.5	

N of cases	933
N of invariant cases	756
Proportion of invariant cases	0.810
N of observations	2,799
N of congruent observations	2,616
Proportion of congruent observations	0.935
Set reliability	0.992

Question text:

1956   In talking to people about the election we find that a lot of people weren't able to vote because they weren't registered or they were sick or they just didn't have time. How about you, did you vote this time? <If yes> Whom did you vote for for President?

1958   Two years ago, in 1956, you remember that Mr. Eisenhower ran against Mr. Stevenson for the second time. Do you remember for sure whether or not you voted in that election. <If 'yes, did vote'> Did you vote for Stevenson or Eisenhower?

1960   Now in 1956, you remember that Mr. Eisenhower ran against Mr. Stevenson . Do you remember for sure whether or not you voted in that election. <If 'yes, did vote'> Which one did you vote for?

Latent Class and Observed Variables:

    1 Did not vote
    2 Stevenson (Dem.)
    3 Eisenhower (Rep.)

Specifically, this table presents the 3^3 table that cross-classifies three trichotomous indicators and their relationship to the latent classes of the model, i.e., the *probability of latent class membership*.[4] The table also gives the *predicted latent class* and the *expected frequencies* under the model of mutual independence (see Clogg and Manning, 1996, pp. 174–175). Recall that if the latent class variable (X)—in this case, voted for Stevenson, voted for Eisenhower, or did not vote—could be observed, then the relationships could be arranged in a four-way contingency table cross-classifying A, B, C, and X. As above, in this model $\pi_{A_1A_2A_3X}(ijkt)$, denotes the probability of cell (i,j,k,t) in this indirectly observed contingency table. The expected frequencies in this table are simply a function of the observed frequencies for the response patterns multiplied by the probabilities of latent class membership.

Following the example discussed above, we can calculate some of the non-model-based indices of reliability in the case of categorical variables. Specifically, we can calculate the proportion of cases that produce consistent scores across all waves, i.e., those "invariant" cells. In Table 11.3, there are 756 cases (101 + 239 + 416) representing .81 (or 756 ÷ 933) cases that had the same value across the three waves. Again, it can be seen that the level of consistency reflected in this measure is dependent on both the number of categories and the number of over-time observations. As above, this computation does not take the latent class model into account, and we can use another approach that assesses the proportion of overtime observations in which the response is "congruent" (or in agreement) with the predicted latent class. For the data in Table 11.3, the proportion of congruent observations is .94 (or 2,616 ÷ 2,799), and because it is based on a model for measurement (i.e., the latent class model), this is probably a more sensitive measure of consistency than the "proportion invariant" measure given above.

Table 11.4 presents the probabilities of levels of the response variables conditional on latent class membership, along with the estimates of reliability as defined above. Specifically, these results present $\pi_{A_1|X=t}^{(i)}$, $\pi_{A_2|X=t}^{(j)}$, and $\pi_{A_3|X=t}^{(k)}$ where A_1, A_2, and A_3 are replicates of the same measure at each of three waves of the study—wave 1 (A_1), wave 2 (A_2) and wave 3 (A_3). As noted above, since the latent class membership does not change, these may be considered multiple measures of the same variable. The numbers in the table are the conditional probabilities that particular levels of the measures are observed given the latent classes, i.e., $\pi_{A_1|X=t}^{(i)}$ = Pr ($A_1 = i \mid X = t$), $\pi_{A_2|X=t}^{(j)}$ = Pr ($A_2 = j \mid X = t$), and $\pi_{A_3|X=t}^{(k)}$ = Pr ($A_3 = k \mid X = t$). The indices i, j, and k for measures A_1, A_2, and A_3 take on the values 1, 2, or 3 at each wave. For example, the conditional probability that a response of "1" (Did not vote) is reported given the latent class is "1" (Did not vote) is .980 at wave 1, .980 at wave 2, and .795 at wave 3. Similarly, the conditional probability that a response of "2" (voted for Stevenson) is reported given the latent class is "2" (voted for Stevenson) is .871 at wave 1, .953 at wave 2, and .901 at wave 3. Finally, the conditional probability that a response of "3" (voted for Eisenhower) given the latent class is "3" (voted for Eisenhower) is .909 at wave 1, .971 at wave 2 and .977 at wave 3.

---

[4]Note that some of these cells contain no cases and are therefore not presented here.

Table 11.4. Estimated parameter values and reliability estimates for a three latent class model of three-wave measurement of electoral behavior in the 1950s NES Panel Study

Item Response	Probability of Item Response Conditional on Class Membership			$\hat{\Theta}$	Item Reliability $\hat{Q}$	Probability of Latent Class Conditional on Item Response		
	pr(item|X=1)	pr(item|X=2)	pr(item|X=3)			pr(X=1|item)	pr(X=2|item)	pr(X=3|item)
A_1 (1)	0.980	0.065	0.047	3,304.8	0.999	0.748	0.120	0.132
A_1 (2)	0.011	0.871	0.044			0.005	0.925	0.070
A_1 (3)	0.009	0.064	0.909			0.003	0.044	0.953
A_2 (1)	0.980	0.008	0.004	33,266.9	1.000	0.966	0.019	0.015
A_2 (2)	0.011	0.953	0.025			0.005	0.958	0.037
A_2 (3)	0.009	0.039	0.971			0.002	0.026	0.972
A_3 (1)	0.795	0.000	0.000	1,499.2	0.999	1.000	0.000	0.000
A_3 (2)	0.045	0.901	0.023			0.020	0.944	0.036
A_3 (3)	0.160	0.099	0.977			0.040	0.060	0.900
X	0.142	0.341	0.517					

Notes:

$\hat{\Theta}$ = odds ratio expressing likelihood of item response congruent with latent class membership.

$\hat{Q}$ = Yule's Q transform of $\hat{\Theta}$. This expresses the odds ratio in a correlational metric, and is an index of item reliability.

As reported in Table 11.4, these item-specific level probabilities combine to contribute to item-specific reliabilities of .999, 1.000, and .999, as defined by the Yule's Q estimate given above for the measurement of electoral behavior in the 1956, 1968, and 1960 waves of the NES panel. These are extremely high levels of reliability, suggesting that there is hardly any error in reports of electoral behavior in this sample over this period.

We sought additional examples in these panel surveys for the measurement of latent classes that do not change over time. A serious limitation exists in most panel data sets, however, in that since these qualities are assumed not to change, it is therefore considered a time-saving strategy to ask or observe them *only once*. We are fortunate in having identified seven additional such variables in our data where these models apply, and for purposes of illustration we implement the Clogg-Manning latent class approach for these variables as well. These include multiple over-time reports of categorical variables such as sex and race, as well as other things that do not change, e.g., parental nativity and maternal employment. These results are presented in Table 11.5.

As with the estimates for electoral behavior given above, these results reveal without exception very high levels of reliability, at or near the theoretical maximum in all cases. We conclude that this is likely due to several factors: (1) as these variables all assess factual content, they are very easy to specify in the form of survey questions with relatively clear answers; (2) they involve a small number of response categories; and (3) the production of a response to these questions does not involve a great deal of information processing or place cognitive demands on the respondent in coming up with a correct response. One can easily imagine that the level of assessed reliability might be considerably lower for categorical latent variables in which these conditions do not obtain. It might also be the case that variables for which change is likely may be less reliably measured.

11.4   THE LATENT MARKOV MODEL

Except for the classic writings by Coleman (1964, 1968) and a pioneering book by Wiggins (1973), which formulated the basic ideas behind the longitudinal approach to reliability estimation (see Chapter 5), little attention has been devoted to reliability and its estimation for measures of categorical latent variables. The work of Wiggins (1973) combines features of the latent class model and the simple Markov chain; and although this model shares some features with the latent class model, as developed by Clogg and Manning (1996), it is not the same. Indeed, the latent Markov model (LMM) addresses an important deficiency in the static latent class model, namely, the possibility that membership in the latent classes can change over time. As we will show, however, one may think of the Clogg-Manning latent class model as a special case of the latent Markov model.

This latent class Markov chain model has existed in the sociological literature for several decades (see Coleman, 1968; Wiggins, 1973; van de Pol and de Leeuw, 1986; van de Pol and Langeheine, 1990; van de Pol, Langeheine and de Jong, 1991;

Table 11.5. Reliability estimates of selected questions involving three-wave measurement in the 1950s and 1970s NES Panel Studies

Source	Topic of Question	Nominal classes	Number of Classes	Set Reliability	Mean Item Reliability	Proportion Invariant	Proportion Congruent
NES 50s-5028	Sex	Male, Female	2	1.000	1.000	0.992	0.997
NES 50s-5080	Race	White, Black	2	1.000	1.000	1.000	1.000
NES 50s-5085	Parental nativity	One, Both, or Neither parent native-born	3	0.994	0.999	0.907	0.968
NES 50s-5092	1956 Presidential election behavior	Did Not Vote, Stevenson, Eisenhower	3	0.991	0.999	0.810	0.935
NES 70s7015	Parental nativity	Both parents native-born, At least one parent foreign-born	2	1.000	0.999	0.972	0.991
NES 70s-7020	Sex	Male, Female	2	1.000	1.000	0.999	1.000
NES 70s-7037	Maternal employment	Mother worked/did not work during respondent's youth	2	1.000	0.985	0.875	0.958

Vermunt, Langeheine, and Böckenholt, 1999), but to our knowledge it has not been applied on a mass scale to the assessment of reliability. The main limitations appear to be the flexibility of statistical software that can place the necessary restrictions on the model. Our preliminary analyses indicate that the kind of single-variable models we wish to estimate for dynamic latent class models for the purposes of reliability estimation can be estimated for three or more waves. There are several available software packages for estimating these models, including the PANMARK computer program for latent class analysis (van de Pol, Langeheine and de Jong, 1991), M*plus* (Muthén and Muthén, 2004), and *WinLTA* (Collins, Flaherty, Hyatt, and Schafer, 1999).

11.4.1 The Model

The latent Markov model depicted in Figure 11.3 is a dynamic version of the latent class model. It is an analog to the quasi-Markov simplex model discussed in Chapter 5 (above) (see Wiggins, 1973, pp. 97–106). In this case, the model retains the property of a one-to-one correspondence between categories of observed and latent classes, which is an essential ingredient to models that provide information on reliability. At the same time, an important component of this model is its allowance for individuals to move in and out of the latent categories over time. Marital status and employment status are examples of variables for which such a model would apply. In such a case, there is an obvious unitary correspondence between the response classes and the latent classes, and over time people can move in and out of the latent classes. Thus, any case in which the categories of the observed variable can be assumed to be identical to the categories of the latent variable, and where the unit of analysis can change membership in the latent classes over time, this is an appropriate candidate for this model.

There are two sets of parameters of interest in this model: (1) the matrices of conditional probabilities linking the observed and latent classes can be designated as $\Pi_{A_1X_1}$, $\Pi_{A_2X_2}$, and $\Pi_{A_3X_3}$; and (2) the matrices of transition probabilities linking the latent categories over time, designated as $\Pi_{X_2X_1}$ and $\Pi_{X_3X_2}$. The latent transition probabilities in this model embody the same intent of the quasi-Markov simplex model, as developed by Heise (1969), Jöreskog (1970), and Wiley and Wiley (1970) for the

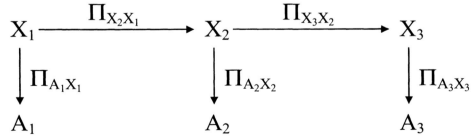

Figure 11.3. Latent Markov model for a single measure over three occasions of measurement.

continuous case, in that the model allows for change at the unit of observation level while examining the reliability of survey measures. Indeed, Wiggins (1973) developed this as an approach to reliability estimation for the case in which the latent variables were categories, i.e., latent classes; and the unobserved processes that entailed movement between latent classes were conceptualized in terms of a Markov process (van de Pol and de Leeuw, 1986; Langeheine and van de Pol, 2002). Just as the quasi-Markov simplex model allows for change in the latent variable, here the model allows change in the latent distribution and expresses change in the form of a latent transition matrix. At the time Wiggins (1973) published his work, approaches to estimation were still relatively underdeveloped, and as a result these models have not seen wide application. More recent developments in estimation have made the application of these models possible (see van de Pol and de Leeuw, 1986; Langeheine and van de Pol, 2002).[5] As we noted in the discussion of the latent class model above, the utility of these models for reliability estimation stems from the fact that the latent classes are put into a one-to-one correspondence with the manifest classes (Wiggins, 1973, p. 26). I briefly review this model because it is likely to play an important role in reliability estimation of categorical measures in future research.

Consistent with the above development, Wiggins' latent Markov model for three points in time can be written as follows:

$$\pi_{A_1 A_2 A_3}(ijk) = \sum_{t_1 = 1}^{T} \sum_{t_2 = 1}^{T} \sum_{t_3 = 1}^{T} \pi_{X_1}(t_1) \pi_{A_1 | X_1 = t_1}(i) \pi_{X_2 | X_1 = t_1}(t_2) \pi_{A_2 | X_2 = t_2}(j)$$
$$\pi_{X_3 | X_2 = t_2}(t_3) \pi_{A_3 | X_3 = t_3}(k)$$

where $X_1 - X_3$ represent a latent variable at three timepoints, whose values are denoted by the index t (see Langeheine and van de Pol, 2002). Each X contains T discrete classes, and A_1, A_2, and A_3 denote replicate manifest measures of membership in these categories at three timepoints, having observed levels i, j, and k, respectively, corresponding to the levels of t. For three occasions of measurement the model parameters cannot be identified. Wiggins suggests that two sets of assumptions be employed to make the model identified: (1) the homogeneity of the latent transition probabilities, $\Pi_{X_2 X_1} = \Pi_{X_3 X_2}$ and (2) the equality of reliabilities over time: $\Pi_{A_1 X_1} = \Pi_{A_2 X_2} = \Pi_{A_3 X_3}$. It can be shown that when membership in the latent classes does not change, i.e., $\Pi_{X_2 X_1} = \Pi_{X_3 X_2} = I$, the latent Markov model reduces to the static latent class model for replicate measures, as depicted in Figure 11.2.

11.4.2   An Example

The example we employ to illustrate the features of the LMM involves the same set of three dichotomous measures of social trust from the 1972–1974–1976 National

---

[5]Wiggins' (1973) work was based on a 1955 Ph.D. dissertation completed at Columbia University under the direction of Paul Lazarsfeld. Langeheine and van de Pol (2002, p. 325) note that "it took nearly 30 years until problems of parameter estimation were solved for the latent Markov model."

Election Study panel study (NES70) discussed above with respect to the latent class model: TRUST (Generally speaking, would you say that most people can be trusted or that you can't be too careful in dealing with people?), HELP (Would you say that most of the time people try to be helpful, or that they are mostly just looking out for themselves?), and FAIR (Do you think most people would try to take advantage of you if they got a chance, or would they try to be fair?). In the present application, however, rather than thinking of these measures as multiple indicators of a set of latent classes representing the concept of social trust, we examine each of them separately, as multiple replications of the same measure within a longitudinal design that allows change in class membership from time to time. This particular example involves observations of these variables across three (i.e., $G = 3$) occasions of measurement, specifically, 1972, 1974, and 1976, referred to here as time 1, time 2, and time 3, respectively. We designate TRUST as A, measured as A_1, A_2, and A_3, corresponding to the three occasions of measurement; we designate HELP as B, measured as B_1, B_2, and B_3, and FAIR as C, measured as C_1, C_2, and C_3. The latent variables assessed by A, B, and C are designated X, Y, and Z, respectively. The depiction in Figure 11.3 corresponds directly to the latent Markov model we estimate for the variable A, and similar representations can be given for B and C. The point is that, in contrast to the latent class model discussed above, rather than three measures of the same concept (as in Figures 11.1 and 11.2), there is no assumption in these models that this is the case. Rather, we model each variable separately, where each is measured longitudinally across three waves (as in Figure 11.3).

We represent the distribution across the categories of a given observed measure using the (T × 1) column vector π_{Ag} (where $g = 1, 2$, or 3), and, similarly, the latent distribution across latent classes is represented as the column vector π_{Xg} (where $g = 1, 2$, or 3). Recall that T is the number of latent classes and G is the number of measures. In the present application of the model, as a way of identifying the model, we constrain the within-time matrix of response probabilities, which contain probabilities of levels of response conditional on latent class membership, to be equal across waves. In the case of the TRUST variable, for example, we have placed the constraint that $\Pi_{A_1X_1} = \Pi_{A_2X_2} = \Pi_{A_3X_3}$ on the model. Similarly, for the HELP and FAIR variables we assume $\Pi_{B_1Y_1} = \Pi_{B_2Y_2} = \Pi_{B_3Y_3}$ and $\Pi_{C_1Z_1} = \Pi_{C_2Z_2} = \Pi_{C_3Z_3}$.

As noted above, another feature of this model that makes it interesting is the fact that latent class membership can change over time; that is, it allows us to estimate the latent transition probabilities, $\Pi_{X_2X_1}$ and $\Pi_{X_3X_2}$, which are potentially of substantive interest. It is also possible to place constraints on these parameters, but in the present estimation of this model, we have not done so. This parallels the quasi-Markov simplex model we discussed in Chapter 5, which similarly separates unreliability from true change. Like our reporting of results from the simplex models, we focus entirely on the reliability of measurement and do not report the results involving the between-occasion transition processes.

In Tables 11.6, 11.7, and 11.8 we present the estimated parameter values and reliability estimates for the LMM models for the longitudinal measurement of TRUST, HELP, and FAIR measures, respectively. Note that for each variable, the within-time matrix of response probabilities is constrained equal across time, and this produces equivalent reliability estimates across the three occasions of measurement in each

Table 11.6. Estimated parameter values and reliability estimates for the measurement of two latent categories and probabilities of latent class membership from a latent Markov model of a measure of social trust in the 1970s National Election Study: "Generally speaking, would you say that most people can be trusted, or that you can't be too careful in dealing with people?" (N = 1,133)

Response	$Pr(A_1)$	Probability of Item Response Conditional on Class Membership		$\hat{\Theta}$	Item Reliability $\hat{Q}$	Probability of Latent Class Conditional on Item Response	
		$Pr(item|X_1=1)$	$Pr(item|X_1=2)$			$Pr(X_1=1|item)$	$Pr(X_1=2|item)$
$A_1=1$	0.509	0.916 (0.887, 0.945)	0.158 (0.117, 0.199)	58.1	0.966	0.834	0.166
$A_1=2$	0.491	0.084 (0.055, 0.113)	0.842 (0.801, 0.883)			0.079	0.920
$Pr(X_1)$		0.464	0.536				

Response	$Pr(A_2)$	$Pr(item|X_2=1)$	$Pr(item|X_2=2)$	$\hat{\Theta}$	$\hat{Q}$	$Pr(X_2=1|item)$	$Pr(X_2=2|item)$
$A_2=1$	0.531	0.916 (0.887, 0.945)	0.158 (0.117, 0.199)	58.1	0.966	0.852	0.147
$A_2=2$	0.469	0.084 (0.055, 0.113)	0.842 (0.801, 0.883)			0.107	0.893
$Pr(X_2)$		0.503	0.497				

Response	$Pr(A_3)$	$Pr(item|X_3=1)$	$Pr(item|X_3=2)$	$\hat{\Theta}$	$\hat{Q}$	$Pr(X_3=1|item)$	$Pr(X_3=2|item)$
$A_3=1$	0.590	0.916 (0.887, 0.945)	0.158 (0.117, 0.199)	58.1	0.966	0.881	0.118
$A_3=2$	0.410	0.084 (0.055, 0.113)	0.842 (0.801, 0.883)			0.094	0.906
$Pr(X_3)$		0.559	0.441				

Notes:

$\hat{\Theta}$ = odds ratio expressing likelihood of item response congruent with latent class membership.

$\hat{Q}$ = Yule's Q transform of $\hat{\Theta}$. This expresses the odds ratio in a correlational metric and is an index of item reliability.

Latent Class and Observed Variables:

1 = People can be trusted

2 = You can't be too careful

Table 11.7. Estimated parameter values and reliability estimates for the measurement of two latent categories and probabilities of latent class membership from a latent Markov model of a measure of social trust in the 1970s National Election Study: "Would you say that most of the time people try to be helpful, or that they are mostly just looking out for themselves?" (N = 1,064)

Response	Pr(B_1)	Probability of Item Response Conditional on Class Membership		Item Reliability		Probability of Latent Class Conditional on Item Response	
		$\Pr(\text{item}|Y_1=1)$	$\Pr(\text{item}|Y_1=2)$	$\hat{\Theta}$	$\hat{Q}$	$\Pr(Y_1=1|\text{item})$	$\Pr(Y_1=2|\text{item})$
$B_1=1$	0.533	0.918 (0.881, 0.955)	0.192 (0.151, 0.233)	47.1	0.958	0.812	0.188
$B_1=2$	0.467	0.082 (0.045, 0.119)	0.808 (0.767, 0.849)			0.082	0.918
$\Pr(Y_1)$		0.471	0.529				

Response	Pr(B_2)	$\Pr(\text{item}|Y_2=1)$	$\Pr(\text{item}|Y_2=2)$	$\hat{\Theta}$	$\hat{Q}$	$\Pr(Y_2=1|\text{item})$	$\Pr(Y_2=2|\text{item})$
$B_2=1$	0.586	0.918 (0.881, 0.955)	0.192 (0.151, 0.233)	47.1	0.958	0.847	0.153
$B_2=2$	0.414	0.082 (0.045, 0.119)	0.808 (0.767, 0.849)			0.107	0.893
$\Pr(Y_2)$		0.541	0.459				

Response	Pr(B_3)	$\Pr(\text{item}|Y_3=1)$	$\Pr(\text{item}|Y_3=2)$	$\hat{\Theta}$	$\hat{Q}$	$\Pr(Y_3=1|\text{item})$	$\Pr(Y_3=2|\text{item})$
$B_3=1$	0.595	0.918 (0.881, 0.955)	0.192 (0.151, 0.233)	47.1	0.958	0.857	0.143
$B_3=2$	0.405	0.082 (0.045, 0.119)	0.808 (0.767, 0.849)			0.112	0.888
$\Pr(Y_3)$		0.555	0.445				

Notes:

$\hat{\Theta}$ = odds ratio expressing likelihood of item response congruent with latent class membership.

$\hat{Q}$ = Yule's Q transform of $\hat{\Theta}$. This expresses the odds ratio in a correlational metric and is an index of item reliability.

Latent Class and Observed Variables:
0 = People are helpful
1 = Look out for themselves

Table 11.8. Estimated parameter values and reliability estimates for the measurement of two latent categories and probabilities of latent class membership from a latent Markov model of a measure of social trust in the 1970s National Election Study: "Do you think most people would try to take advantage of you if they got a chance, or would they try to be fair?" (N = 1,065)

Response	$\Pr(C_1)$	Probability of Item Response Conditional on Class Membership		Item Reliability		Probability of Latent Class Conditional on Item Response	
		$\Pr(item|Z_1=1)$	$\Pr(item|Z_1=2)$	$\hat{\Theta}$	$\hat{Q}$	$\Pr(Z_1=1|item)$	$\Pr(Z_1=2|item)$
$C_1=1$	0.656	0.879 (0.844, 0.914)	0.189 (0.120, 0.258)	31.2	0.938	0.907	0.093
$C_1=2$	0.344	0.121 (0.086, 0.156)	0.811 (0.742, 0.880)			0.237	0.762
$\Pr(Z_1)$		0.677	0.323				
Response	$\Pr(C_2)$	$\Pr(item|Z_2=1)$	$\Pr(item|Z_2=2)$	$\hat{\Theta}$	$\hat{Q}$	$\Pr(Z_2=1|item)$	$\Pr(Z_2=2|item)$
$C_2=1$	0.655	0.879 (0.844, 0.914)	0.189 (0.120, 0.258)	31.2	0.938	0.907	0.093
$C_2=2$	0.345	0.121 (0.086, 0.156)	0.811 (0.742, 0.880)			0.237	0.763
$\Pr(Z_2)$		0.655	0.345				
Response	$\Pr(C_3)$	$\Pr(item|Z_3=1)$	$\Pr(item|Z_3=2)$	$\hat{\Theta}$	$\hat{Q}$	$\Pr(Z_3=1|item)$	$\Pr(Z_3=2|item)$
$C_3=1$	0.663	0.879 (0.844, 0.914)	0.189 (0.120, 0.258)	31.2	0.938	0.911	0.089
$C_3=2$	0.337	0.121 (0.086, 0.156)	0.811 (0.742, 0.880)			0.245	0.754
$\Pr(Z_3)$		0.663	0.337				

Notes:

$\hat{\Theta}$ = odds ratio expressing likelihood of item response congruent with latent class membership.

$\hat{Q}$ = Yule's Q transform of $\hat{\Theta}$. This expresses the odds ratio in a correlational metric and is an index of item reliability.

Latent Class and Observed Variables:

0 = People try to be fair

1 = People take advantage

case. This assumption of equal reliabilities across waves is comparable to the quasi-Markov simplex models employed in the earlier parts of the book and discussed at length in Chapter 5. The estimated reliabilities in these tables are comparable to those presented in Table 11.2 for the multiple indicator latent class model, wherein TRUST, HELP, and FAIR were modeled as alternative indicators of a set of two latent classes representing the concept of social trust.

There are two tests of hypotheses that can be entertained with respect to the Π_{AX}, Π_{BY}, and Π_{CZ} parameter matrices. (We have dropped the subscripts from this notation, given the assumption of equal reliabilities across waves.) First, as noted above, standard errors are available from modern software applications, and it is possible to test specific hypotheses about population parameters. It is possible to address the question of whether a particular parameter estimate is 1.0 or 0.0—the case of perfect reliability. The response probabilities presented in Tables 11.6, 11.7, and 11.8 are accompanied by the 95% confidence intervals (given inside the parentheses), and in no case are the values of 1.0 and 0.0 within these intervals. It is also possible to formulate such hypotheses as a global test of perfect reliability, where the matrices Π_{AX}, Π_{BY}, and Π_{CZ} are set equal to an identity matrix. Although possible, we have not performed this test here, preferring instead to read the tests individually from the confidence intervals provided.

A second set of hypotheses that we test using these model constructions is to examine whether the category-specific reliabilities are equal. We use the term *reliability* in the sense of Clogg and Manning's (1996) "reliability as predictability," and ask whether from a statistical viewpoint there is an equality in the probability of the item response given the latent class. This amounts to the test of the equality of diagonal elements of Π_{AX}, Π_{BY}, and Π_{CZ} parameter matrices. Tests of equality of diagonal elements of these parameter matrices were carried out by placing a linear constraint on the threshold logits for the latent response function. The constraint is analogous to the typical equality constraints used for model testing in SEM; in this case the constraint takes the form $t_{01} = -t_{11}$. The test statistic in this case is distributed as χ^2 with 1 degree of freedom. The application of this test in the present case yields significant differences in the case of TRUST ($\chi^2 = 5.915$, df = 1, p = 0.0150) and HELP ($\chi^2 = 11.153$, df = 1, p = 0.0008), but the results for FAIR ($\chi^2 = 2.913$, df = 1, p = 0.0879) indicate a failure to reject the null hypothesis of no differences in the category-specific reliabilities.

Another interesting feature of this model is that it provides estimates of latent distributions for the latent classes at each point in time. These are labeled as $Pr(X_g)$, $Pr(Y_g)$, and $Pr(Z_g)$ (where g = 1,2,3) in Tables 11.6, 11.7, and 11.8. There are at least three potentially meaningful comparisons of these latent class distributions. First, it may be meaningful to compare the latent distributions for a given measure across time. In the present example, the analyses for TRUST and HELP yield the finding that social trust is increasing between 1972 and 1976. There appears to be no change in the latent distribution for FAIR. Second, it may be useful in some circumstances to compare the latent distributions with the observed distributions, and these differences may interact with time. What we find in the present case, at least for TRUST and HELP, is that the observed distributions overestimate the amount of social trust in the population, expressed in terms of the proportion of respondents providing a social trust response. The observed distributions for FAIR, on the other hand, are virtually identical to the latent distributions across all three occasions of measurement. Finally, it may be meaningful to compare measures with respect to the latent distributions. In the present example, TRUST and HELP provide a comparable level of endorsement of social trust at the latent level, which tracks about the same across time, whereas FAIR yields a much higher level of assessed social trust.

It is important to be aware of the relationship between the observed and latent distributions under the latent Markov model; specifically, they are related by way of the matrix of response probabilities discussed earlier. Expressing this relationship formally produces the following equivalence:

$$\pi_{Ag} = \Pi_{AgXg}\, \pi_{Xg}$$

Note that observed distribution, π_{Ag}, is a $(T \times 1)$ column vector of observed probabilities; Π_{AgXg} is a $(T \times T)$ matrix of response probabilities, arranged so that the observed categories define the columns and the latent classes define the columns; and π_{Xg} is a $(T \times 1)$ column vector of latent probabilities. Of course, these are all model-based parameters, so it is more precise to designate these as estimated probabilities.

11.5 CONCLUSIONS

In the study of reliability methods for survey data, a fundamental distinction must be made between measures of latent classes versus measures of latent continuous variables. Regardless of whether measures are continuous or categorical, if the latent variable is continuous, approaches based on classical true-score models are appropriate; but when the latent variable is a set of latent classes, the CTST models will not work. Although there have been some applications of categorical latent class models, to this point the systematic study of measurement error in survey research has focused primarily on the continuous latent variables. In this chapter, we discussed three types of latent class models for categorical latent variables, which provide an analog to the conception of reliability involved in models based on classical theory. We illustrated that for those event classes that do not change, reliability estimation for categorical variables is a straightforward application of latent class modeling techniques. The more difficult challenge is the development of comparable models permitting change in the composition of latent classes, and we discussed the prospects offered by the latent Markov model. These models have not found wide application to date, in part because they are relatively unknown or poorly understood. We expect these circumstances to change in future studies of the reliability of measurement involving categorical latent variables.

Following from the discussion of the latent class models in this chapter, it is worthwhile to draw several comparisons between the latent class approach for categorical variables and the CTST models for continuous variables. Although we can see parallels between the two sets of models, at least at an abstract level, there are several differences worth noting. First, in the CTST tradition, the observed and latent distributions are different, but they have equivalent means. It follows from the assumption of random measurement error in the model, $Y = T + E$, that $\mu_T = \mu_Y$. Although the observed and true scores have the same central tendency, as a consequence of the random error assumption, they obviously have different variances. In the latent class models, there is no constraint on the central tendency and no constraint on the nature of the latent distribution, except as reflected in the response probabilities. Second, although there is an approach to defining the reliability for a given measure in the latent class model, the essential feature of the latent class model involves a set of probabilities linking the observed and latent categories. In the latent class models, the conditional probabilities express the elements of what one might consider an analog to the CTST's *index of reliability*, but there is a fundamental difference. The probabilities involved in the latent class model are category-specific; that is, there is a different estimate of the relationship between the observed data and the latent level for each category of the variable. This assumption can be tested, but there is no counterpart to this in the CTST approach. If we were to formulate the CTST approach in terms of the definition of a set of conditional probabilities relating the observed measures to the latent variables, the probabilities would not differ across levels of the variable.

Although we see many similarities between the latent class approach and the CTST approach at an abstract level of conceptualization, in the end we must conclude that the concepts of reliability involved are fundamentally different and cannot be compared at the level of empirical estimation. Essential to the formulation of a notion of reliability in both models is the assumption of local independence, i.e., the assumption that the measures are conditionally independent, given the latent variable. This assumption is common to all latent variable models. Despite this similarity, and the analog between the two models with respect to the notion of "reliability as predictability," the fact remains that they are very different formulations, from traditions that begin with very different assumptions. While there may be some understandable desire to "bridge the gap between reliability assessment of continuous variables and reliability assessment of categorical variables" (Clogg and Manning, 1996, p. 181), one must admit that the languages of the two traditions are different, and so are the statistical cultures in which they are embedded.

CHAPTER TWELVE

Final Thoughts and Future Directions

When measurement departs from theory, it is likely to yield mere numbers and their very neutrality makes them particularly sterile as a source of remedial suggestions. But numbers register the departure from theory with an authority and finesse that no qualitative technique can duplicate, and that departure often is enough to start the search.

Thomas Kuhn, "The function of measurement in modern physical science" (1961, p. 180)

There are many spheres of human activity where the concept of *reliability* is a compelling indicator of quality, whether in industrial production, consumer product evaluation, in scientific research, or everyday life. I have argued throughout this book that this concept is an indispensable criterion for evaluating the quality of survey data—one among many. In Chapter 2, I pointed out that there should be little doubt that the potential for random measurement errors exists at each of several stages in the response process—from the selection of the question to the selection of a response. The existence of these likely sources of measurement error poses a serious potential limitation to the validity and usefulness of survey data. In addition to other threats to inferences from surveys (coverage or sampling frame problems, sampling errors, and survey nonresponse) (see Groves, 1989; Groves and Couper, 1998), measurement errors can seriously reduce the usefulness of survey data. At the same time, by studying the sources of unreliability in survey measurement, one can improve the survey enterprise.

The assessment of reliability in survey research is, however, not a routine exercise, and a commitment to obtaining information on reliability poses significant challenges for research design and statistical estimation. In the present book, I have shown how it is possible to meet these challenges, but even if it is possible, assessing the reliability of survey questions is by no means an accepted strategy. One is forced to conclude that there is a widespread lack of an appreciation of the concept among survey researchers, as well as some degree of resistance. While other criteria for data quality in survey research are often justifiably stressed (e.g., the quality of the sample, levels of cooperation and response rates, and the extent of sampling error), the quality of measurement is often somewhat of a lesser concern. No matter how good the sample, or how high the response rates, the presence of measurement error renders survey results less useful. With this in mind, I have suggested that it is possible to define and operationalize one aspect of survey measurement quality—reliability of measures—and I have argued that information gathered on the reliability of measurement, such as that reported in this book, can help improve the methods of data collection and the inferences drawn from survey research. Such information can supplement other experiences that can help evaluate the effectiveness of specific types of survey question, and, in turn, be used to improve survey questionnaires.

The purpose of this chapter is to further make the case for the development of a research strategy that systematically investigates the nature and extent of random measurement errors in survey data and the conditions under which such errors are more (or less) likely to occur. In previous chapters, I articulated an approach that addresses these issues and presented several sets of results that examine some of these questions. In this chapter, I summarize the implications of these results and discuss what I consider to be profitable future directions of research on this topic. First, however, I return briefly to the rationale for focusing on the reliability of measurement and the approach adopted to assess it. I engage these issues here in part because I know that some people who read this book will have fundamental doubts or skepticism about the usefulness and applicability of the concept of reliability to evaluating the quality of survey data and/or my approach to the matter.

12.1   RELIABILITY AS AN OBJECT OF STUDY

Without accurate and consistent measurement, the statistical tabulation and quantitative analysis of survey data hardly makes sense. Yet, there is a general lack of empirical information about potential problems and very little available information on the reliability of measurement from large-scale population surveys for the types of survey measures in common use. Literature on this topic is growing, and the examination of these issues is considered a legitimate topic of investigation. Still, there are several questions that survey methodologists and others will raise about this approach—questions that need to be addressed before I draw conclusions and present some suggestions for next steps.

First, some will suggest that an emphasis on reliability is misplaced. Reliability, some will argue, is nothing more than a measure of the consistency of measurement and therefore *focuses on the wrong thing*. A concern with measurement quality should focus on validity, this perspective would argue, *not* reliability. This is potentially persuasive in that *validity* of measurement is an extremely important consideration—more important than *reliability*, many would argue—however, this perspective is based on a misconception. As I pointed out (see Chapter 3), one cannot assess validity without having a high degree of reliability. The two are intimately related, and in the discussion below I will show formally *how the level of reliability places an upper limit on estimates of validity*. It serves, therefore, as a sine qua non of survey measurement—put simply, *without a high degree of reliability one cannot have validity.*

Second, even if the topic of "errors of measurement" is viewed as a legitimate focus of research on survey quality, the argument is often made that reliability captures only one kind of measurement error—random errors—and there are many more interesting types of measurement error that should be given a greater priority. Again, this argument is specious. By definition, systematic errors are a part of reliable variance, and as I pointed out earlier, they are confounded with the "true" variables we are interested in. There is little question that systematic (or nonrandom) errors are reliably present in survey data, and they are an important focus of research. However, the existence of systematic error in no way removes or prevents the additional possibility of random errors. Indeed, *it is difficult, if not impossible, to identify systematic errors without first taking random errors into account.* These are complementary, not competing, conceptualizations of measurement error; and to focus on reliability in no way denies the existence of nonrandom errors. Both are important topics of examination.

Third, the approach taken in this book has originated from the psychometric perspective on measurement error, and is not formulated from complementary perspectives (e.g., the statistical sampling framework). Although I have made an effort to point out how differing perspectives can be integrated, the present approach will be unsatisfactory from the point of view of sampling perspectives because in classical theory errors are formulated at the population level and the models involved do not incorporate sampling errors. Here I think the issue is a matter of perspective. As Groves (1989) has argued, there are many valuable perspectives on survey errors, and *much can be learned from viewing survey measurement from a psychometric perspective.* Building upon these comments, in the following section I provide a more refined basis for addressing many of these and other potential criticisms of my approach, with a focus on the rationale for the program of research on which it is based. It is neither incompatible with, nor does it seek to replace, other approaches to the study of survey errors.

12.2 WHY STUDY THE RELIABILITY OF SURVEY MEASUREMENT?

The short answer to this question is: *Because almost everything depends on the absence of random errors of measurement.* To explain what this means and to hopefully promote greater understanding of why estimating survey measurement reliability is important, I return to several of the arguments I made earlier. First, nearly everyone would agree that reliability of measurement is a necessary condition for validity in the social sciences. Paraphrasing the words of Lord and Novick (1968, p. 72) the linkage between reliability and validity represents a major justification for the study of the reliability of survey measurement. The simplest form of *validity*, expressed as the correlation of a given variable, X, with another variable, Y, that it is expected to predict—also referred to as *predictive validity* or *criterion-related validity*—can be used as the focus of this discussion.

Consider the following proof. First, it can be shown that COR(X,Y) is always less than or equal to the *index of reliability* of either and both X or Y. We can bring in the notation used in Chapter 3—that ρ_X and ρ_Y are the indices of reliability (i.e., the square root of reliability) for X and Y. These inequalities can be stated more formally as

$$COR(X,Y) \leq \rho_X$$
$$COR(X,Y) \leq \rho_Y$$

where $\rho_X = COR(X,T_X)$ and $\rho_Y = COR(Y,T_Y)$ (see Chapter 3). To understand the basis for these inequalities, note first that it makes sense that a measure correlates at least as highly with its own true score as it does with the true score on a second measurement, i.e., $COR(X,T_X) \geq COR(X,T_Y)$. Second, the absolute level of correlation of a given measure with the true score of a second measure is at least as high as with the observed score on the second measure, i.e., $|COR(X,T_Y)| \geq |COR(X,Y)|$ (see Lord and Novick, 1968, pp. 69–70). This leads to the following inequality:

$$COR(X,T_X) \geq COR(X,T_Y) \geq COR(X,Y)$$
$$COR(Y,T_Y) \geq COR(Y,T_X) \geq COR(X,Y)$$

From these observations, then, it can readily be deduced that the index of reliability of a particular measure places an upper bound of the correlation of that measure with any other variable, i.e., *reliability is a necessary condition for validity*. Of course, as should be clear, it does not follow that a high level of reliability implies (or even suggests) high validity, so it is important to add that *while reliability is a necessary condition for validity, it is not a sufficient one*.

These inequalities reinforce (and clarify further) the point made in Chapter 3 that *criterion validity* of measurement cannot exceed the *index of reliability* (the square root of reliability) (see Lord and Novick, 1968, pp. 72–74). Recall that, in this case, the term "criterion validity" is simply defined as the correlation of the measure, X, with some other variable, Y, presumably a criterion linked to the purpose of measurement. Similar proofs can be developed for more sophisticated notions of validity, e.g., construct validity, but these are unnecessary for present purposes.

From these considerations one can readily conclude that (1) *if reliability is a necessary condition for validity* and (2) *if the failure to obtain valid measurement presents a major limitation on the contribution of survey measurement to social scientific knowledge*, then (3) *a priority should be placed on finding out how reliable are our measures*, and (4) *making improvements where possible*.

12.2.1 Effects of Unreliability on Statistical Inference

Another way to justify the importance of a strategy for assessing the reliability of measurement is to examine the effects of random errors of measurement on statistical inference. It is well known, for example, that unreliability of measurement inflates sample estimates of population variances, and it can therefore readily be seen to have a biasing effect on estimates of standard errors of sample estimates of population means, among other things. This bias artificially inflates confidence intervals; and it is well known that statistical analyses ignoring unreliability of variables underestimate the strength and statistical significance of the statistical association between those variables. Although this underestimation tends to make analyses more conservative from a scientific perspective, it also increases the probability of type II error and consequent rejection of correct, scientifically productive hypotheses about the effects of independent variables on the dependent variables that measure phenomena of interest.

We can see the influence of unreliability on statistical inference by examining a concrete example. Consider the example of testing the null hypothesis that there are no differences between two groups. This is a standard test that is used routinely in research employing survey data intended to evaluate whether the two groups involved come from the same population. I present the p-values associated with the *critical ratios* for two cases where two factors are allowed to vary—the sample size per group and the reliability of measurement—in Tables 12.1 and 12.2. I have simplified the examples by constraining the sample size per group to be the same and the population response variance to be equal across groups. The examples are given for variables in standard form, and the true standard deviations in each group are equal to one, conditions set in order to make clear the effects of reliability/unreliability. Results are presented for two cases: (1) where the observed sample mean difference is equal to one-quarter of a standard deviation and (2) where the mean difference is equal to one-half of a standard deviation.

These tables reproduce two results for the critical ratios and their associated p-values involved in these tests—the effect of sample size and the effect of unreliability of measurement. Specifically, as one moves down the columns for a given level of reliability, i.e., as the sample size per group increases, the null hypothesis becomes easier to reject. However, the gains from sample size are tempered by unreliability of measurement, as the p-values associated with increasing sample sizes decline at a decreasing rate with lower levels of reliability. In addition, as one moves across the rows, for a given sample size the ability to reject the null hypothesis is made more difficult with declining reliability. The first of these results—the effect of sample size on statistical inference—is much better appreciated than the effect of unreliability of measurement in affecting statistical inferences, but the implications of these results are clear. *Unreliability of measurement places limitations on tests of simple hypotheses.*

Interestingly, the negative consequences of lower reliability can be offset by increasing sample size, but this is a principle that cannot be exploited to any serious lengths. For example, with a sample size of 125 in each group, a difference of .25 standard deviation reaches significance at the p < .05 level when there is perfect measurement. As reliability declines—reading across the row—the p-value increases and the ability to detect such a difference also declines. In order to find such a difference as significant when reliability of measurement is .50, one would need to double the sample size. This is a very expensive solution to the problem of unreliability of measurement, and I do not recommend this as a solution to the problem.

In addition to making it more difficult to reject the null hypothesis in such cases, unreliability of measurement has an effect on statistical power, that is, it increases

Table 12.1.   Probability of type I error for a two-tailed difference of means test as a function of sample size and reliability of measurement where the true variance equals 1.0 and the true difference of means is .25

Sample Size per Group	Reliability									
	1	0.9	0.8	0.7	0.6	0.5	0.4	0.3	0.2	0.1
30	0.337	0.362	0.390	0.421	0.456	0.496	0.543	0.598	0.667	0.761
35	0.299	0.325	0.353	0.385	0.421	0.462	0.511	0.569	0.641	0.742
40	0.267	0.292	0.320	0.352	0.389	0.432	0.482	0.542	0.618	0.725
45	0.239	0.264	0.292	0.324	0.361	0.404	0.455	0.518	0.597	0.709
50	0.214	0.239	0.266	0.298	0.335	0.379	0.431	0.495	0.577	0.693
55	0.193	0.216	0.244	0.275	0.312	0.356	0.409	0.474	0.559	0.679
60	0.174	0.196	0.223	0.254	0.291	0.335	0.388	0.455	0.541	0.666
65	0.157	0.179	0.205	0.235	0.272	0.315	0.369	0.436	0.525	0.653
70	0.141	0.163	0.188	0.218	0.254	0.297	0.351	0.419	0.509	0.641
75	0.128	0.149	0.173	0.202	0.238	0.281	0.335	0.403	0.495	0.629
80	0.116	0.136	0.159	0.188	0.222	0.265	0.319	0.388	0.481	0.618
85	0.105	0.124	0.147	0.175	0.209	0.251	0.304	0.373	0.467	0.607
90	0.095	0.113	0.135	0.162	0.196	0.237	0.290	0.360	0.454	0.597
95	0.087	0.104	0.125	0.151	0.184	0.225	0.277	0.347	0.442	0.586
100	0.079	0.095	0.115	0.141	0.172	0.213	0.265	0.334	0.430	0.577
125	0.049	0.062	0.078	0.099	0.127	0.163	0.212	0.280	0.378	0.533
150	0.031	0.041	0.054	0.071	0.095	0.127	0.172	0.237	0.334	0.494
175	0.020	0.027	0.037	0.051	0.071	0.099	0.140	0.201	0.296	0.460
200	0.013	0.018	0.026	0.037	0.054	0.078	0.115	0.172	0.264	0.430
225	0.008	0.012	0.018	0.027	0.041	0.061	0.094	0.147	0.236	0.402
250	0.005	0.008	0.013	0.020	0.031	0.049	0.078	0.126	0.212	0.377
275	0.004	0.006	0.009	0.014	0.024	0.039	0.064	0.109	0.190	0.354
300	0.002	0.004	0.006	0.011	0.018	0.031	0.053	0.094	0.171	0.333
325	0.002	0.003	0.005	0.008	0.014	0.025	0.044	0.081	0.155	0.314
350	0.001	0.002	0.003	0.006	0.011	0.020	0.037	0.071	0.140	0.296
375	0.001	0.001	0.002	0.004	0.008	0.016	0.031	0.061	0.126	0.279
400	0.000	0.001	0.002	0.003	0.006	0.013	0.026	0.053	0.114	0.264

the likelihood of failing to reject the null hypothesis when there is in fact a true effect. Although researchers have long wanted to set low probabilities of making type I errors (the probability of rejecting the null hypothesis when it is true) they have historically not given equal consideration to type II errors, i.e., accepting the null hypothesis when there is in fact a real effect. A test has *statistical power* when it leads the researcher to correctly reject the null hypothesis (Murphy and Myors, 2004). In this context, it can readily be shown that unreliability of measurement reduces statistical power (Cleary, Linn, and Walster, 1970).

This conclusion regarding statistical power can be shown by considering a concrete example. Consider again the example of testing the null hypothesis that there are no differences between the means of two groups, i.e., that they come from the same population. Standard power calculations in this case require knowledge of an effect size, the standard error, and the number of cases per group (see Castelloe,

Table 12.2. Probability of type I error for a two-tailed difference of means test as a function of sample size and reliability of measurement where the true variance equals 1.0 and the true difference of means is .50

Sample Size per Group	Reliability									
	1	0.9	0.8	0.7	0.6	0.5	0.4	0.3	0.2	0.1
30	0.058	0.071	0.089	0.111	0.139	0.176	0.226	0.293	0.390	0.543
35	0.040	0.051	0.066	0.085	0.110	0.144	0.190	0.256	0.353	0.511
40	0.028	0.037	0.049	0.065	0.087	0.118	0.161	0.224	0.320	0.482
45	0.020	0.027	0.037	0.050	0.070	0.097	0.137	0.197	0.292	0.455
50	0.014	0.020	0.028	0.039	0.056	0.080	0.117	0.174	0.266	0.431
55	0.010	0.014	0.021	0.030	0.045	0.066	0.100	0.154	0.244	0.409
60	0.007	0.011	0.016	0.024	0.036	0.055	0.086	0.136	0.223	0.388
65	0.005	0.008	0.012	0.019	0.029	0.046	0.074	0.121	0.205	0.369
70	0.004	0.006	0.009	0.015	0.023	0.038	0.063	0.107	0.188	0.351
75	0.003	0.004	0.007	0.011	0.019	0.032	0.055	0.096	0.173	0.335
80	0.002	0.003	0.005	0.009	0.015	0.027	0.047	0.085	0.159	0.319
85	0.001	0.002	0.004	0.007	0.013	0.022	0.041	0.076	0.147	0.304
90	0.001	0.002	0.003	0.006	0.010	0.019	0.035	0.068	0.135	0.290
95	0.001	0.001	0.002	0.004	0.008	0.016	0.031	0.061	0.125	0.277
100	0.001	0.001	0.002	0.003	0.007	0.013	0.026	0.054	0.115	0.265
125	0.000	0.000	0.000	0.001	0.002	0.006	0.013	0.031	0.078	0.212
150	0.000	0.000	0.000	0.000	0.001	0.002	0.007	0.018	0.054	0.172
175	0.000	0.000	0.000	0.000	0.000	0.001	0.003	0.011	0.037	0.140
200	0.000	0.000	0.000	0.000	0.000	0.000	0.002	0.006	0.026	0.115
225	0.000	0.000	0.000	0.000	0.000	0.000	0.001	0.004	0.018	0.094
250	0.000	0.000	0.000	0.000	0.000	0.000	0.000	0.002	0.013	0.078
275	0.000	0.000	0.000	0.000	0.000	0.000	0.000	0.001	0.009	0.064
300	0.000	0.000	0.000	0.000	0.000	0.000	0.000	0.001	0.006	0.053
325	0.000	0.000	0.000	0.000	0.000	0.000	0.000	0.001	0.005	0.044
350	0.000	0.000	0.000	0.000	0.000	0.000	0.000	0.000	0.003	0.037
375	0.000	0.000	0.000	0.000	0.000	0.000	0.000	0.000	0.002	0.031
400	0.000	0.000	0.000	0.000	0.000	0.000	0.000	0.000	0.002	0.026

2000). In this example we calculate power for two cases of effect size: (1) one where the effect size is equal to one-quarter of a standard deviation (see Table 12.3) and (2) one where the effect size is equal to one-half of a standard deviation (Table 12.4). Within each of these cases, we set the type I error at three standard levels: .05, .01 and .001. Then we examine the consequences of sample size and levels of reliability (via its effect on the standard error) for the calculated level of statistical power.

The results in Tables 12.3 and 12.4 reinforce several facts about the factors affecting statistical power: (1) tests involving smaller effect sizes have less power, for a given sample size (compare results in any given cell across the two tables); (2) for a particular effect size, power decreases as the type I error is lowered (compare results in any given cell across the three panels of either table); (3) increases in sample size are associated with greater statistical power (compare

Table 12.3. Power calculations for a two-tailed difference of means test as a function of sample size and reliability of measurement where the true variance equals 1.0 and the true difference of means is .25

Sample Size per Group	Reliability									
	1	0.9	0.8	0.7	0.6	0.5	0.4	0.3	0.2	0.1
Type I Error = .05										
30	0.159	0.147	0.136	0.125	0.114	0.103	0.092	0.082	0.071	0.060
60	0.274	0.252	0.229	0.206	0.183	0.161	0.138	0.115	0.093	0.071
90	0.385	0.353	0.320	0.287	0.253	0.218	0.184	0.150	0.116	0.082
125	0.504	0.463	0.421	0.377	0.332	0.285	0.238	0.190	0.142	0.095
175	0.645	0.600	0.550	0.497	0.439	0.378	0.314	0.248	0.181	0.114
225	0.754	0.709	0.658	0.600	0.536	0.465	0.387	0.305	0.220	0.133
300	0.864	0.826	0.781	0.725	0.658	0.580	0.489	0.388	0.277	0.162
375	0.928	0.900	0.864	0.816	0.754	0.676	0.580	0.465	0.333	0.191
425	0.954	0.932	0.903	0.861	0.805	0.730	0.634	0.514	0.370	0.210
Type I Error = .01										
30	0.051	0.046	0.042	0.037	0.033	0.029	0.025	0.021	0.017	0.013
60	0.110	0.098	0.086	0.074	0.063	0.053	0.043	0.034	0.025	0.017
90	0.180	0.159	0.138	0.118	0.099	0.081	0.064	0.048	0.034	0.021
125	0.270	0.238	0.206	0.175	0.146	0.118	0.091	0.067	0.045	0.026
175	0.402	0.356	0.311	0.265	0.220	0.176	0.135	0.097	0.063	0.033
225	0.526	0.472	0.416	0.357	0.298	0.240	0.183	0.130	0.082	0.041
300	0.684	0.626	0.562	0.492	0.417	0.338	0.260	0.183	0.113	0.054
375	0.800	0.747	0.684	0.611	0.528	0.436	0.339	0.240	0.147	0.067
425	0.856	0.809	0.751	0.680	0.595	0.498	0.391	0.280	0.171	0.077
Type I Error = .001										
30	0.009	0.008	0.007	0.006	0.005	0.004	0.003	0.003	0.002	0.002
60	0.025	0.022	0.018	0.015	0.012	0.010	0.007	0.005	0.004	0.002
90	0.051	0.042	0.035	0.028	0.022	0.017	0.012	0.009	0.005	0.003
125	0.091	0.076	0.062	0.049	0.038	0.028	0.020	0.013	0.008	0.004
175	0.166	0.138	0.112	0.089	0.068	0.049	0.034	0.022	0.012	0.005
225	0.256	0.215	0.175	0.139	0.106	0.077	0.052	0.032	0.017	0.007
300	0.404	0.345	0.286	0.230	0.176	0.128	0.086	0.053	0.027	0.010
375	0.548	0.478	0.405	0.331	0.258	0.190	0.129	0.077	0.039	0.013
425	0.634	0.562	0.484	0.401	0.317	0.235	0.160	0.097	0.048	0.016

power calculations within any column of any of the panels in either table); and (4) statistical power is reduced by unreliability (compare values across any row of any of the panels in either table). Of course, the point of these tables is to introduce the idea that lower levels of measurement reliability reduce statistical power, a result that may not be as well known as the others. This can be seen, for example, in Table 12.4, where the effect size is .50 of a standard deviation and where the level of type I

Table 12.4.   Power calculations for a two-tailed difference of means test as a function of sample size and reliability of measurement where the true variance equals 1.0 and the true difference of means is .50

Sample Size per Group	Reliability									
	1	0.9	0.8	0.7	0.6	0.5	0.4	0.3	0.2	0.1
					Type I Error = .05					
30	0.478	0.439	0.399	0.357	0.314	0.270	0.226	0.181	0.136	0.092
60	0.775	0.731	0.681	0.623	0.557	0.484	0.405	0.319	0.229	0.138
90	0.916	0.886	0.847	0.797	0.734	0.655	0.560	0.447	0.320	0.184
125	0.976	0.962	0.941	0.909	0.862	0.795	0.702	0.578	0.421	0.238
175	0.997	0.993	0.987	0.974	0.951	0.910	0.839	0.724	0.550	0.314
225	1.000	0.999	0.997	0.993	0.984	0.963	0.917	0.826	0.658	0.387
300	1.000	1.000	1.000	0.999	0.997	0.991	0.972	0.918	0.781	0.489
375	1.000	1.000	1.000	1.000	1.000	0.998	0.991	0.963	0.864	0.580
425	1.000	1.000	1.000	1.000	1.000	0.999	0.996	0.979	0.903	0.634
					Type I Error = .01					
30	0.244	0.215	0.186	0.158	0.132	0.107	0.083	0.061	0.042	0.025
60	0.549	0.494	0.436	0.376	0.314	0.253	0.193	0.136	0.086	0.043
90	0.772	0.718	0.654	0.581	0.499	0.411	0.318	0.225	0.138	0.064
125	0.912	0.875	0.825	0.761	0.679	0.579	0.463	0.335	0.206	0.091
175	0.981	0.967	0.944	0.906	0.848	0.763	0.644	0.490	0.311	0.135
225	0.997	0.993	0.984	0.967	0.935	0.877	0.778	0.625	0.416	0.183
300	1.000	0.999	0.998	0.994	0.984	0.959	0.901	0.779	0.562	0.260
375	1.000	1.000	1.000	0.999	0.997	0.988	0.959	0.878	0.684	0.339
425	1.000	1.000	1.000	1.000	0.999	0.995	0.979	0.921	0.751	0.391
					Type I Error = .001					
30	0.075	0.062	0.051	0.041	0.032	0.024	0.017	0.011	0.007	0.003
60	0.269	0.226	0.185	0.147	0.112	0.081	0.055	0.034	0.018	0.007
90	0.505	0.438	0.368	0.299	0.232	0.170	0.115	0.069	0.035	0.012
125	0.732	0.662	0.582	0.492	0.397	0.299	0.207	0.125	0.062	0.020
175	0.912	0.867	0.805	0.723	0.619	0.496	0.361	0.227	0.112	0.034
225	0.976	0.956	0.923	0.869	0.786	0.669	0.517	0.343	0.175	0.052
300	0.997	0.994	0.985	0.965	0.924	0.846	0.714	0.519	0.286	0.086
375	1.000	0.999	0.998	0.992	0.977	0.937	0.847	0.672	0.405	0.129
425	1.000	1.000	0.999	0.997	0.990	0.968	0.904	0.755	0.484	0.160

error is set at .01. Acceptable levels of statistical power (above .9) are achieved with a sample size of 125 in each group when there is perfect measurement, but when reliability declines, so does the calculated level of power, and it does not take much loss in reliability to reduce the power to below acceptable levels. As noted above, the negative consequences of unreliability are offset by increasing sample sizes, but while true, this is not a recommended strategy in general.

12.2.2   Effects of Unreliability on Estimates of Regression Coefficients

Perhaps the best-known fact about the effects of unreliability on statistical inference involves its attenuation of bivariate statistical associations and regression coefficients. It can be shown in the simplest regression models that unreliability of measurement in predictor variables bias regression coefficients downward, making it
more difficult to reject the null hypothesis, and unreliability in both dependent and
independent variables attenuate estimates of statistical associations.

Taking the first assertion, consider the formula for the simple regression of T_Y on
T_X, the true scores for observed variables Y and X, as follows:

$$b(T_Y, T_X) = COV(T_X, T_Y) / VAR(T_X)$$
$$= COV(X, Y) / VAR(T_X).$$

This shows an established result of classical true-score theory that random errors
of measurement do not affect the covariances among variables (see Chapter 3). The
use of the simple regression of the observed variables Y on X, $b(X,Y) = COV(X,Y) /
VAR(X)$, will, therefore, produce an underestimate of this true regression due to
the bias in the estimate of the denominator, that is, $b(X,Y) \leq b(T_Y, T_X)$. This comes
about because of the fact that in the presence of measurement error in X, VAR(X)
overestimates $VAR(T_X)$, that is, $VAR(X) = VAR(T_X) + VAR(E_X)$. It is less straightforward an exercise to show the biases happening in multivariate regression models,
but suffice it to say that to the extent that the predictor variables have substantial
amounts of measurement error, the regression coefficients will be biased. It is a particularly serious matter when predictor variables have different levels of reliability,
given that the purpose of such models is often to draw inferences regarding the relative importance of variables. In short, variables with greater amounts of measurement error are relatively disadvantaged in such endeavors.

With appropriate estimates of reliability of measurement it is possible using
some statistical software packages (e.g., Stata) to isolate some types of error statistically and therefore control for them in the analysis of data.[1] The work of Bielby,
Hauser and their colleagues (Bielby and Hauser, 1977; Bielby, Hauser and Featherman, 1977a, 1977b; Hauser, Tsai and Sewell, 1983), discussed below, illustrate how
inferences about the relative effects of variables can be significantly altered by taking measurement error into account.

12.2.3   Effects of Unreliability on Estimates of Statistical Association

It is a relatively straightforward exercise to show how Pearson product-moment correlations are attenuated by random measurement errors in either X or Y. It is well

---

[1]More specifically, the Stata statistical program includes an "errors in variables" routine (eivreg)
that permits the importation of reliability estimates to improve the precision of the estimates of the
regression model. Although more cumbersome, the same objectives can be accomplished using SEM
programs, such as LISREL, AMOS, or M*plus*.

known that, if reliability is known, there is a standard tool that can be employed to disattenuate the observed correlations. In order to show this, note, first, that the correlation between the true scores, T_X and T_Y, is equal to

$$COR(T_X,T_Y) = COR(X,Y) / \rho_X \rho_Y$$

where ρ_X and ρ_Y are defined as the indexes of reliability for X and Y, respectively, that is, $\rho_X = COR(X,T_X)$ and $\rho_Y = COR(Y,T_Y)$ (see Chapter 3). This is a standard attenuation correction formula given in any basic textbook on test theory (see, e.g., Lord and Novick, 1968, p. 69). One can see from this that if the reliability of measurement for X and Y is known, the true correlation, that is, the correlation between T_X and T_Y, can be inferred given the observed correlation.

From here let us rewrite the above equation as

$$COR(T_X,T_Y) \rho_X \rho_Y = COR(X,Y) .$$

Imagine in the extreme case where $COR(T_X,T_Y) = 1.0$ and $\rho_X = 1.0$—here $\rho_Y = COR(X,Y)$. So, by logic, as $COR(T_X,T_Y)$ and ρ_X drop from unity, so will the correlation between X and Y, and ρ_Y is clearly seen as an upper bound on $COR(X,Y)$. Figure 12.1 plots the observed correlation as a function of the true correlation and reliability. From these graphs it is clear that the only instance in which the observed correlation equals the true correlation is when reliability of both variables, X and Y, is 1.0, i.e., there is no random measurement error in either X or Y. A reliability < 1.0 attenuates the true correlation, resulting in systematic bias in the observed correlation. For example, reading from the plot in Figure 12.1, when the true correlation is .40 and reliability of both variables is .50, the observed correlation is .20. Or, taking another example, when the true correlation is .70 and reliability for both variables is also .70, the observed correlation is .49. These identities should give some pause to researchers who study statistical associations among variables and who draw inferences about the extent of true association in populations of interest.

12.3   THE LONGITUDINAL APPROACH

Assuming that reliability of measurement is a legitimate approach to evaluating (one aspect of) the quality of survey data, how then should we obtain estimates of reliability of survey measures? In Chapter 5, I argued in favor of adopting a longitudinal approach. While there is no need to reproduce the entire argument here, the key elements of the argument hinge in part on the limitations of the use of cross-sectional survey designs to generate estimates of measurement reliability. One major disadvantage of the cross-sectional approach derives from the inability to design replicate measures that can be used in cross-sectional studies that meet the requirements of the statistical models for estimating reliability. Specifically, given the substantial potential for sources of specific variance when "nonreplicate" measures are used, it is difficult to make the assumption of *univocity*, that is, that the measures measure one and only one thing. As we argued, this assumption is not required in the longitudinal case—measures can be multidimensional. In addition, given the fact

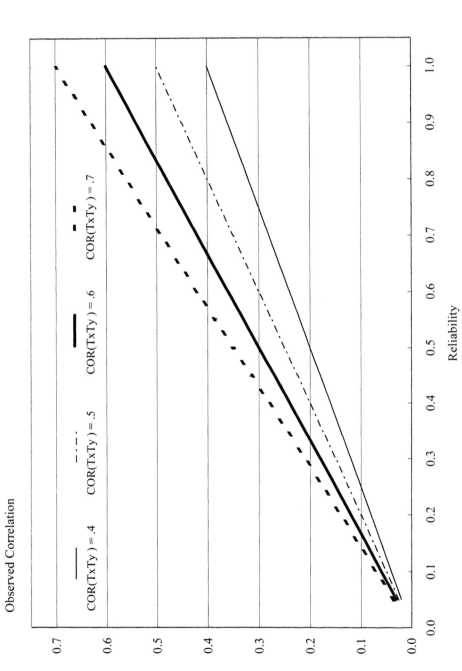

Figure 12.1. Observed correlation as a function of true correlation and reliability.

that reliability estimation is based on the idea of replicate measures, it is difficult to assume in cross-sectional designs that their errors are independent, an assumption that is critical to the statistical models for estimating reliability.

In making these observations, I do not intend to imply that there is anything new here. The limitations of cross-sectional designs for reliability estimation have been recognized for some time and the problems with interpretations of reliability placed on "multiple indicators" models are increasingly becoming better known (e.g., see Alwin and Jackson, 1979; Goldstein, 1995). Even when one is able to employ the multitrait-multimethod (MTMM) approach in cross-sectional designs (i.e., measures of the same trait or concept that vary key features of the measures), and even if one is willing to extend the logic of the CTST model to include multiple sources of "true variance," the problems do not go away (see Chapter 4). The assumption of the independence of errors continues to be problematic. Specifically, in order to interpret the parameters of such models in terms of reliability of measurement, it is essential to be able to rule out the operation of memory, or other forms of consistency motivation, in the organization of responses to multiple measures of the same construct.

Thus, as I suggested in Chapter 5, a strong alternative to the cross-sectional designs for reliability estimation is the multiple-wave reinterview design, a design that goes beyond the traditional test-retest design (see Moser and Kalton, 1972, pp. 353–354). In that chapter, I subjected such test-retest designs to careful scrutiny with respect to estimating reliability, but argued that the limitations of the test-retest design can be overcome by incorporating three or more waves of data separated by lengthy periods of time and by modeling the change in the underlying true score using the quasi-Markov simplex approach (Alwin, 1989; Heise, 1969; Jöreskog, 1970; Saris and Andrews, 1991). I argued specifically that the problem of consistency due to retest effects or memory could be remedied, or at least minimized, through the use of design strategies with relatively distant reinterview intervals of more than 2 years, thereby reducing the likelihood of correlated errors due to memory.

12.4   ASSEMBLING KNOWLEDGE OF SURVEY MEASUREMENT RELIABILITY

Although these issues have been around for many years, knowledge has only recently been cumulating regarding the levels of reliability for the typical kinds of information gathered by surveys. We know even less about the factors linked to the quality of measurement. While there is a general lack of information about the level of reporting reliability in the standard survey interview, there is a small and growing cadre of investigators addressing these important questions. In this section we briefly describe the convergence of our estimates of reliability with other studies and introduce the idea for the development of an archive of reliability estimates that can be drawn upon for information about measurement quality. I argue that the development of such an archive of known reliability estimates for routinely used survey measures will not only in the long run improve the quality of our inferences but will also serve as a basis ultimately for the evaluation of the quality of survey data.

12.4.1 Some Comparisons of Reliability Estimates

The potential of the simplex models used here to produce meaningful estimates of reliability can be seen by comparing our estimates to other empirical applications of similar designs. I limit the focus here to the measurement of variables reflecting socioeconomic position. In Table 12.5, we present estimates for reliability from three well-known studies of the reliability of standard measures of socioeconomic variables—Siegel and Hodge (1968), Bielby and colleagues (Bielby and Hauser, 1977; Bielby, Hauser and Featherman, 1977a, 1977b), and Hauser and colleagues (Hauser, Tsai and Sewell, 1983). This table reports reliability estimates for three frequently used measures of socioeconomic status: education, occupational standing, and income. These three studies are classic studies in reliability estimation and have served as the basis of statistical adjustment of relationships to remove bias in regression coefficients.

The Siegel and Hodge (1968) study was one of the earliest systematic studies of reliability of census reports of socioeconomic variables. The data for their study were taken from cross-tabulations from the *1960 Census of Population*, the March 1960 *Current Population Survey* (CPS), and the *1960 Census Post-Enumeration Survey* (PES). The correlations on which the reliability estimates are derived are based on matched cases from census reports and one of the two surveys—the CPS in the case of years of schooling and the PES for occupation and income. This design utilized two independent reports of the variable in each case—and, as such, this design closely approximates a *multiple-measures* design (see Chapter 3), although the measures were obtained at different times. Except for the potential problem of correlated errors in such designs, the simple correlations among these two reports can be interpreted as estimates of the reliability of the census reports. The resulting estimates for the number of years of schooling attained (reported in the CPS), the Socioeconomic Index (SEI) scores for detailed occupation (reported in the PES), and personal income (also reported in the PES) are given in Table 12.5.

The Bielby et al. studies (Bielby and Hauser, 1977; Bielby, Hauser and Featherman, 1977a, 1977b) are based on a similar methodology involving the March 1973 *Current Population Survey.* The March 1973 CPS questionnaire was supplemented by a mail questionnaire (OCGQ) conducted by Featherman and Hauser (1975) in conjunction with their Occupational Changes in a Generation study (see also Hauser and Featherman, 1977; Featherman and Hauser, 1978). The OCGQ questionnaire gathered data on social background and occupational careers not typically collected in the CPS. In order to examine response errors in the measurement of socioeconomic variables, a random subsample of respondents to the OCGQ was interviewed principally by telephone about three weeks subsequent to the return of the mail questionnaire. Again, apart from the potential for correlated errors in such designs (see Chapter 5), the "test-retest" correlations among comparable variables in these data can be interpreted as estimates of reliability of the CPS and OCGQ data. Resulting estimates for years of schooling, the occupational standing of first and current jobs, and earnings for nonblack males are presented in Table 12.5.

Table 12.5. Comparison of reliability estimates of socioeconomic measures from the present study with estimates from other large-scale projects

Education

Study	Concept	Population	Survey	Design	Estimate
Siegel and Hodge (1968)	Yrs. of schooling	Persons in U.S. households	Census-PES	Test-retest	0.933
Bielby, Hauser & Featherman (1977)	Yrs. of schooling	Nonblack males in U.S. households	CPS-OCG	Test-retest	0.840
Hauser, Tsai & Sewell (1983)	Yrs. of schooling	Wisconsin 1957 high school grads	WLS	Test-retest	0.845
Present study	Yrs. of schooling	U.S. households	NES50	Three-wave simplex	0.909
Present study	Yrs. of schooling	U.S. households	NES70	Three-wave simplex	0.972
Present study	Yrs. of schooling	Detroit families -- mothers	SAFMO	Three-wave simplex	0.948
Present study	Yrs. of schooling	Detroit families -- children	SAFCH	Three-wave simplex	0.938
				Average	0.912

Occupational Standing

Study	Concept	Population	Survey	Design	Estimate
Siegel and Hodge (1968)	Current job --SEI	Persons in U.S. households	Census-CPS	Test-retest	0.873
Bielby, Hauser & Featherman (1977)	First job -- SEI	Nonblack males in U.S. households	CPS-OCG	Test-retest	0.850
Bielby, Hauser & Featherman (1977)	Current job --SEI	Nonblack males in U.S. households	CPS-OCG	Test-retest	0.800
Hauser, Tsai & Sewell (1983)	Early job -- SEI	Wisconsin 1957 high school grads	WLS	Test-retest	0.685
Hauser, Tsai & Sewell (1983)	Current job --SEI	Wisconsin 1957 high school grads	WLS	Test-retest	0.817
Present study	Current job --SEI	U.S. households	NES70	Three-wave simplex	0.827
Present study	Current job --SEI	U.S. households	NES90	Three-wave simplex	0.804
				Average	0.808

Income

Study	Concept	Population	Survey	Design	Estimate
Siegel and Hodge (1968)	Personal income	Persons in U.S. households	Census-CPS	Test-retest	0.847
Bielby & Hauser (1977)	Earnings	Nonblack males in U.S. households	CPS-OCG	Test-retest	0.904
Present study	Job income	U.S. households	ACL	Three-wave simplex	0.955
Present study	Family income	U.S. households	ACL	Three-wave simplex	0.882
Present study	Family income	U.S. households	NES50	Three-wave simplex	0.895
Present study	Family income	U.S. households	NES70	Three-wave simplex	0.869
Present study	Personal income	U.S. households	NES90	Three-wave simplex	0.953
Present study	Family income	Detroit families -- mothers	SAFMO	Three-wave simplex	0.769
				Average	0.884

Footnote: Reliability estimates recorded here from the present study are the Heise reliability estimates based on Pearson correlations.

A third study providing estimates of reliability for socioeconomic variables, conducted by Hauser, Tsai and Sewell (1983), used as its basis the Wisconsin Longitudinal Study, which has followed a large sample of 1957 high school graduates from the state of Wisconsin, assessed on several successive occasions since high school (Sewell, Hauser, Springer, and Hauser, 2003). The Hauser et al. (1983) study reports "test-retest" correlations for years of schooling completed, and the occupational standing of early and current jobs from a parental report obtained in 1964 and a respondent self-report in 1975, which are presented in Table 12.5. One could argue that these correlations confound validity and reliability, since they are not relying on multiple reports by the same individuals (see discussion below). For present purposes it can be argued that they provide very close estimates of reliability.[2]

The comparison of reliability estimates for education, occupational standing, and income from the present study (also displayed in Table 12.5) with results of reliability estimation from other studies reveals several interesting observations.[3] First, overall the level of reliability across studies within each of these domains is relatively consistent, and with a few exceptions, quite close to the average. Second, the estimates from the present study are quite comparable to those from other studies. For example, our two estimates of the reliability of measuring occupational standing for current job are almost identical to those three estimates obtained from these other studies—an average of .816 for our two studies versus an average of .830 for the other three studies. Third, there is obvious variation in levels of estimated reliability across these various studies, but there are some amazing consistencies. For example, the average of the reliability of education for the four simplex-based estimates from the present study, .94, is very close to the .93 estimate given by Siegel and Hodge (1968).

There are a number of details of these individual studies that we cannot take into account here, and there may be some very good reasons to expect differences in reliability estimates. Reliability can potentially be affected by a number of aspects of data collection, for example, by the nature of the question and response categories, the quality of the interviewing and interviewer training, the mode of administration, although we do not try to capture any of these differences here. Additionally, reliability can also be affected by the nature of the population (see Chapter 10) and by the design employed to gather multiple measures. I have emphasized the convergences here, and leave an account of the differences for further study.

12.4.2 Considerations in Archiving Reliability Estimates

One part of the future agenda for a program of research on reliability of survey measurement is a proposal for developing an archive of reliability estimates. Such

---

[2]The Hauser et al. (1983) analysis provides a much more detailed examination of question-specific reliability than is conveyed by the test-retest correlations I present here.

[3]In order to maximize comparability with the results from other studies in this table, I have presented estimates based on the Heise (1969) estimation strategy applied to Pearson product-moment correlations (see Chapter 6 and the Appendix).

an archive could serve a number of different purposes. First, when combined with information on the formal properties of the questions (e.g., fact versus nonfact, number of response categories, etc.) estimates of reliability could be analyzed in the same manner we have done in the present research, with the potential for there being a much broader base of question content and many more numbers of questions on a particular topic. Second, given the variability in reliability estimates across studies, with a greater number of estimates of reliability for a given question topic, it would be possible to get a better handle on the role of data collection features of different studies and the role of population differences in accounting for this variability. Third, for analysts wishing to take unreliability of measurement into account in their own analyses, an archive involving a greater number of "point estimates" of reliability can provide a stronger basis upon which to build an average for exporting information on reliability into a data set that contains no re-measurement design internal to the study. Fourth, if reliability estimates from cross-sectional studies, particularly those assessing nonfactual content (e.g., MTMM studies), are included with estimates from longitudinal studies, it may be possible to sort out any systematic bias in reliability estimation that could result from either approach.

With the increasing availability of longitudinal data for secondary analysis, particularly the types of data sets meeting the requirements of the present project, it is possible to generate many more estimates of reliability of the types of questions included here. Combined with these results, as well as those from previous studies using a variety of designs (as in Table 12.5), the results from future analyses can be used as the basis for the type of archival information proposed here. There are some additional examples of this from past studies that we can mention here (assembled in Table 12.6), which illustrate some of the potential problems with defining reliability. These examples involve assessments of socioeconomic background, specifically father's and mother's years of schooling completed, father's occupational standing, and parental income, from four studies—Bielby, Hauser and Featherman (1977a), Hauser, Tsai and Sewell (1983), Corcoran (1980), and Mare and Mason (1980).

The Bielby et al. (1977a) and Hauser et al. (1983) studies were described above. The Corcoran (1980) study uses data from the Panel Study of Income Dynamics (PSID), specifically white noninstitutionalized male household heads, female household heads, and wives aged 23–30 in 1976 who were living with both parents in 1968. The Mare and Mason (1980) study relied on male school students and their families from Ft. Wayne, Indiana, in which both parents and children were asked to report on parental statuses. The Mare and Mason (1980) report presents a detailed examination of measurement models of status attainment processes (see also Mason, Hauser, Kerckhoff, Poss and Manton, 1976), and I do not adequately summarize their results here. In order to maximize comparability to the other studies reported upon in Table 12.6, I present reliability results only for 12th-grade boys.

In both the Bielby et al. (1977a) and Hauser et al. (1983) studies the design for reliability estimation is classified in Table 12.6 as involving "multiple measures," inasmuch as in both studies the measurement design involved two independent efforts to obtain replicate measures of parental statuses. The Bielby et al. (1977a)

Table 12.6. Reliability estimates from existing literature on social background variable

Father's Education

Study	Concept	Population	Survey	Design	Estimate
Bielby, Hauser & Featherman (1977)	Father's yrs. of schooling	Nonblack males in U.S. households	CPS-OCG	Multiple measures	0.940
Hauser, Tsai & Sewell (1983)	Father's yrs. of schooling	Wisconsin 1957 high school grads	WLS	Multiple measures	0.767
Corcoran (1980)	Father's yrs. of schooling	U.S. households	PSID	Multiple indicators	0.801
Mare & Mason (1980)	Father's yrs. of schooling	Ft. Wayne, Indiana male 12th graders	Local	Multiple indicators	0.944
				Average	0.863

Mother's Education

Study	Concept	Population	Survey	Design	Estimate
Hauser, Tsai & Sewell (1983)	Mother's yrs. of schooling	Wisconsin 1957 high school grads	WLS	Multiple measures	0.733
Corcoran (1980)	Mother's yrs. of schooling	U.S. households	PSID	Multiple indicators	0.887
Mare & Mason (1980)	Mother's yrs. of schooling	Ft. Wayne, Indiana male 12th graders	Local	Multiple indicators	0.870
				Average	0.830

Father's Occupational Standing

Study	Concept	Population	Survey	Design	Estimate
Bielby, Hauser & Featherman (1977)	Father's job at R's age 16 -- SEI	Nonblack males in U.S. households	CPS-OCG	Multiple measures	0.870
Hauser, Tsai & Sewell (1983)	Father's job at R's age 16 -- SEI	Wisconsin 1957 high school grads	WLS	Multiple indicators	0.701
Corcoran (1980)	Father's job at R's age 16 -- SEI	U.S. households	PSID	Multiple indicators	0.817
Mare & Mason (1980)	Father's job at R's age 16 -- SEI	Ft. Wayne, Indiana male 12th graders	Local	Multiple indicators	0.937
				Average	0.831

Parental Income

Study	Concept	Population	Survey	Design	Estimate
Bielby, Hauser & Featherman (1977)	Family income	Nonblack males in U.S. households	CPS-OCG	Multiple measures	0.910

measures occurred within 3 weeks of one another, whereas the Hauser et al. (1983) measures were separated by 18 years (1957 and 1975). This possibly accounts for the substantial differences in the levels of estimated reliability across the two studies. In the two additional studies included in the table—by Corcoran (1980) and Mare and Mason (1980)—the authors present these values as estimates of reliability, but they in fact confound reliability and validity. For this reason I refer to the designs involved as "multiple-indicators" studies. This usage is consistent with the distinction I introduced in Chapter 3 to refer to multiple attempts to measure the same concept involving different questions (as distinguished from "multiple measures" that involve exact replicates). This is true as well in the case of the Hauser et al. (1983) assessments of occupational standing. In these examples the designs involved reports from different persons, specifically children and parents, so these are clearly not multiple measures in the sense I have used that concept in this book. Although reports from different persons may be getting at the same concept, it is harder to argue they are measuring the same true score or that specific variance in the measure is absent. Considering them as multiple indicators is a more secure position to take.

These results illustrate the complexity of reliability estimation and the difficulties of drawing interpretations about measurement error from such a variety of designs. Several factors contribute to an appropriate interpretation of these results. First, as noted, most of the reliability estimates reported in Table 12.6 are not precise estimates of reliability; that is, they do not estimate the same true score. Clearly, the measures are tapping the same underlying common factor, but the measures are not truly univocal in the sense required by CTST. The result of this, as shown in Chapter 3, is that the correlation will underestimate reliability. In this case, a statistical adjustment (in the sense of a correction for attenuation) will overadjust for reliability. One should be aware that this is the case when using such types of reliability estimates. Second, I already mentioned that the time interval between multiple measures, i.e., independent reports of the same true score, can affect the reliability estimates (see also Alwin, 1992). Generally, greater intervals of time are expected to reduce reliability due to a number of different factors, e.g., greater forgetting, decreased salience, and consequently greater amounts of guessing. I noted that the correlations among measures obtained over many years, e.g., in the Hauser et al. (1983) reports of parental statuses, the overall estimates of reliability were considerably lower than in the case of the estimates obtained in the Bielby et al. (1977a, 1977b) studies. Note that in both cases the estimates were based on reports of the same true scores—the key difference was in the amount of time between independent reports. Of course, in the Bielby et al. studies the short time interval may contribute to upward bias due to correlated errors in the multiple reports (see Chapter 5, Section 5.1).

One factor that should not be lost in the consideration of the results in both Tables 12.5 and 12.6 is that in virtually all cases the reliabilities are at relatively high levels—.91 for self-reports of schooling, .81 for self-reports of occupational standing, .88 for reports of personal and/or family income, .86 for reports of father's education,

.83 for reports of mother's education, .83 for father's occupational standing, and .91 for parental income. These patterns are consistent with the overall conclusion in Chapter 7 that surveys can obtain relatively reliable reports of factual information. Although it may seem otherwise, there is little suggestion in these results that social background variables are measured less reliably than are contemporaneous achievement variables (see Bielby et al. 1977a, p. 734). However, even though the levels of reliability seem to be high, adjustment for levels of measurement error can make a difference for such predictors. In one of the most systematic analyses to date of the impact of survey measurement error on the nature of causal inferences, Bielby et al. (1977a, pp. 733–734) concluded that the effects of family socioeconomic factors and schooling are underestimated by analyses that do not take measurement error into account.

12.5   COMPENSATING FOR MEASUREMENT ERROR USING COMPOSITE VARIABLES

Earlier in the chapter, I argued that if one has an appropriate estimate of reliability for a given variable, it is possible to employ "errors in variables" approaches to adjust for the relative effects of measurement error. In order to use this approach, one must have access to good estimates of reliability, and as the previous discussion illustrates, the choice of a reliability estimate is not unproblematic. Ideally, this information would be based on the same sample used for one's analysis—this was true in the studies by Bielby et al. (1977a, 1977b), Bielby and Hauser (1977), Mare and Mason (1980), and Hauser et al. (1983) discussed above. In other cases, where one is exporting a reliability estimate into an independent sample, one must give serious consideration to the population on which the estimate is based, the design used to estimate reliability, and the extent to which the assumptions of the CTST model for reliability are warranted. I discussed some of the problems with drawing such inferences in the above section.

   This kind of approach is less likely to be useful as a way of compensating for unreliability for nonfactual measures, which are less often considered as independent variables, and that generally have much lower levels of reliability. It is well known that reliability can be improved by using multiple measures, or lacking that, multiple indicators. Recall from the discussion in Chapter 3 that a great deal is known about the reliability of composite variables. Using the notation developed there, let Y be a linear composite made up of G measures of the same true score, that is, $Y = Y_1 + Y_2 + \cdots + Y_g + \cdots + Y_G$, i.e., $\Sigma_g(Y_g)$. Recall further that the purpose of reliability estimation for composite variables is to estimate $VAR(T)/VAR(Y)$, where Y is as defined above and $T = T_1 + T_2 + \cdots + T_g + \cdots + T_G$, i.e., $\Sigma_g(T_g)$, where the T's are the true scores for each of the Y terms. In Chapter 3 we indicated that the quantities derived from the application of the algebra of writing linear composites can be manipulated to form an internal consistency measure of composite reliability, as follows:

$$ICR = \frac{\sum_j \sum_i \Sigma_{YY} - \sum_j \sum_i \Theta^2}{\sum_j \sum_i \Sigma_{YY}}$$

In Chapter 3 we showed how Cronbach's α (1951), computed as follows,

$$\alpha = \frac{G}{G-1}\left[1 - \frac{\sum_g VAR(Y_g)}{VAR(Y)}\right]$$

can be derived from this general formula under the assumption of G unit-weighted (or equally weighted) *tau-equivalent* measures.

We can simplify this formula even further if we assume that the Ys are all standardized with mean 0 and standard deviation 1, in which case the numerator of the fraction to the right simply equals G and the denominator of that fraction is made up of $G + G(G-1) COR(Y_i, Y_j)$ (see Bohrnstedt, 1969). Recall that under the CTST model for *tau-equivalent* measures, the covariances among the measures are by definition equal. These covariances in the present case are correlations, given the assumption of a constant standard deviation of 1. Recall also that in the CTST model the true covariances and observed covariances are equal, given that random measurement error does not affect the covariances among variables. Let r symbolize the typical off-diagonal element in the covariance matrix (in this case the correlation matrix) for the observed measures, i.e., $COR(Y_i, Y_j) = r$. In this special case, the formula for alpha reduces to

$$\alpha = \frac{G}{G-1}\left[1 - \frac{G}{G + (G^2 - G)r}\right]$$

which can be rewritten as

$$\alpha = \frac{G}{G-1}\left[1 - \frac{1}{1 + (G-1)r}\right]$$

Suppose, using this simplified formula, that we wanted to calculate how many individual questions one would need to combine to form a composite score that would achieve a given level of reliability. As we will see shortly, levels of reliability for composite variables are a function of three factors: (1) the number of items, (2) the reliability of the individual items, and (3) the average level of intercorrelation among the true scores of items in the composite. In order to show this, recall that an observed correlation between any two items in a scale, Y_i and Y_j, can be written as follows:

$$COR(Y_i, Y_j) = COR(T_{Yi}, T_{Yj}) \, \rho_{Yi} \, \rho_{Yj}$$

where $COR(T_{Yi}, T_{Yj})$ is the correlation between their true scores, and ρ_{Yi} and ρ_{Yj} are the indices of reliability for Y_i and Y_j, respectively.[4] In calculating the reliability of a composite score, we can disaggregate the effects on composite reliability, i.e., coefficient α, of the three factors mentioned above.

Before showing how this works, let me clarify the simplifying assumptions that will enable us to see the role of these factors more clearly. Let us first assume for the measures that all items in a composite have the same reliability, and second, that all the items in a composite have the same level of true-score intercorrelation. Then we can calculate composite reliability, i.e., coefficient α, for scales that differ in item reliability, level of true-score correlation, and the number of items in the composite (i.e., G). The results of these computations are presented graphically here—the individual graphs differ with respect to the level of reliability at the item level, beginning with an item reliability of .40, then advancing through reliabilities of .60, .80, and 1.0 in Figures 12.2 through 12.5 respectively. Within each graph the level of intercorrelation among true scores is varied systematically, from .40 through 1.0, and the composite reliability is calculated in each case for composites containing numbers of items from two to 25.

As noted above, generally composite reliability can be improved by increasing the number of items, by choosing highly reliable items, and by choosing items that measure the same true scores, i.e., $COR(T_{Yi}, T_{Yj}) = 1.0$. When item reliability is low, e.g., .40 (see Figure 12.2), large numbers of measures must be combined to achieve acceptably high levels of composite reliability (say, .90). This is true even when *multiple measures*, i.e., those measures having the same (or linearly related) true scores and that therefore have high true score intercorrelations, are used. To achieve a composite reliability of .90 for such items with true-score correlations of 1.0, one would need at least 12 such items. This is obviously impractical, unless substantial amounts of survey interview time can be devoted to the measurement of a particular construct, and even if this were the case, true-score correlations of unity would amount to asking the same exact question multiple times.

It is obviously good to have items with highly correlated true scores, but it is even better to have higher levels of reliability. Even slightly more reliable items, e.g., reliabilites of .60 (see Figure 12.3), would require only five such items (those with true-score correlations of 1.0) for a composite reliability of .90. And in this case, one can see the benefits of this slightly higher level of reliability, because when the true-score correlations are .70 for items measured at .60 reliability, 10 measures are needed to achieve a composite reliability of .90. From the examples given so far, it is clear that lower levels of true-score intercorrelations produce relatively hopeless scenarios. Even in the case of perfect item reliabilities (see Figure 12.5)—an extremely rare case—the only way to counteract low levels of true score intercorrelation is to build composites with many items.

---

[4]Note that this is simply a manipulation of the standard formula for the correction for attenuation, and that the denominator of this formula contains the product of the square root of reliability (i.e., the index of reliability) for the two variables, Y_i and Y_j'.

Coefficient Alpha

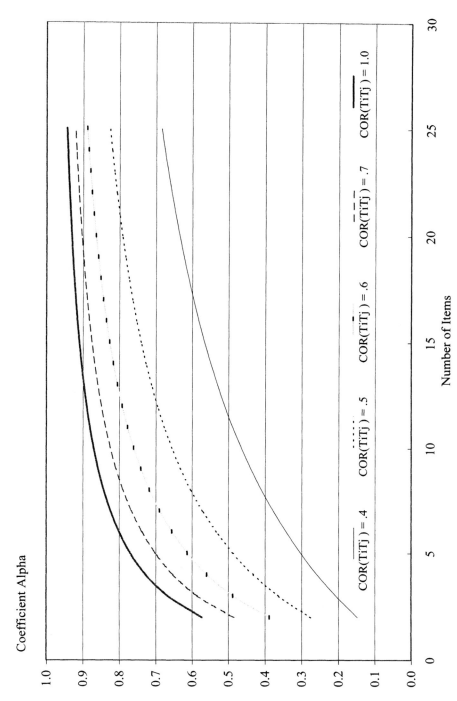

COR(TiTj) = .4   COR(TiTj) = .5   COR(TiTj) = .6   COR(TiTj) = .7   COR(TiTj) = 1.0

Figure 12.2. Coefficient alpha as a function of the true correlation among items and the number of items where reliability = .40.

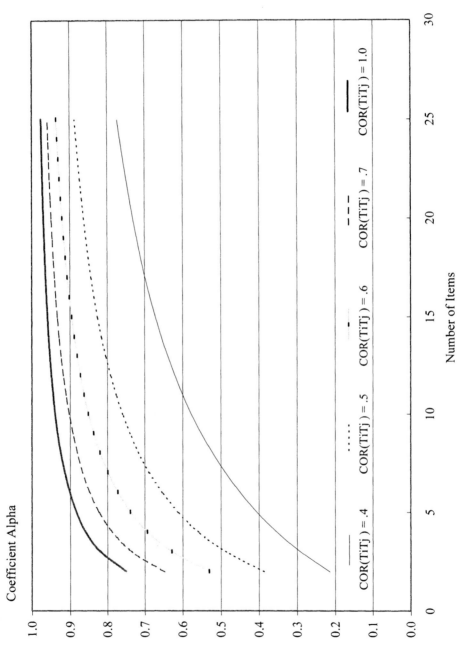

Figure 12.3. Coefficient alpha as a function of the true correlation among items and the number of items where reliability = .60.

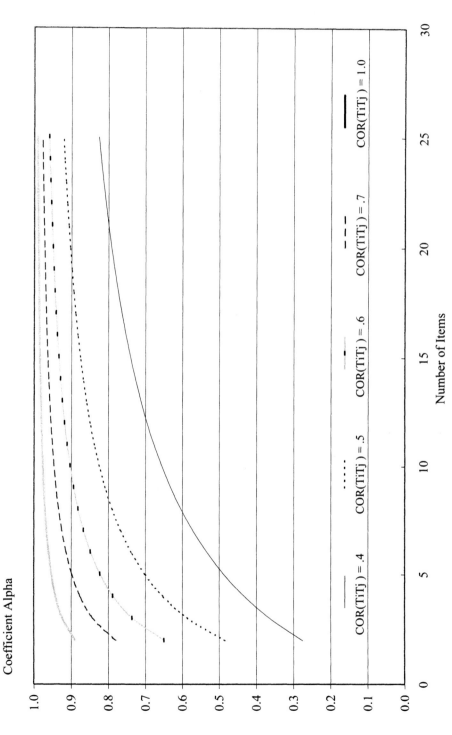

Coefficient Alpha

COR(TiTj) = .4   COR(TiTj) = .5   COR(TiTj) = .6   COR(TiTj) = .7   COR(TiTj) = 1.0

Number of Items

Figure 12.4. Coefficient alpha as a function of the true correlation among items and the number of items where reliability = .80.

Coefficient Alpha

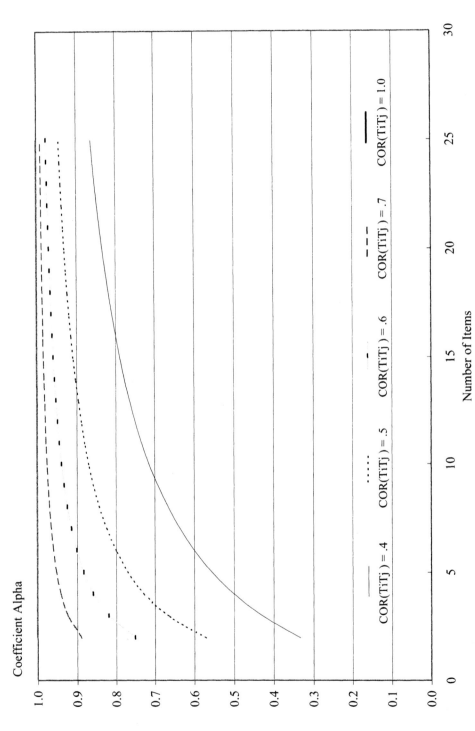

Figure 12.5. Coefficient alpha as a function of the true correlation among items and the number of items where reliability = 1.0.

Obviously, from these examples, one is much better off with highly reliable measures involving high true-score intercorrelations. One can see what a premium should be placed on such properties by considering the case where the item reliabilities are .80 (see Figure 12.4). Such high levels of reliability coupled with perfect correlations among true scores, i.e., tau-equivalent or congeneric measures, yields the welcome result that only two such items would be necessary to achieve a composite reliability of .90. Some five questions would need to be included to achieve this level of composite reliability when the true-score correlations are .70, and one would need 10 such items when the true-scores correlate .60. Again, this is impractical, as rarely can this much survey interview time be devoted to the measurement of one concept.

The story that is emerging from this exercise should be clear—little can be gained by using composite scores unless one or more of the following conditions hold: (1) item reliability exceeds .80, (2) true item correlations are above .70, or (3) it is possible to include the responses of at least five or more questions into a composite assessing a given concept. This conclusion runs somewhat counter to the common practice in survey research of adding the responses together to achieve a more reliable measure. That sentiment is pointed in the right direction, as it *is* the case that summing similar questions will result in more reliable measures. The point I wish to make here is that there is no real solution to poor measurement, and throwing more questions at an ill-defined concept will not necessarily help. Moreover, summing similar indicators with low reliabilities that do not have high true-score intercorrelations will require far more questionnaire space than is likely to be justified. If there is a solution to the problem, it probably lies in clearer theoretical specification of the concepts of interest, a much more careful approach to measurement, and extensive pretesting that will improve the comprehension of the questions and their utility for measuring the concept of interest.

12.6 CONCLUSIONS

Despite the infrequency with which survey measures are evaluated with respect to the criterion of measurement reliability, there should be little doubt that this is an important consideration in evaluating the quality of survey data. Earlier in the chapter I posed the question: Why study the reliability of survey measurement? In this chapter I have provided several statistical rationales—the biases in simple tests of significance, the decrements in statistical power, the biases in estimates of simple regression coefficients, and the attenuation in estimates of statistical association—but as I have argued throughout this book, there are other reasons as well. One involves the benefits to be derived from the evaluation of survey questions and questionnaires in terms of the quality of data they produce. In the foregoing chapters, I considered some of the ways in which the extent of measurement errors can be detected and estimated in order to better understand their consequences. The major vehicle for achieving these purposes involved a study of nearly 500 survey measures obtained in surveys conducted at the University of Michigan over the past several years. Assembling information on reliability from these data sources provides a benchmark against which we can further improve knowledge about the strengths and weaknesses of survey data. In addressing these issues our analyses focused on estimating the effects on reliability of content and source of information (Chapter 7), the design of questionnaires (Chapter 8) and the formal attributes of questions (Chapter 9). These analyses have covered terrain that is familiar to survey methodologists, albeit from what may perhaps be an unfamiliar perspective.

12.6.1 Future Agenda

In keeping with the goals of this chapter, I want to mention several areas where future research might profitably focus. I remarked at the outset that we have seen a number of very positive developments over the past two decades in the research literature on survey methodology, especially that part of the literature focusing on problems of measurement. Over the past two or more decades, we have seen a virtual explosion of interest in the application of theoretical ideas from linguistic, communication, and cognitive sciences to problems of survey measurement. As a field, we probably better understand the sources of measurement errors in surveys than ever before, particularly those involving cognitive processes and the effects of question wording. Despite these important theoretical developments, there has been little effort to link these discussions to problems of modeling measurement errors and develop equational representations (or models) of these theoretical ideas permitting the examination of specific types of measurement errors suggested in these applications. We can improve our knowledge about the nature and extent of measurement errors in surveys, and can better understand the factors contributing to errors of measurement to the extent that we can provide empirical estimates of indicators of data quality. The consideration of the presence and extent of measurement errors in survey data will ultimately lead to improvements in the overall collection and analysis of survey data. Ultimately studies of measurement errors can identify, for example, which types of questions and which types of interviewer practices produce the most valid and reliable data.

Another important issue on the agenda, one that has been relatively neglected, involves the potential lack of homogeneity in the behavior of measurement errors across relevant subpopulations. Although studies have examined how to adapt survey design to attributes of survey participants (Conrad, 1999, pp. 308–310), we know too little about how most of our measures perform in different segments of the population. In Chapter 10, I introduced procedures to test these issues, and continued systematic efforts to further investigate these issues is necessary. These efforts will need to grapple with the paradox (discussed in Chapter 10) that our models for measurement error are written for individuals, but our indicators of measurement quality are instead linked fundamentally to attributes of populations (and not individuals). The idea that reliability, or validity, or some other indicator of data quality depends on the characteristics of the subpopulation investigated may have implications for survey design.

A third research priority that will require renewed emphasis in future research is the issue of reliability estimation for categorical latent variables, an issue covered in a preliminary fashion in Chapter 11. Although models for measurement error are relatively well developed for continuous latent variables, it is increasingly the case that researchers prefer to consider survey responses as inherently categorical rather than assume the response categories map to underlying interval scale or an approximation, e.g., one that is at least ordinal in character. The assumptions of the classical true-score models that underlie approaches to estimating reliability (as discussed in Chapters 4 and 5) are problematic for categorical latent variables. Although appropriate models have existed for specifying parameters expressing the epistemic relationships among categorical latent and observed variables, only recently has it been possible to estimate these model parameters. As pointed out in Chapter 11, the challenge to the development of models permitting change in the composition of latent classes has been met by relatively new estimation strategies for discrete-time Markov chain latent class models (see Langeheine and van de Pol, 2002). These tools are now available, which provide workable solutions to the estimation problems, and preliminary evidence suggests that these categorical models provide a valuable analog to the conception of reliability involved in the structural equation modeling approaches for panel data discussed throughout this book.

A final future research priority involves the creative use of design strategies linking within- and between-person models for measurement error. In Chapter 3, I highlighted the differences between the experimental or "between subjects" approach to measurement error and the "within subjects" approach used here, and discussed ways to combine them. Between-subjects or experimental designs typically compare two or more conditions (e.g., question forms, question sequence, or interview modes), with the object of comparing the responses of members of the sample assigned to these conditions. Such designs cannot assess reliability of measures unless a within-subjects design is embedded in the experimental conditions. By combining the two types of designs, one can examine differences in the marginal distributions or means across conditions and study the variance/covariance properties across experimental forms. Traditionally, split-ballot experiments do not include a within-subject replication of measures and therefore cannot assess measurement reliability as it is defined classically. With a few exceptions (e.g., McClendon and Alwin, 1993), research on question wording, question context, and mode effects does not focus on measurement errors per se. The implementation of such an experimental approach within the longitudinal framework employed here may, for example, take advantage of existing longitudinal studies that implement mode comparisons. For example, Rodgers, McCammon, and Alwin (2004) designed a study that assesses error variance in questions from the Health and Retirement Study asked in personal interviews and self-administered mail questionnaires. To permit such estimates to be made, at least three waves of data are needed within each mode that meet the design requirements spelled out in Chapter 5. For a substantial number of questions that were asked using both modes, we can compare the estimated error variances in the answers given to the mail questionnaire to those in the answers given in the personal interviews.

12.6.2 Closing

In drawing this discussion to a close, I want to return to the issues engaged throughout this book. How reliable are standard types of survey measures in general use in surveys of the general population? Does survey measurement reliability depend on the content being measured? Specifically, is factual information gathered more precisely than attitudinal and/or other subjective data? Also, does the measurement reliability of nonfactual questions (attitudes, beliefs and self-assessments) depend on their specific content? Does reliability of measurement vary as a function of the source of the information? In the case of factual data, are proxy reports as reliable as self-reports? How reliable are interviewer observations? Is reliability of measurement affected by the context in which the questions are framed? Specifically, does the location of the question in the questionnaire, or the use of series or batteries of questions, produce detectable differences in levels of measurement error? Do the formal properties of survey questions affect the quality of data that results from their use? Specifically, how is reliability affected by (1) the form of the question, (2) the length of the question, (3) the number of response categories, (4) the extent of verbal labeling, and (5) other properties of the question response format? Are measurement errors linked to attributes of the respondent population? Specifically, how are education and age related to reliability of measurement?

The overarching questions are (1) How reliable are survey data? and (2) Why are some measures more reliable than others? In Chapter 2, I argued that it is possible to better understand the potential errors in survey measurement if we make explicit the assumptions that are often made in the collection of survey data. These included the following key elements: that the question asked is appropriate, relevant, and has an answer; that the question is posed in such a way that the respondent or informant understands the information requested; that the respondent or informant has access to the information requested; that the respondent or informant can retrieve the information from memory; that the respondent or informant is motivated to make the effort to provide an accurate account of the information retrieved; and that the respondent or informant can communicate this information into the response categories provided by the survey question.

We looked at two issues that relate to the element of *access*—the *topic of the questions* and the *source of the information requested*—that have received little attention in the survey methods literature on measurement error. I hypothesized that the content of the information requested from the respondent would be more reliably reported if it had a factual basis, and thus could be retrieved more easily. Few survey questions are perfectly reliable—the typical factual question in this study has a reliability of .80 (see Table 7.4) and the typical nonfactual question has a reliability of slightly more than .60. There is little difference in reliability estimates across types of nonfactual content in average reliability (see Table 7.5). Although true in only a minority of cases, some factual questions produce highly reliable data, e.g., the number of children a woman reports she has had (see Alwin and Thornton, 1984). The HRS measure of "estimated weight" is another example of a variable that is measured with a great deal of precision (see Chapter 10). Regardless of what many researchers may believe, however, factual survey content is measured with error, although it is perhaps less vulnerable to sources of error than are nonfactual questions. In this regard, I have confirmed the commonly held view that factual material can be more precisely measured than content that is essentially subjective, although there is considerable overlap. For a number of reasons, variables that involve subjective content (including perceptions of facts) are expected to have lower reliabilities because of the difficulties of translating internal cues related to such content into the response framework offered to the respondent.

Despite the potential limitations of self-reports, *respondents are better at reporting information about themselves than about others.* Self-reports are on average more reliable than are proxy reports. I reported a controlled comparison of a small set of variables involving the same or similar content reported for self and others obtained from the same respondents. Our results reinforced the conclusion that self-reports are likely to be more reliable than information that can be provided second hand by a proxy informant. However, proxy reports of the types of factual information included were not all that bad (see Table 7.2). In addition, although the main focus of this study was on self-reports, we also reported on a small amount of data on organizational and interviewer reports of factual material, and drew on data assessing the reliability of interviewer beliefs about respondents and their reactions to the interview. Although facts reported by the survey organization are without question the most reliable, interviewer reports of factual aspects of the household are also highly reliable. Interviewer judgments regarding respondent characteristics, however, are substantially less reliable (see Table 7.1).

Many guidelines for developing good survey questionnaires represent the "tried and true" and aim to codify a workable set of rules for question construction. It is hard to argue against these guidelines, but there is little hard evidence to confirm that following these rules improves survey data quality. The argument set forth in this book is that we can assess these issues and evaluate the link between question/questionnaire characteristics and the quality of data by using levels of measurement error, or reliability, associated with a particular question or question type as a major criterion of evaluation.

When survey researchers build questionnaires, there are considerations that affect the quality of data other than the formal properties of the questions themselves. In other words, developing survey measures involves more than writing the questions; it also involves the organization of questions into subunits larger than the question. Ordinarily, a questionnaire is divided into *sections*, subquestionnaires focusing on similar general content, and within sections there are often *series* of questions that pertain to the same specific topic. Further, sometimes a series of questions will include questions that not only deal with the same topic but also involve the exact same response formats. The latter are referred to as *batteries* of questions.

Thus, we began the examination of the reliability of survey questions with a focus on the *architecture of survey questionnaires*, focusing specifically on whether a given question is part of a series or battery of questions, the position of the question in the series or battery, and the position of the question in the larger questionnaire.[5] Our results provide relatively strong evidence that questions in what we referred to as a "topical series" are less reliable than stand-alone questions (among factual questions), and questions in batteries are less reliable than questions in series (among nonfactual questions) (see Table 8.3). In contrast, we find little or no effect on reliability of question location in the questionnaire or the position of the question within a series or battery (see Tables 8.6 and 8.8). Question context interacts to a slight degree with questionnaire position, in that questions in batteries appearing later in the questionnaire are somewhat less reliable than those appearing earlier. In addition, the introduction length in both series or batteries seems to matter for reliability of measurement—although it operates somewhat differently in the two cases. Among series, those having long introductions (16 or more words) appear to have lower reliability, whereas questions in batteries having *any* introduction appear to have lower reliability (see Table 8.9).

It is not completely clear what accounts for these patterns. The "stand-alone > series > battery" ordering with respect to reliability is certainly intriguing. Perhaps the best explanation is that the same factors motivating the researcher to group questions together—contextual similarity—are the same factors that promote measurement errors (see Andrews, 1984). Similarity of question content and response format may actually distract the respondent from fully considering what information is being asked for, and this may reduce the respondent's attention to the specificity of questions. Thus, measurement errors may be generated in response to the efficiency features of the questionnaire. Unfortunately, it appears that the respondents may also be more likely to streamline their answers when the investigator streamlines the questionnaire.

On the other hand, it may be the type of content measured using different ways of organizing questionnaires that accounts for these results—that is, the measurement of types of survey content for which batteries are used are measured less reliably than those types for which series are used. Or, the types of content assessed in stand-alone questions are simply the kinds of things that are measured more reliably. Because I anticipated this argument, I controlled for survey content—fact versus nonfact—where possible in our analyses of survey context. I noted the inherent limitations, given that batteries are hardly ever used to measure facts and nonfactual questions are much less likely to be measured using stand-alone questions than with series and batteries. While I believe this argument is therefore less powerful in successfully accounting for the effects of question context, it remains an issue that must be investigated in future research.

---

[5]I refer to this as "question context" although this is generally not the way survey methodologists use this term (see Tourangeau, 1999). In the present study, I use the term "context" to refer to the following classification: (1) stand-alone questions (not in topical series or in batteries); (2) questions that are part of a topical series, which has an introduction; (3) questions that are part of a topical series, which has no introduction; (4) Questions that are part of a battery, which has an introduction; and (5) questions that are part of battery, which has no introduction. A "battery" is a type of series where the response scale is identical for all questions in the battery (see Chapter 8).

Another possibility is that the *context effects* result from other aspects of the questions, particularly question form (e.g., type of rating scale used, length of the question, number of response categories, labeling of response categories, or explicit provision of a Don't Know option). In other words, measurement reliability may have less to do with question context (stand-alone, series, or batteries) per se, and more to do with the formal properties of questions appearing in one or the other type of questionnaire unit. If we can account for the observed differences in reliability by reference to the formal attributes of questions rather than the context in which they are embedded, then the focus of our attention can be simplified. That is, we can pay less attention to the way in which the survey questionnaire is organized and more attention to the internal characteristics of the questions themselves.

Following up on this possibility, my analysis then turned to the examination of the questions themselves, focusing specifically on the formal characteristics of survey questions that are believed to affect the quality of measurement. Using measurement reliability as a criterion for evaluating these issues, I focused on several features of survey questions that potentially impact measurement quality. I begin the investigation with a focus on the topic of question form, and moved from there to a consideration of several additional attributes of questions, specifically the number of response options provided in closed-form questions, the use of unipolar versus bipolar response formats, the use of visual aids, particularly the use of verbal labeling of response categories, the provision of an explicit Don't Know options, and the length of questions. As noted throughout this study, isolating the effects of formal attributes of questions is challenging, due to the confounding of question content, context, and question characteristics. However, in most cases it is possible to isolate critical comparisons of specific forms of questions by carefully controlling for other pertinent features of question content and context.

One of the earliest areas of debate in the design of questionnaires originated during World War II involving whether "open" versus "closed" question forms should be used (Converse, 1987). Although the "closed question" or "forced choice" approach won out for most survey purposes, even in the early 1980s there was enough concern that such questions might ignore unconsidered options to respondents to warrant further study (Schuman and Presser, 1981). There are clearly many uses for open-ended questions in survey research, but the standard approach in modern survey research is to formulate questions within a fixed-format set of response categories. In most instances, carefully pretested closed form questions are preferred to open-ended questions because of their greater specificity.

Open-ended questions are, however, employed in a relatively selective way. For example, few nonfactual questions are ever open-ended. By contrast, closed-form questions are almost exclusively contained within series and batteries. Our results with respect to reliability demonstrates that information assessed in *open-ended* response formats tends to be more reliable than questions with *closed-form* response formats (see Table 9.1). Although this may be due in part to question content, these conclusions apply even when controlling for the fact-versus-nonfact distinction. I find substantial differences favoring the open-ended format within categories of facts and nonfacts, although there are too few cases in which the measurement of nonfacts employs the open-ended format, so we do not have the data to establish this empirically. My discussion of this issue (see Chapter 9), however, suggested that researchers might more profitably exploit the uses of the open-ended approach, relying more on open-ended formats, at least when measuring factual content.

With respect to fixed-form or closed questions, we usually consider the question and response format as one package. There is, however, some value in disentangling the role of the response format per se in its contribution to the likelihood of measurement error. There are several types of closed-form questions, distinguished by the type of response format: Likert-type or agree-disagree questions, forced-choice questions, feeling thermometers, and various kinds of *other* rating scales. The results of the present study suggest that there is little apparent difference in estimated reliability associated with these several types of closed-form question formats, but we are not in a position to draw significant import from these results (see Table 9.2).

Previous comparisons of various types of closed-form response formats have ignored the complexity of other variations in the formal characteristics of survey questions. One of the most serious of these is the fact that rating scales differ with respect to whether they are assessing unipolar versus bipolar concepts. The key distinction I make here is between unipolar versus bipolar rating scales. Typically, the ends of a bipolar scale are of the same intensity but opposite valence, while the ends of a unipolar scale tend to differ in amount or intensity, but not valence. Our discussion of this issue noted that one often finds that some types of content, such as attitudes, are always measured using bipolar scales, whereas others, such as behavioral frequencies, are always measured using unipolar scales. Unipolar scales rarely use more than five categories, and it can be readily seen when viewed from this perspective that three-, four-, or five-category scales used to measure unipolar concepts is quite different from similar scales intended to measure bipolar concepts. Clearly, the issue of what the middle category means for three- and five-category scales is quite different depending on whether the concept being measured is unipolar or bipolar.

One clearly cannot evaluate the use of bipolar and unipolar scales separately from the issue of the number of categories and vice versa. Although the early research on question form essentially ignored the question of the number of response options used in survey measurement, there is a small body of literature developing around the issue of how many response categories to use in the gathering of information in survey interviews. The overall expectation regarding the number of response categories was based on information theory, which argues that scales with a greater number of categories can carry more information and thereby enhance measurement accuracy. Several conclusions emerge from our examination of these issues. First, there seems to be little if any support for the information-theoretic view that more categories produce more reliability. There is *no* monotonic increase in estimated reliability with increases in the number of categories in the present sample of measures (see Table 9.4). Indeed, if one were to leave aside the nine-category scales, the conclusion would be that more categories produce systematically less reliability. Similarly, I do not find any support for the suggestion that five-category response formats are less reliable than four- and six-category scales, nor that seven-category scales are superior to all others. One aspect of previous analyses which is upheld is the conclusion that nine-category scales are superior to seven-point scales in levels of assessed reliability. Finally, there are relatively few differences between unipolar and bipolar measures, net of number of response categories, that cannot be attributed to the content measured. Evaluating the differences in reliability of measurement across categories of different lengths reveals the superiority of four-and five-category scales for unipolar concepts. For bipolar measures, the two-category scale continues to show the highest levels of measurement reliability, and following this the three- and five-category scales show an improved level of reliability relative to all others. There are many fewer substantive differences among the bipolar measures, although it is relatively clear that seven-category scales achieve the poorest results (see Table 9.4)

A problem with assessing the reliability of measurement of questions with different numbers of response categories is that the results depend on the approach used to estimate reliability (see Chapter 3). With few exceptions, past results have been based on the use of Pearson-based correlation coefficients and have ignored the problem addressed in earlier chapters concerning the role of number of response categories in the calculation of reliability estimates. Following discussions in the statistical literature, I argued that when continuous latent variables are measured with a small number of response categories, the estimated level of association is attenuated with respect to what would be estimated with direct measurement of an "ideal" continuous variable. Given that the degree of attenuation due to crude categorization is directly related to the number of response categories, this aspect of our methods of estimation must be controlled in the analysis of this issue (see Table 9.3).

Due to the vagueness of many nonfactual questions and response categories and the perceived pressure on the part of respondents to answer such questions (even when they have little knowledge of the issues or have given little thought to what is being asked), there is also concern respondents will behave randomly, producing measurement unreliability. The explicit offering of a Don't Know option may forestall such contributions to error. Numerous split-ballot experiments have found when respondents are asked if they have an opinion, the number of Don't Know responses is significantly greater than when they must volunteer a no-opinion response. Our analysis compared measures of nonfactual content using questions without an explicit Don't Know option and comparable questions that did provide such an option—within three-, five-, and seven-category bipolar rating scales—and found no significant differences (see Table 9.5). This result is consistent with the majority of studies on this topic [see review by Krosnick (2002)].

Labels of response options reduce ambiguity in translating subjective responses into the categories of the response scales. Based on simple notions of communication and information transmission, the better labeled response categories may be more reliable. We therefore hypothesized that the more verbal labeling that is used in the measurement of subjective variables, the greater will be the estimated reliability. Our analysis of the available data on this issue suggests that a significant difference in reliability exists for fully vs. partially labeled response categories, such that measures with fully labeled categories were more reliable (see Table 9.6). This supports the conclusions of prior research which found that among seven-point scales, those that were fully labeled were significantly more reliable.

One element of survey question writing to which the majority of researchers subscribes is that questions should be as short as possible. I examined the relationship of question length to the reliability of measurement controlling for question content, question context and length of introductions. For stand-alone questions and questions in series, there is consistently a negative relationship between question length and the level of estimated measurement reliability. In the case of batteries the results indicate, except for the case of batteries with introductions of medium length, the relationship of reliability and question length is also negative for questions in batteries (see Table 9.11). The exception, though not significant, does pose an interesting puzzle with respect to the possible interaction between question context, question length, and reliability of measurement. However, the bulk of these results present a relatively convincing case for the conclusion that there are declining levels of reliability associated with greater numbers of words in questions and further support for the typical advice given to survey researchers that their questions should be as short as possible.

In examining these several hypotheses regarding the sources of survey measurement error, the present results may simply confirm for many what is all too obvious about survey measurement. For example, the finding that facts can be measured more reliably than nonfacts may not be all that surprising (see Kalton and Schuman, 1982). Similarly, the finding that proxy reports are less dependable than self-reports is unlikely to be controversial, although this does tend to disagree with common advice on the subject (see Sudman, Bradburn and Schwarz, 1996, p. 243). Similarly, our finding that the location of the question in the questionnaire bears little, if any, relationship to the extent of measurement error may not be one of our more profound results. Admittedly, in some cases we may have done little more than quantify a pattern that virtually all survey experts would have expected, supported what others have already documented, or confirmed what might otherwise have been known on the basis of other types of data.

Of course, in the scientific enterprise the replication of others' findings is a valuable exercise, and quantifying conventional wisdom is a worthy goal. At the same time, some of our reliability estimates depart from theoretical expectations. But unlike much of the lore that exists in the area of survey research, we have quantitative results to support our conclusions. I agree with Kuhn (1961, p. 180) that our "numbers register the departure from theory with an authority and finesse that no qualitative technique can duplicate," and departures from theory that can be quantified provide at least a starting point for detecting the "reality" involved.

Indeed, it is when the findings of prior research cannot be replicated, or when "what everyone believes" cannot be substantiated that questions are raised. In several instances we produced results that either seriously questioned the received wisdom, or we have documented a pattern that others have heretofore only speculated about. For example, our finding that offering a Don't Know option bears little on the extent of measurement error questions strongly held beliefs about the ways questionnaires are designed in many organizations (see e.g., Converse, 1964). Similarly, our findings that batteries of questions are more prone to measurement errors is somewhat novel and one that deserves a great deal more attention, especially given the pervasiveness of this approach. Further, our findings that longer questions (among stand-alone and questions in series) are systematically less reliable is one that fits with many of the desiderata frequently given for writing survey questions, namely, to keep questions short. However, these are the first results that actually confirm the hypothesized relationship between question length and the quality of the resulting data. The results of this research should be relevant to the general task of uncovering the sources of measurement error in surveys and improving survey data collection methods. The application of this knowledge is justified given the substantial social and economic resources invested each year in survey data collection. Clearly, in all of these areas further research is called for, and in future years we can continue to generate comparable information on many more such measures and build upon these results in ways that will clarify the nature of the influences of attributes of questions and questionnaires on the level of measurement error present in survey data.

Reliability of Survey Measures Used in the Present Study

Source	Variable Number	Question Content	Note	Sample Size	Reliability[a]
I. Respondent Self-reports					
A. Facts					
1. Personal Characteristics					
ACL	2	Age of respondent		2,220	0.999
ACL	40	Year of most recent widowhood		353	0.967
ACL	125	Respondent's weight		2,184	0.956
ACL	244	Year in which respondent was married		882	0.983
NES50	5030	Age of respondent, bracketed	DE,1	1,128	1.002[b]
NES50	5033	Age of youngest child		591	0.921
NES50	5055	Age of respondent, continuous		1,128	0.995
NES70	7001	Age of respondent		1,276	0.994
NES90	9121	Month of respondent's birth	E,1	593	1.000
NES90	9122	Year of respondent's birth	E,1	592	0.997
NES90	9123	Age of respondent	S	596	0.997
SAFMO	8074	Number of children respondent has had		947	1.113[b]
2. Household Characteristics					
NES50	5032	Number of children under 18 years old in family		1,093	1.024[b]
3. Place of Residence					
NES50	5091	Type and size of place respondent grew up		889	0.919
NES70	7042	How long respondent has lived in community		1,211	0.856
NES70	7043	Type of place where respondent grew up		1,194	0.862
NES90	9185	How long respondent has lived in community		595	0.724

Source	Variable Number	Question Content	Note	Sample Size	Reliability[a]
NES90	9186	How long respondent has lived in current home		594	0.845
4. Education					
NES50	5034	Respondent's education	S,2	1,117	0.929
NES70	7002	Respondent's years of schooling		1,274	0.972
SAFCH	8150	Respondent's years of schooling		447	0.938
SAFMO	8032	Respondent's education		966	0.948
5. Employment					
ACL	57	Hours per week respondent works on job		943	0.823
ACL	143	Weeks during the past 12 months respondent was employed, including paid vacation and sick leave		823	0.398
ACL	155	Year respondent stopped working on last regular job		398	0.922
NES50	5039	How much respondent does farming		72	1.057[b]
NES50	5040	How much land respondent farms		69	0.918
NES50	5043	How long respondent has belonged to union		195	0.808
NES70	7024	Hours per week respondent works/ worked on job		648	0.754
NES90	9134	When respondent retired		77	0.968
NES90	9143	Hours per week respondent works on job (R working/temp laid off)	E,4	346	0.756
NES90	9162	Hours per week respondent works/ worked on job (all Rs)		362	0.744
SAFMO	8033	Hours per week respondent works on main job (Not Working = 0)		864	0.706
SAFMO	8034	Hours per week respondent works on main job (Working all 3 waves)	E,4	378	0.720
6. Occupation					
NES70	7022	Respondent's present/former Duncan SEI		853	0.827
NES90	9139	Respondent's occupation—prestige score		521	0.804
7. Income					
ACL	245	Respondent's job income		798	0.955
ACL	247	Family income for the last 12 months		1,841	0.882
NES50	5035	Family income for this year		1,072	0.917
NES70	7018	Family income for the previous year		1,177	0.869
NES90	9176	Family income for the previous year	E,6	490	0.917
NES90	9177	Respondent's income for the previous year		529	0.953

Source	Variable Number	Question Content	Note	Sample Size	Reliability[a]
SAFMO	8029	Family income last year (1976, 1979, 1984)		802	0.769
SAFMO	8030	Family income last year (1979, 1984, 1992)	E,7	705	0.779

8. Other Economic Characteristics

Source	Variable Number	Question Content	Note	Sample Size	Reliability[a]
ACL	249	Assets of respondent and spouse		1,912	0.880
SAFMO	8055	Amount of mortgage or debt on home		477	0.742
SAFMO	8057	Number of different types of family assets	S,2	1,113	0.718
SAFMO	8058	Present value of family assets	S,2	877	0.799

9. Social Interactions

Source	Variable Number	Question Content	Note	Sample Size	Reliability[a]
ACL	4	How often respondent talks on the telephone with friends, neighbors, or relatives		2,216	0.588
ACL	5	How often respondent gets together with friends, neighbors, or relatives		2,214	0.472
ACL	23	Frequency of contact with child/children who are 16 years old or older and do not live at home		1,085	0.766
ACL	36	Number of family and friends with whom respondent can share very private feelings and concerns		2,190	0.448
ACL	58	Amount of unpaid help given in the last year to friends, neighbors, or relatives who did not live with respondent		1,439	0.397
ACL	71	Hours per week spent caring for children who live with respondent		452	0.814
ACL	175	Hours in the past year respondent spent caring for friends and relatives who had trouble taking care of themselves		2,214	0.693

10. Activities

Source	Variable Number	Question Content	Note	Sample Size	Reliability[a]
ACL	6	How often respondent attends meetings or programs of groups, clubs, or organizations respondent belongs to		2,220	0.634
ACL	7	How often respondent works in the garden or yard		2,214	0.825
ACL	8	How often respondent engages in active sports or exercise other than taking walks		2,217	0.656
ACL	9	How often respondent takes walks		2,211	0.514
ACL	59	Hours respondent spent in the past 12 months doing paid work that was not part of a regular job		2,216	0.813

Source	Variable Number	Question Content	Note	Sample Size	Reliability[a]
ACL	60	Hours respondent spent in the past 12 months doing volunteer work		619	0.723
ACL	78	Hours respondent spent in the past 12 months doing home maintenance or improvement work		1,586	0.574
ACL	86	Hours respondent spends in an average week doing housework, including cooking and cleaning		1,986	0.690
NES90	9025	How many days in the past week respondent watched the news on television		593	0.615
NES90	9026	How many days in the past week respondent read a daily newspaper		592	0.682

11. Religious Activities

Source	Variable Number	Question Content	Note	Sample Size	Reliability[a]
ACL	201	How often respondent usually attends religious services		2,217	0.875
NES50	5027	How often respondent goes to church		1,045	0.814
NES70	7019	How often respondent goes to church or synagogue		1,142	0.914
NES90	9101	How often respondent prays, outside of attending religious services		587	0.886
NES90	9102	How often respondent reads the Bible, outside of attending religious services		590	0.830
SAFCH	8133	How often respondent usually attends religious services		865	0.746
SAFMO	8035	How often respondent usually attends religious services		874	0.835

12. Political Behavior

Source	Variable Number	Question Content	Note	Sample Size	Reliability[a]
NES50	5023	Frequency with which respondent has voted in elections for President since old enough to vote		1,056	0.907
NES50	5067	If there was ever a time when respondent had a different party identification, and which party that was		1,022	0.435
NES70	7137	How much respondent follows what's going on in government and public affairs		1,176	0.635
NES70	7144	Frequency with which respondent has voted in elections for President since old enough to vote		1,135	0.893

13. Health Behavior

Source	Variable Number	Question Content	Note	Sample Size	Reliability[a]
ACL	53	Average number of cigarettes respondent smokes in a day		2,204	0.873

Source	Variable Number	Question Content	Note	Sample Size	Reliability[a]
ACL	54	Number of days during the last month respondent drank beer, wine, or liquor		887	0.773
ACL	56	On days respondent drinks, how many cans of beer, glasses of wine, or drinks of liquor does respondent usually have		771	0.535

B. Values
1. Morality and Spirituality

ACL	202	How important religious or spiritual beliefs are in respondent's day-to-day life		2,216	0.808
NES90	9100	Would respondent say her/his religion provides some guidance in her/his day-to-day living, quite a bit of guidance, or a great deal of guidance in her/his day-to-day life		411	0.614
NES90	9323	Extent to which respondent agrees/disagrees that the world is always changing and we should adjust our view of moral behavior to those changes		537	0.661
NES90	9324	Extent to which respondent agrees/disagrees that we should be more tolerant of people who choose to live according to their own moral standards, even if they are very different from our own		538	0.700

2. Social Groups

NES50	5008	Extent to which respondent agrees/disagrees that if Negroes are not getting fair treatment in jobs and housing, the government should see to it that they do		795	0.588
NES50	5016	Extent to which respondent agrees/disagrees that the government in Washington should stay out of the question of whether white and colored children go to the same school		816	0.717
NES70	7104	Where respondent would place herself/himself on the help to minority groups scale (the government in Washington should make every possible effort to improve the social and economic position of Blacks and other minority groups versus the government should not make any special effort to help minorities because they should help themselves)		948	0.657

Source	Variable Number	Question Content	Note	Sample Size	Reliability[a]
NES70	7107	Where respondent would place herself/ himself on the school busing scale (achieving racial integration of schools is so important that it justifies busing children to schools out of their own neighborhoods versus letting children go to their neighborhood schools is so important that people oppose busing)		1,042	0.754
NES90	9092	Where respondent would place herself/ himself on the help to Blacks scale (the government in Washington should make every effort to improve the social and economic position of Blacks versus the government should not make any special effort to help Blacks because they should help themselves)		452	0.777
NES90	9331	Does respondent think the number of immigrants from foreign countries who are permitted to come to the United States to live should be increased a little, increased a lot, decreased a little, decreased a lot, or left the same as it is now		514	0.597

3. Government

Source	Variable Number	Question Content	Note	Sample Size	Reliability[a]
NES50	5002	Extent to which respondent agrees/ disagrees that the government in Washington ought to see to it that everybody who wants to work can find a job		868	0.594
NES50	5006	Extent to which respondent agrees/ disagrees that the United States should give economic help to the poorer countries of the world even if those countries can't pay for it		753	0.461
NES50	5010	Extent to which respondent agrees/ disagrees that if the cities and towns around the country need help to build more schools, the government in Washington ought to give them the money they need		871	0.610
NES50	5012	Extent to which respondent agrees/ disagrees that the United States should keep soldiers overseas where they can help countries that are against Communism		688	0.615
NES50	5014	Extent to which respondent agrees/ disagrees that the government should leave things like electric power and housing for private businessmen to handle		596	0.456

Source	Variable Number	Question Content	Note	Sample Size	Reliability[a]
NES70	7097	Where respondent would place herself/ himself on the job assurance scale (the government in Washington should see to it that every person has a job and a good standard of living versus the government should just let each person get ahead on his own)		908	0.577
NES70	7100	Where respondent would place herself/ himself on the urban unrest and rioting scale (use all available force to maintain law and order no matter what results versus correct the problems of poverty and unemployment that give rise to the disturbances)		400	0.386
NES70	7114	Where respondent would place herself/ himself on the rights of the accused scale (do everything possible to protect the legal rights of those accused of committing crimes versus it is more important to stop criminal activity even at the risk of reducing the rights of the accused)		925	0.516
NES90	9084	Where respondent would place herself/ himself on the government spending scale (the government should pro- vide fewer services in order to reduce spending versus the government should provide many more services even if it means an increase in spending)		464	0.654
NES90	9088	Where respondent would place herself/ himself on the defense spending scale (we should spend much less money for defense versus defense spending should be greatly increased)		487	0.581
NES90	9089	Where respondent would place herself/ himself on the medical and hospital costs scale (there should be a govern- ment insurance plan which would cover all medical expenses for everyone versus all medical expenses should be paid by individuals, and through private insurance plans like Blue Cross or other company paid plans)		465	0.575

Source	Variable Number	Question Content	Note	Sample Size	Reliability[a]
NES90	9090	Where respondent would place herself/ himself on the job assurance scale (the government in Washington should see to it that every person has a job and a good standard of living versus the government should just let each person get ahead on their own)		497	0.632
NES90	9093	Whether respondent thinks federal spending on food stamps should be increased, decreased, or kept about the same		562	0.648
NES90	9094	Whether respondent thinks federal spending on welfare programs should be increased, decreased, or kept about the same		569	0.752
NES90	9095	Whether respondent thinks federal spending on AIDS research should be increased, decreased, or kept about the same		576	0.796

4. Family

Source	Variable Number	Question Content	Note	Sample Size	Reliability
NES70	7113	Where respondent would place herself/ himself on the women's equal role scale (women should have an equal role with men in running business, industry, and government versus women's place is in the home)		1,016	0.730
NES90	9098	Where respondent would place herself/ himself on the women's equal role scale (women should have an equal role with men in running business, industry, and government versus a woman's place is in the home)		544	0.657
SAFCH	8102	Extent to which respondent agrees/ disagrees that a young couple should not live together unless they are married		863	0.751
SAFCH	8103	Extent to which respondent agrees/ disagrees that young people should not have sex before marriage		848	0.860
SAFCH	8105	Extent to which respondent agrees/ disagrees that when there are children in the family, parents should stay together even if they don't get along		818	0.595
SAFCH	8107	Extent to which respondent agrees/ disagrees that most of the important decisions in the life of the family should be made by the man of the house		833	0.647

Source	Variable Number	Question Content	Note	Sample Size	Reliability[a]
SAFCH	8109	Extent to which respondent agrees/ disagrees that there is some work that is men's and some that is women's, and they should not be doing each others'		831	0.567
SAFCH	8110	Extent to which respondent agrees/ disagrees that a wife should not expect her husband to help around the house after he comes home from a hard days work		833	0.818
SAFCH	8114	Extent to which respondent agrees/ disagrees that it is more important for a wife to help her husband's career than to have one herself		829	0.586
SAFMO	8002	Extent to which respondent agrees/ disagrees that a young couple should not live together unless they are married		833	0.706
SAFMO	8003	Extent to which respondent agrees/ disagrees that young people should not have sex before marriage		813	0.582
SAFMO	8005	Extent to which respondent agrees/ disagrees that when there are children in the family, parents should stay together even if they don't get along		821	0.714
SAFMO	8007	Extent to which respondent agrees/ disagrees that most of the important decisions in the life of the family should be made by the man of the house		858	0.718
SAFMO	8009	Extent to which respondent agrees/ disagrees that there is some work that is men's and some that is women's, and they should not be doing each others'		826	0.635
SAFMO	8010	Extent to which respondent agrees/ disagrees that a wife should not expect her husband to help around the house after he comes home from a hard days work		746	0.609
SAFMO	8014	Extent to which respondent agrees/ disagrees that it is more important for a wife to help her husband's career than to have one herself		781	0.914
SAFMO	8070	What does respondent think is the ideal number of children for the average American family		903	0.517
SAFMO	8079	Does respondent feel that almost all married couples who can ought to have children		871	0.742

Source	Variable Number	Question Content	Note	Sample Size	Reliability[a]
C. Beliefs					
1. Relationships					
ACL	19	How much respondent's husband/wife/ partner makes respondent feel loved and cared for		1,142	0.749
ACL	20	How much respondent's husband/ wife/partner is willing to listen when respondent needs to talk about worries or problems		1,142	0.719
ACL	21	How much respondent's son/daughter/ children make respondent feel loved and cared for		1,289	0.683
ACL	22	How much respondent's son/daughter/ children is/are willing to listen when respondent needs to talk about worries or problems		1,263	0.690
ACL	29	How much respondent's mother/mother figure makes respondent feel loved and cared for		392	0.719
ACL	30	How much respondent's mother/mother figure is willing to listen when respondent needs to talk about worries or problems		384	0.673
ACL	31	How much respondent's father/father figure makes respondent feel loved and cared for		86	0.951
ACL	32	How much respondent's father/father figure is willing to listen when respondent needs to talk about worries or problems		87	0.668
ACL	33	How much respondent's friends and other relatives make respondent feel loved and cared for		2,211	0.531
ACL	34	How much respondent's friends and other relatives are willing to listen when respondent needs to talk about worries or problems		2,174	0.518
SAFCH	8119	How often respondent can respect mother's ideas and opinions about the important things in life		809	0.626
SAFCH	8120	How often mother respects respondent's ideas and opinions about the important things in life		810	0.552
SAFCH	8121	How often mother accepts and understands respondent as a person		809	0.538
SAFCH	8123	How often mother makes it easy for respondent to confide in her		808	0.718

Source	Variable Number	Question Content	Note	Sample Size	Reliability[a]
SAFCH	8124	How often mother gives respondent the right amount of affection		810	0.586
SAFCH	8141	How often respondent can respect father's ideas and opinions about the important things in life		732	0.560
SAFCH	8142	How often father respects respondent's ideas and opinions about the important things in life		729	0.417
SAFCH	8143	How often father accepts and understands respondent as a person		728	0.535
SAFCH	8145	How often father makes it easy for respondent to confide in him/her		731	0.600
SAFCH	8146	How often father gives respondent the right amount of affection		732	0.682
SAFCH	8148	How happy respondent thinks her/his parent's marriage is, taking things all together		521	0.919
SAFCH	8149	How often respondent thinks her/his parents have problems getting along with each other		517	0.704
SAFMO	8019	How often respondent can respect her child's ideas and opinions about the important things in life		845	0.566
SAFMO	8020	How often child respects respondent's ideas and opinions about the important things in life		842	0.393
SAFMO	8024	How well respondent thinks her husband understands her, her feelings, likes and dislikes, and problems		627	0.713
SAFMO	8069	Are most of the people respondent knows really well and feels close to friends or relatives		854	0.473

2. *Economy and Society*

Source	Variable Number	Question Content	Note	Sample Size	Reliability[a]
ACL	199	Extent to which respondent agrees/ disagrees that people die when it is their time to die, and nothing can change that		2,189	0.659
ACL	200	Extent to which respondent agrees/ disagrees that if bad things happen, it is because they were meant to be		2,165	0.587
ACL	221	How much money respondent would get for her/his house/apartment/farm, after paying off the mortgage, if respondent sold it today		972	0.876

Source	Variable Number	Question Content	Note	Sample Size	Reliability[a]
NES70	7075	Comparing how much people in respondent's line of work earn with how much money people in other occupations earn, does respondent feel that the money people in respondent's line of work make is much less than their fair share, somewhat less than their fair share, about their fair share, or more than their fair share		266	0.592
NES70	7080	Generally speaking, would respondent say that most people can be trusted, or that you can't be too careful in dealing with people		1,133	0.705
NES70	7081	Would respondent say that most of the time people try to be helpful, or that they are mostly just looking out for themselves		1,064	0.746
NES70	7082	Does respondent think most people would try to take advantage of her/him if they got a chance, or would they try to be fair		1,065	0.676
NES90	9318	Extent to which respondent agrees/disagrees that if people were treated more equally in this country we would have many fewer problems		537	0.481
NES90	9325	Extent to which respondent agrees/disagrees that this country would have many fewer problems if there were more emphasis on traditional family ties		534	0.636
NES90	9326	Extent to which respondent agrees/disagrees that the newer lifestyles are contributing to the breakdown of our society		530	0.553
NES90	9387	Whether respondent expects the economy to get better, get worse, or stay about the same in the next 12 months	E,7	507	0.552
SAFMO	8054	Compared with other men respondent's husband's age with about the same education who are in his general line of work, is respondent's husband's income above average, about average, or below average		776	0.463
SAFMO	8056	What the present value of respondent's house is, or what it would bring if respondent sold it today		534	0.972

Source	Variable Number	Question Content	Note	Sample Size	Reliability[a]
3. Influence of Social Groups					
NES70	7045	Whether respondent thinks big business has too much influence, just about the right amount of influence, or too little influence in American life and politics		1,067	0.742
NES70	7046	Whether respondent thinks Blacks have too much influence, just about the right amount of influence, or too little influence in American life and politics		1,011	0.710
NES70	7048	Whether respondent thinks labor unions have too much influence, just about the right amount of influence, or too little influence in American life and politics		1,047	0.623
NES70	7049	Whether respondent thinks liberals have too much influence, just about the right amount of influence, or too little influence in American life and politics		780	0.639
NES70	7050	Whether respondent thinks old people have too much influence, just about the right amount of influence, or too little influence in American life and politics		1,078	0.458
NES70	7051	Whether respondent thinks people on welfare have too much influence, just about the right amount of influence, or too little influence in American life and politics		942	0.643
NES70	7052	Whether respondent thinks poor people have too much influence, just about the right amount of influence, or too little influence in American life and politics		1,071	0.546
NES70	7054	Whether respondent thinks women have too much influence, just about the right amount of influence, or too little influence in American life and politics		1,054	0.369
NES70	7055	Whether respondent thinks young people have too much influence, just about the right amount of influence, or too little influence in American life and politics		1,039	0.410
NES70	7103	Whether respondent thinks that civil rights leaders are trying to push too fast, are going too slowly, or are moving at about the right speed		1,104	0.613

Source	Variable Number	Question Content	Note	Sample Size	Reliability[a]
4. Government					
NES50	5042	How much effect does respondent think government farm policies have had on what respondent gets for crops		65	0.294
NES70	7120	How much of the time respondent thinks you can trust the government in Washington to do what is right		1,215	0.581
NES70	7121	Would respondent say the government is pretty much run by a few big interests looking out for themselves or that it is run for the benefit of all the people		1,009	0.745
NES70	7124	Does respondent agree or disagree that people like her/him don't have any say about what the government does		1,151	0.639
NES70	7125	Does respondent agree or disagree that voting is the only way that people like her/him can have any say about how the government runs things		1,145	0.690
NES70	7126	Does respondent agree or disagree that sometimes politics and government seem so complicated that a person like her/him can't really understand what's going on		1,157	0.746
NES70	7130	How much attention does respondent feel the government pays to what the people think when it decides what to do		1,139	0.382
NES70	7132	How much does respondent feel that having elections makes the government pay attention to what the people think		1,141	0.421
NES90	9074	Would respondent say that during the past year the United States' position in the world has grown weaker, stayed about the same, or grown stronger		576	0.697
NES90	9320	Extent to which respondent agrees/ disagrees that people like her/him don't have any say about what the government does		591	0.500
NES90	9322	Extent to which respondent agrees/ disagrees that sometimes politics and government seem so complicated that a person like her/him can't really understand what's going on		538	0.551
NES90	9327	How much of the time respondent thinks you can trust the government in Washington to do what is right		537	0.372
NES90	9329	Would respondent say the government is pretty much run by a few big interests looking out for themselves or that it is run for the benefit of all the people		473	0.712

Source	Variable Number	Question Content	Note	Sample Size	Reliability[a]
5. Public Officials					
NES70	7094	Where respondent would place George Wallace on the liberal-to-conservative scale		542	0.634
NES70	7119	Does respondent think that people in the government waste a lot of money we pay in taxes, waste some of it, or don't waste very much of it		1,237	0.602
NES70	7122	Does respondent feel that almost all of the people running the government are smart people, or that quite a few of them don't seem to know what they are doing		1,091	0.504
NES70	7123	Does respondent think that quite a few of the people running the government are crooked, not very many are, or hardly any of them are crooked		1,133	0.611
NES70	7127	Does respondent agree or disagree that public officials don't care much what people like her/him think		1,089	0.639
NES70	7128	Does respondent agree or disagree that those we elect to Congress in Washington lose touch with the people pretty quickly		1,082	0.711
NES70	7133	How much attention does respondent think most Congressmen pay to the people who elect them when they decide what to do in Congress		1,126	0.581
NES90	9063	Where respondent would place Bill Clinton on the liberal-to-conservative scale		493	0.680
NES90	9079	In respondent's opinion, how well does "moral" describe Bill Clinton		536	0.755
NES90	9080	In respondent's opinion, how well does "provides strong leadership" describe Bill Clinton		524	0.694
NES90	9081	In respondent's opinion, how well does "really cares about people like you" describe Bill Clinton		534	0.772
NES90	9082	In respondent's opinion, how well does "knowledgeable" describe Bill Clinton		559	0.683
NES90	9083	In respondent's opinion, how well does "gets things done" describe Bill Clinton		464	0.725
NES90	9085	Where respondent would place Bill Clinton on the government spending scale (the government should provide fewer services in order to reduce spending versus the government should provide many more services even if it means an increase in spending)		433	0.463

Source	Variable Number	Question Content	Note	Sample Size	Reliability[a]
NES90	9091	Where respondent would place Bill Clinton on the job assurance scale (the government in Washington should see to it that every person has a job and a good standard of living versus the government should just let each person get ahead on their own)		452	0.502
NES90	9321	Extent to which respondent agrees/ disagrees that public officials don't care much what people like her/him think		535	0.424
NES90	9328	Does respondent think that people in government waste a lot of the money we pay in taxes, waste some of it, or don't waste very much of it		536	0.505
NES90	9330	Does respondent think that quite a few of the people running the government are crooked, not very many are, or hardly any of them are crooked		524	0.751

6. Political Parties

Source	Variable Number	Question Content	Note	Sample Size	Reliability[a]
NES50	5003	Whether respondent thinks the Democrats are closer to what respondent wants on the issue of the government seeing to it that everyone who wants to work can find a job, or are the Republicans closer, or wouldn't there be any difference between them on this		353	0.653
NES50	5007	Whether respondent thinks the Democrats or the Republicans are closer to what respondent wants on the issue of the government giving economic help to the poorer countries of the world even if they can't pay for it, or isn't there any difference between them		271	0.633
NES50	5009	Whether respondent thinks the Democrats or the Republicans are closer to what respondent wants on the issue of the government seeing to it that Negroes get fair treatment in jobs and housing, or wouldn't there be any difference		341	0.559
NES50	5011	Whether respondent thinks the Democrats or the Republicans are closer to what respondent wants on the issue of the government in Washington giving money to cities and towns around the country who need help to build more schools, or wouldn't there be any difference between them on this		301	0.535

Source	Variable Number	Question Content	Note	Sample Size	Reliability[a]
NES50	5013	Whether respondent thinks the Democrats or the Republicans are closer to what respondent wants on the issue of the U.S. keeping soldiers overseas where they can help countries that are against Communism, or wouldn't there be any difference		348	0.649
NES50	5015	Whether respondent thinks the Democrats or the Republicans are closer to what respondent wants on the issue of the government leaving things like electric power and housing for private businessmen to handle, or isn't there be any difference		242	0.784
NES70	7047	Whether respondent thinks Democrats have too much influence, just about the right amount of influence, or too little influence in American life and politics		948	0.644
NES70	7053	Whether respondent thinks Republicans have too much influence, just about the right amount of influence, or too little influence in American life and politics		945	0.681
NES70	7095	Where respondent would place the Democratic Party on the liberal-to-conservative scale		643	0.506
NES70	7096	Where respondent would place the Republican Party on the liberal-to-conservative scale		638	0.516
NES70	7098	Where respondent would place the Democratic Party on the job assurance scale (the government in Washington should see to it that every person has a job and a good standard of living versus the government should just let each person get ahead on his own)		668	0.388
NES70	7099	Where respondent would place the Republican Party on the job assurance scale (the government in Washington should see to it that every person has a job and a good standard of living versus the government should just let each person get ahead on his own)		665	0.420
NES70	7101	Where respondent would place the Democratic Party on the urban unrest and rioting scale (use all available force to maintain law and order no matter what results versus correct the problems of poverty and unemployment that give rise to the disturbances)		284	0.355

Source	Variable Number	Question Content	Note	Sample Size	Reliability[a]
NES70	7102	Where respondent would place the Republican Party on the urban unrest and rioting scale (use all available force to maintain law and order no matter what results versus correct the problems of poverty and unemployment that give rise to the disturbances)		280	0.409
NES70	7105	Where respondent would place the Democratic Party on the help to minority groups scale (the government in Washington should make every possible effort to improve the social and economic position of Blacks and other minority groups versus the government should not make any special effort to help minorities because they should help themselves)		621	0.328
NES70	7106	Where respondent would place the Republican Party on the help to minority groups scale (the government in Washington should make every possible effort to improve the social and economic position of Blacks and other minority groups versus the government should not make any special effort to help minorities because they should help themselves)		621	0.387
NES70	7108	Where respondent would place the Democratic Party on the school busing scale (achieving racial integration of schools is so important that it justifies busing children to schools out of their own neighborhoods versus letting children go to their neighborhood schools is so important that people oppose busing)		540	0.451
NES70	7109	Where respondent would place the Republican Party on the school busing scale (achieving racial integration of schools is so important that it justifies busing children to schools out of their own neighborhoods versus letting children go to their neighborhood schools is so important that people oppose busing)		559	0.378
NES70	7115	Where respondent would place the Democratic Party on the rights of the accused scale (do everything possible to protect the legal rights of those accused of committing crimes versus it is more important to stop criminal activity even at the risk of reducing the rights of the accused)		538	0.451

Source	Variable Number	Question Content	Note	Sample Size	Reliability[a]
NES70	7116	Where respondent would place the Republican Party on the rights of the accused scale (do everything possible to protect the legal rights of those accused of committing crimes versus it is more important to stop criminal activity even at the risk of reducing the rights of the accused)		553	0.313
NES70	7129	Does respondent agree or disagree that parties are only interested in people's votes but not in their opinions		1,075	0.730
NES70	7131	How much does respondent feel that political parties help to make the government pay attention to what the people think		1,106	0.363
NES90	9064	Where respondent would place the Republican Party on the liberal-to-conservative scale		456	0.631
NES90	9065	Where respondent would place the Democratic Party on the liberal-to-conservative scale		461	0.742
NES90	9086	Where respondent would place the Republican Party on the government spending scale (the government should provide fewer services in order to reduce spending versus the government should provide many more services even if it means an increase in spending)		411	0.290
NES90	9087	Where respondent would place the Democratic Party on the government spending scale (the government should provide fewer services in order to reduce spending versus the government should provide many more services even if it means an increase in spending)		420	0.387
NES90	9286	Which political party respondent thinks would be most likely to get the government to do a better job in dealing with the single most important problem the country faces—the Republicans, the Democrats, or wouldn't there be much difference between them		248	0.676

7. Family

Source	Variable Number	Question Content	Note	Sample Size	Reliability[a]
SAFCH	8111	Extent to which respondent agrees/disagrees that a working mother can establish as warm and secure a relationship with her children as a mother who does not work		838	0.528

Source	Variable Number	Question Content	Note	Sample Size	Reliability[a]
SAFCH	8113	Extent to which respondent agrees/ disagrees that women are much happier if they stay at home and take care of their children		744	0.515
SAFCH	8115	Extent to which respondent agrees/ disagrees that married people are usually happier than those who go through life without getting married		841	0.540
SAFCH	8116	Extent to which respondent agrees/ disagrees that there are few good or happy marriages these days		860	0.438
SAFCH	8118	Extent to which respondent agrees/ disagrees that all in all, there are more advantages to being single than to being married		840	0.393
SAFMO	8011	Extent to which respondent agrees/ disagrees that a working mother can establish as warm and secure a relationship with her children as a mother who does not work		823	0.688
SAFMO	8013	Extent to which respondent agrees/ disagrees that women are much happier if they stay at home and take care of their children		744	0.616
SAFMO	8015	Extent to which respondent agrees/ disagrees that married people are usually happier than those who go through life without getting married		812	0.604
SAFMO	8016	Extent to which respondent agrees/ disagrees that there are few good or happy marriages these days		851	0.632
SAFMO	8018	Extent to which respondent agrees/ disagrees that all in all, there are more advantages to being single than to being married		805	0.589

D. Attitudes
1. Social Groups

Source	Variable Number	Question Content	Note	Sample Size	Reliability[a]
NES70	7056	Respondent's feeling thermometer rating of Black militants		1,082	0.791
NES70	7057	Respondent's feeling thermometer rating of Blacks		1,120	0.565
NES70	7058	Respondent's feeling thermometer rating of big business		1,033	0.492
NES70	7059	Respondent's feeling thermometer rating of civil rights leaders		1,064	0.773

Source	Variable Number	Question Content	Note	Sample Size	Reliability[a]
NES70	7060	Respondent's feeling thermometer rating of conservatives		940	0.582
NES70	7062	Respondent's feeling thermometer rating of labor unions		1,051	0.694
NES70	7063	Respondent's feeling thermometer rating of liberals		920	0.643
NES70	7064	Respondent's feeling thermometer rating of middle-class people		1,120	0.452
NES70	7065	Respondent's feeling thermometer rating of the military		1,116	0.690
NES70	7066	Respondent's feeling thermometer rating of people who use marijuana		1,113	0.729
NES70	7067	Respondent's feeling thermometer rating of policemen		1,152	0.648
NES70	7068	Respondent's feeling thermometer rating of poor people		1,112	0.589
NES70	7069	Respondent's feeling thermometer rating of radical students		1,055	0.567
NES70	7071	Respondent's feeling thermometer rating of Whites		1,129	0.534
NES70	7072	Respondent's feeling thermometer rating of the women's liberation movement		1,057	0.650
NES70	7073	Respondent's feeling thermometer rating of young people		1,133	0.510
NES90	9230	Respondent's feeling thermometer rating of labor unions		501	0.735
NES90	9231	Respondent's feeling thermometer rating of people on welfare		504	0.425
NES90	9232	Respondent's feeling thermometer rating of conservatives		493	0.694
NES90	9233	Respondent's feeling thermometer rating of poor people		501	0.701
NES90	9234	Respondent's feeling thermometer rating of big business		520	0.488
NES90	9235	Respondent's feeling thermometer rating of Blacks		504	0.566
NES90	9236	Respondent's feeling thermometer rating of the women's movement		519	0.771
NES90	9237	Respondent's feeling thermometer rating of liberals		504	0.748
NES90	9238	Respondent's feeling thermometer rating of Hispanic-Americans/Hispanics or Latinos		482	0.563
NES90	9239	Respondent's feeling thermometer rating of people seeking to protect the environment		512	0.697

Source	Variable Number	Question Content	Note	Sample Size	Reliability[a]
NES90	9240	Respondent's feeling thermometer rating of Whites		496	0.583
NES90	9241	Respondent's feeling thermometer rating of gay men and lesbians		498	0.781
NES90	9242	Respondent's feeling thermometer rating of Christian fundamentalists		469	0.856

2. Civil Rights

Source	Variable Number	Question Content	Note	Sample Size	Reliability[a]
NES90	9307	Whether respondent is for or against preferential hiring and promotion of Blacks		478	0.962
NES90	9309	Given respondent's view on the issue of school prayer, does respondent favor that choice strongly or not strongly		506	0.594
NES90	9315	Extent to which respondent agrees/ disagrees that we have gone too far in pushing equal rights in this country		537	0.605
NES90	9316	Extent to which respondent agrees/ disagrees that this country would be better off if we worried less about how equal people are		537	0.581
NES90	9317	Extent to which respondent agrees/ disagrees that it is not really that big a problem if some people have more of a chance in life than others		537	0.463
NES90	9319	Extent to which respondent agrees/ disagrees that one of the big problems in this country is that we don't give every- one an equal chance		539	0.459

3. Government

Source	Variable Number	Question Content	Note	Sample Size	Reliability[a]
NES50	5004	Extent to which respondent agrees/ disagrees that this country would be bet- ter off if we just stayed home and did not concern ourselves with problems in other parts of the world		846	0.592
NES70	7112	Whether respondent would say the gov- ernment is doing a good job, only a fair job, or a poor job in fighting inflation or unemployment		504	0.389
NES70	7134	If all other methods have failed and the person decides to try to stop the government from going about its usual activities with sit-ins, mass meetings, demonstrations, and things like that, would respondent approve of that, disapprove, or would it depend on the circumstances		1,135	0.533

Source	Variable Number	Question Content	Note	Sample Size	Reliability[a]
NES90	9305	Does respondent favor or oppose the death penalty for persons convicted of murder		469	0.910
NES90	9306	Does respondent favor/oppose the death penalty for persons convicted of murder strongly or not strongly	S,3	467	0.843

4. Public Officials

NES90	9027	Respondent's feeling thermometer rating of Al Gore		492	0.666
NES90	9028	Respondent's feeling thermometer rating of Hillary Clinton		548	0.811
NES90	9097	Does respondent favor or oppose a proposed law that would limit members of Congress to no more than 12 consecutive years of service in that office		534	0.776
NES90	9310	Does respondent approve or disapprove of the way the U.S. Congress has been handling its job		498	0.712
NES90	9311	Does respondent approve/disapprove strongly or not strongly of the way the U.S. Congress has been handling its job	S,3	496	0.809
NES90	9380	Respondent's feeling thermometer rating of Bill Clinton (Pre Post Pre)	8	525	0.791
NES90	9381	Respondent's feeling thermometer rating of Bill Clinton (Post Post Post)	8	525	0.824

5. Political Parties

NES70	7061	Respondent's feeling thermometer rating of Democrats		1,126	0.656
NES70	7070	Respondent's feeling thermometer rating of Republicans		1,107	0.552
NES90	9030	Respondent's feeling thermometer rating of the Democratic Party		566	0.756
NES90	9031	Respondent's feeling thermometer rating of the Republican Party		562	0.808

6. Political Candidates

NES70	7088	Respondent's feeling thermometer rating of Hubert Humphrey		1,160	0.691
NES70	7089	Respondent's feeling thermometer rating of Henry "Scoop" Jackson		482	0.634
NES70	7090	Respondent's feeling thermometer rating of Edward "Ted" Kennedy		1,188	0.807
NES70	7091	Respondent's feeling thermometer rating of Richard Nixon		1,223	0.766
NES70	7092	Respondent's feeling thermometer rating of George Wallace		1,140	0.777

Source	Variable Number	Question Content	Note	Sample Size	Reliability[a]
NES90	9029	Respondent's feeling thermometer rating of Jesse Jackson		555	0.772
NES90	9228	Respondent's feeling thermometer rating of Democratic House candidate		245	0.741
NES90	9229	Respondent's feeling thermometer rating of Republican House candidate		242	0.867
NES90	9382	Respondent's feeling thermometer rating of Ross Perot (Pre Post Pre)	8	483	0.796
NES90	9383	Respondent's feeling thermometer rating of Ross Perot (Post Post Post)	8	483	0.815
7. Family					
ACL	26	How happy respondent is with the way her/his son has/daughter has/children have turned out to this point		1,780	0.637
SAFCH	8101	Extent to which respondent agrees/ disagrees that it's alright for a couple to live together without planning to get married		863	0.714
SAFCH	8104	Extent to which respondent agrees/ disagrees that premarital sex is alright for a young couple planning to get married		852	0.792
SAFCH	8106	Extent to which respondent agrees/ disagrees that divorce is usually the best solution when a couple can't seem to work out their marriage problems		855	0.376
SAFCH	8108	Extent to which respondent agrees/ disagrees that it's perfectly alright for women to be very active in clubs, politics, and other outside activities before the children are grown up		821	0.587
SAFCH	8112	Extent to which respondent agrees/ disagrees that it is much better for everyone if the man earns the main living and the woman takes care of the home and family		823	0.657
SAFCH	8117	Extent to which respondent agrees/ disagrees that it's better for a person to get married than to go through life being single		839	0.665
SAFMO	8001	Extent to which respondent agrees/ disagrees that it's alright for a couple to live together without planning to get married		842	0.818
SAFMO	8004	Extent to which respondent agrees/ disagrees that premarital sex is alright for a young couple planning to get married		835	0.828

Source	Variable Number	Question Content	Note	Sample Size	Reliability[a]
SAFMO	8006	Extent to which respondent agrees/ disagrees that divorce is usually the best solution when a couple can't seem to work out their marriage problems		825	0.741
SAFMO	8008	Extent to which respondent agrees/ disagrees that it's perfectly alright for women to be very active in clubs, politics, and other outside activities before the children are grown up		806	0.733
SAFMO	8012	Extent to which respondent agrees/ disagrees that it is much better for everyone if the man earns the main living and the woman takes care of the home and the family		801	0.737
SAFMO	8017	Extent to which respondent agrees/ disagrees that it's better for a person to get married than to go through life being single		759	0.513
SAFMO	8082	How much it would bother respondent if things turn out so that respondent's child does not marry		202	0.585
SAFMO	8083	How much it would bother respondent if it turns out that respondent's child does not have any children		345	0.732

E. Self-perception
1. Self-efficacy

Source	Variable Number	Question Content	Note	Sample Size	Reliability[a]
ACL	14	Extent to which respondent agrees/ disagrees that "I can do just about anything I really set my mind to do."		2,186	0.522
ACL	15	Extent to which respondent agrees/ disagrees that "Sometimes I feel that I am being pushed around in life."		2,181	0.482
ACL	16	Extent to which respondent agrees/ disagrees that "There is really no way I can solve the problems I have."		2,160	0.498
NES50	5053	Does respondent feel that she/he is the kind of person that gets her/his share of bad luck, or does respondent feel that she/he has mostly good luck		225	0.775
NES70	7076	Whether respondent thinks it's better to plan her/his life a good way ahead, or that life is too much a matter of luck to plan ahead very far		1,088	0.763

Source	Variable Number	Question Content	Note	Sample Size	Reliability[a]
NES70	7077	Whether respondent usually gets to carry out things as planned, or do things usually come up to make respondent change plans		1,091	0.695
NES70	7078	Has respondent usually felt pretty sure her/his life would work out the way she/he wants it to, or have there been times when respondent hasn't been sure about it		1,137	0.623
NES70	7079	Is respondent most like people who feel they can run their lives pretty much the way they want to, or most like people who feel the problems of life are sometimes too big for them		1,086	0.739

2. Stress and Mental Health

Source	Variable Number	Question Content	Note	Sample Size	Reliability[a]
ACL	25	How often respondent feels bothered or upset as a parent		1,792	0.596
ACL	128	How often respondent felt depressed during the past week		2,205	0.497
ACL	129	How often respondent felt that everything she/he did was an effort during the past week		2,193	0.497
ACL	130	How often respondent felt her/his sleep was restless during the past week		2,196	0.509
ACL	131	How often respondent felt she/he was happy during the past week		2,190	0.504
ACL	132	How often respondent felt lonely during the past week		2,203	0.597
ACL	133	How often respondent felt people were unfriendly during the past week		2,199	0.531
ACL	135	How often respondent did not feel like eating during the past week		2,197	0.410
ACL	136	How often respondent felt sad during the past week		2,201	0.360
ACL	137	How often respondent felt that people disliked her/him during the past week		2,193	0.550
ACL	138	How often respondent felt she/he could not get going during the past week		2,201	0.511
ACL	150	How often respondent feels bothered or upset in her/his work		949	0.408
ACL	176	How stressful is it for respondent to take care of a friend or relative who has trouble taking care of herself/himself because of physical or mental illness, or disability, or how stressful is it for respondent to arrange for her/his care		95	0.600

Source	Variable Number	Question Content	Note	Sample Size	Reliability[a]
3. Quality of Life					
ACL	72	How much respondent enjoys caring for the child/children who lives/live with respondent		452	0.815
ACL	79	How much respondent enjoyed doing work to maintain or improve her/his home, yard, or automobile		1,599	0.598
ACL	87	How much respondent enjoys doing housework, including cooking, cleaning, and other work around the house		2,081	0.666
ACL	149	How much respondent enjoys doing the work on her/his main job		954	0.719
ACL	160	How much respondent enjoyed doing work or chores that people paid her/him to do instead of going to regular businesses		54	0.734
ACL	167	How much respondent enjoyed doing volunteer work		623	0.436
ACL	173	How much respondent enjoyed helping friends, neighbors, and relatives who did not live with respondent		1,445	0.506
SAFCH	8122	How often respondent enjoys doing things together with her/his mother		810	0.581
SAFCH	8144	How often respondent enjoys doing things together with her/his father		729	0.678
SAFMO	8022	How often respondent enjoys doing things together with son/daughter		845	0.531
SAFMO	8023	How often respondent enjoys talking to son/daughter		847	0.550
4. Functional Status					
ACL	106	How much difficulty respondent has bathing by herself/himself		2,222	0.897
ACL	109	How much difficulty respondent has climbing a few flights of stairs		1,989	0.729
ACL	111	How much difficulty respondent has walking several blocks		1,991	0.800
ACL	114	How much difficulty would respondent have doing heavy work around the house such as shoveling snow or washing walls		1,764	0.701
ACL	116	How much are respondent's daily activities limited in any way by her/his health or health-related problems		2,217	0.725
5. Financial Situation					
ACL	218	How difficult is it for respondent/ respondent's family to meet the monthly payments on respondent's/respondent's family's bills		2,187	0.780

Source	Variable Number	Question Content	Note	Sample Size	Reliability[a]
NES50	5019	During the last few years, has respondent's financial situation been getting better, getting worse, or has it stayed about the same		1,092	0.496
NES50	5020	Looking ahead and thinking about the next few years, does respondent expect her/his financial situation will stay about the way it is now, get better, or get worse		937	0.596
NES70	7074	Comparing how much respondent earns on her/his job with what other people in respondent's line of work get, does respondent feel the money respondent makes is much less than her/his fair share, somewhat less than her/his fair share, about her/his fair share, or more than her/his fair share		272	0.599
NES70	7111	Does respondent think that a year from now she/he (and family) will be better off financially, or worse off, or just about the same as now		455	0.402
NES90	9054	Would respondent say that she/he (and family living here) are better off or worse off financially than respondent was a year ago		591	0.484
NES90	9055	Would respondent say that she/he (and family living here) are much better off or somewhat better off /much worse off or somewhat worse off than respondent was a year ago	S,3	588	0.693
NES90	9056	Does respondent think that a year from now she/he (and family living here) will be better off financially, worse off, or just about the same as now		572	0.494
NES90	9057	Does respondent think that a year from now she/he (and family living here) will be much better off or somewhat better off financially/much worse off or somewhat worse off financially	S,3	570	0.489
SAFMO	8053	When respondent thinks of what her family needs to get along on comfortably right now, is her husband's income high enough, a little low, or quite a bit too low		825	0.468

6. Political Interest

Source	Variable Number	Question Content	Note	Sample Size	Reliability[a]
NES50	5024	Would respondent say that she/he was very much interested, somewhat interested, or not much interested in following the political campaigns this year		1,101	0.559

Source	Variable Number	Question Content	Note	Sample Size	Reliability[a]
NES90	9211	Would respondent say that she/he was very much interested, somewhat interested, or not much interested in following the political campaigns this year		538	0.523
NES90	9214	How much would respondent say that she/he personally cared about the way the election to the U.S. House of Representatives came out		584	0.561
NES90	9280	Would respondent say she/he follows what's going on in government and public affairs most of the time, some of the time, only now and then, or hardly at all		536	0.805

7. Political Identification

Source	Variable Number	Question Content	Note	Sample Size	Reliability[a]
NES50	5021	Generally speaking, does respondent usually think of herself/himself as a Republican, a Democrat, an Independent, or what. (If Republican or Democrat) Would respondent call herself/himself a strong Republican/Democrat or a not very strong Republican/Democrat. (If Independent or Other) Does respondent think of herself/himself as closer to the Republican or Democratic Party. [7-Level Party Identification]	S,3	1,045	0.899
NES50	5109	Generally speaking, does respondent usually think of herself/himself as a Republican, a Democrat, an Independent, or what [3-Level Party Identification]		1,045	0.950
NES50	5110	(If Republican or Democrat) Would respondent call herself/himself a strong Republican/Democrat or a not very strong Republican/Democrat	E,1	677	0.696
NES50	5111	(If Independent or Other) Does respondent think of herself/himself as closer to the Republican or Democratic Party	E,1	101	0.530
NES50	5112	(If Republican or Democrat) Would respondent call herself/himself a strong Republican/Democrat or a not very strong Republican/Democrat. (If Independent, No Preference, or Other) Does respondent think of herself/himself as closer to the Republican or to the Democratic party. [4-Level Party Identification]	S,3	1,045	0.729

Source	Variable Number	Question Content	Note	Sample Size	Reliability[a]
NES70	7083	Generally speaking, does respondent usually think of herself/himself as a Republican, a Democrat, an Independent, or what. (If Republican or Democrat) Would respondent call herself/ himself a strong Republican/Democrat or a not very strong Republican/ Democrat. (If Independent, No Preference, or Other) Does respondent think of herself/himself as closer to the Republican or to the Democratic party. [7-Level Party Identification]	S,3	1,237	0.863
NES70	7084	(If Republican or Democrat) Would respondent call herself/himself a strong Republican/Democrat or a not very strong Republican/Democrat	E,1	582	0.665
NES70	7086	(If Independent, No Preference, or Other) Does respondent think of herself/ himself as closer to the Republican or to the Democratic party	E,1	250	0.856
NES70	7093	Where respondent would place herself/ himself on the liberal-to-conservative scale		761	0.708
NES70	7212	Generally speaking, does respondent usually think of herself/himself as a Republican, a Democrat, an Independent, or what [3-Level Party Identification]		1,263	0.901
NES70	7213	(If Republican) Was there ever a time when respondent thought of herself/ himself as a Democrat or an Independent rather than a Republican. (If Yes) Did respondent think of herself/himself as a Democrat or an Independent. (If Democrat) Was there ever a time when respondent thought of herself/himself as a Republican or an Independent rather than a Democrat. (If Yes) Did respondent think of herself/himself as a Republican or an Independent. [Former Party Alignment for Current Party Identifiers]	S,2	586	0.712

Source	Variable Number	Question Content	Note	Sample Size	Reliability[a]
NES70	7214	(If closer to the Republican party) Was there ever a time when respondent thought of herself/himself as closer to the Democratic party instead of the Republican party. (If Neither) Was there ever a time when respondent thought of herself/himself as a Democrat or as a Republican. (Which party was that?) (If closer to the Democratic party) Was there ever a time when respondent thought of herself/himself as closer to the Republican party instead of the Democratic party. [Former Party Alignment for Current Leaners and Independents]	S,2	245	0.617
NES70	7229	(If Republican or Democrat) Would respondent call herself/himself a strong Republican/Democrat or a not very strong Republican/Democrat. (If Independent, No Preference, or Other) Does respondent think of herself/himself as closer to the Republican or to the Democratic party. [4-Level Party Identification]	S,3	1,237	0.695
NES90	9062	Where respondent would place herself/himself on the liberal-to-conservative scale		407	0.824
NES90	9075	Generally speaking, does respondent usually think of herself/himself as a Republican, a Democrat, an Independent or what [3-Level Party Identification]		585	0.978
NES90	9076	(If Republican or Democrat) Would respondent call herself/himself a strong Republican/Democrat or a not very strong Republican/Democrat	E,1	272	0.596
NES90	9077	(If Independent, No Preference, or Other Party) Does respondent think of herself/himself as closer to the Republican Party or to the Democratic Party	E,1	114	0.807

Source	Variable Number	Question Content	Note	Sample Size	Reliability[a]
NES90	9078	Generally speaking, does respondent usually think of herself/himself as a Republican, a Democrat, an Independent, or what. (If Republican or Democrat) Would respondent call herself/himself a strong Republican/Democrat or a not very strong Republican/Democrat. (If Independent, No Preference, or Other Party) Does respondent think of herself/himself as closer to the Republican Party or to the Democratic Party. [7-Level Party Identification]	S,3	584	0.900
NES90	9418	(If Republican or Democrat) Would respondent call herself/himself a strong Republican/Democrat or a not very strong Republican/Democrat. (If Independent, No Preference, or Other) Does respondent think of herself/himself as closer to the Republican or to the Democratic party. [4-Level Party Identification]	S,3	584	0.737

8. Relationship to Social Groups

Source	Variable Number	Question Content	Note	Sample Size	Reliability[a]
NES50	5044	Would respondent say she/he (or head) feels pretty close to labor union members in general or that she/he doesn't feel much closer to them than she/he does to other kinds of people		197	0.811
NES50	5045	How much interest would respondent say she/he (or head) has in how union people as a whole are getting along in this country		203	0.579
NES50	5046	Would respondent say she/he feels pretty close to Negroes in general or that she/he doesn't feel much closer to them than she/he does to other people		83	0.267
NES50	5048	Would respondent say she/he feels pretty close to Catholics in general or that she/he doesn't feel much closer to them than she/he does to other kinds of people		213	0.465
NES50	5049	How much interest would respondent say she/he has in how Catholic people as a whole are getting along in this country		211	0.458
NES50	5108	Whether respondent thinks of herself/himself as being in the middle class or the working class		1,002	0.781

Source	Variable Number	Question Content	Note	Sample Size	Reliability[a]
9. Family					
ACL	239	How often respondent and spouse/partner typically have unpleasant disagreements or conflicts		1,134	0.674
ACL	240	How often respondent feels bothered or upset by her/his marriage/relationship with partner		1,127	0.721
SAFCH	8125	When something is bothering respondent, how often is she/he able to talk it over with her/his mother		811	0.566
SAFCH	8137	Does respondent think she/he will get married		208	1.131[b]
SAFCH	8147	When something is bothering respondent, how often is she/he able to talk it over with her/his father		730	0.632
SAFMO	8021	How often respondent finds it easy to understand son/daughter		848	0.546
SAFMO	8025	How well does respondent think she understands her husband		626	0.644
SAFMO	8026	Generally speaking, would respondent say that the time she spends together with her husband is extremely enjoyable, very enjoyable, enjoyable, or not too enjoyable		624	0.655
SAFMO	8061	Which category best describes how respondent and her husband decide which couples they see most often: husband usually decides, husband decides a little more often than respondent does, respondent decides a little more often than he does, or respondent usually decides		641	0.402
SAFMO	8062	Which category best describes how respondent and her husband decide how much should be spent on major purchases, like furniture and appliances: husband usually decides, husband decides a little more often than respondent does, respondent decides a little more often than he does, or respondent usually decides		648	0.481
SAFMO	8063	Which category best describes how respondent and her husband decide how often they go out for an evening: husband usually decides, husband decides a little more often than respondent does, respondent decides a little more often than he does, or respondent usually decides		640	0.411

Source	Variable Number	Question Content	Note	Sample Size	Reliability[a]
SAFMO	8064	Which category best describes how grocery shopping is divided up in respondent's family: husband usually does it, husband does it a little more often than respondent does, respondent does it a little more often than he does, or respondent usually does it		650	0.543
SAFMO	8065	Which category best describes how doing the evening dishes is divided up in respondent's family: husband usually does it, husband does it a little more often than respondent does, respondent does it a little more often than he does, or respondent usually does it		621	0.819
SAFMO	8066	Which category best describes how repairing things around the house is divided up in respondent's family: husband usually does it, husband does it a little more often than respondent does, respondent does it a little more often than he does, or respondent usually does it		625	0.688
SAFMO	8067	Which category best describes how straightening up before company comes is divided up in respondent's family: husband usually does it, husband does it a little more often than respondent does, respondent does it a little more often than he does, or respondent usually does it		646	0.539
SAFMO	8068	Which category best describes how handling the money and bills is divided up in respondent's family: husband usually does it, husband does it a little more often than respondent does, respondent does it a little more often than he does, or respondent usually does it		647	0.810
SAFMO	8078	How often respondent and her husband have problems getting along with each other		591	0.597

F. Self-assessment
1. *Self-esteem*

Source	Variable Number	Question Content	Note	Sample Size	Reliability[a]
ACL	11	Extent to which respondent agrees/ disagrees that "I take a positive attitude toward myself."		2,164	0.694

Source	Variable Number	Question Content	Note	Sample Size	Reliability[a]
ACL	12	Extent to which respondent agrees/disagrees that "At times I think I am no good at all."		2,168	0.540
ACL	13	Extent to which respondent agrees/disagrees that "All in all, I am inclined to feel that I am a failure."		2,170	0.541
SAFCH	8126	How often for respondent is it true that "I take a positive attitude toward myself."		852	0.633
SAFCH	8127	How often for respondent is it true that "I feel I do not have much to be proud of."		853	0.344
SAFCH	8128	How often for respondent is it true that "I am able to do things as well as most other people."		855	0.474
SAFCH	8129	How often for respondent is it true that "I feel that I can't do anything right."		854	0.425
SAFCH	8130	How often for respondent is it true that "As a person I do a good job these days."		852	0.420
SAFCH	8131	How often for respondent is it true that "I feel that I have a number of good qualities."		855	0.470
SAFCH	8132	How often for respondent is it true that "I feel that I'm a person of worth, at least on an equal level with others."		855	0.534

2. *Quality of Life*

Source	Variable Number	Question Content	Note	Sample Size	Reliability[a]
ACL	24	How satisfied respondent is with being a parent		1,786	0.605
ACL	88	How satisfied respondent is with her/his health		2,210	0.630
ACL	115	How respondent would rate her/his health at the present time		2,218	0.688
ACL	151	How satisfied respondent is with her/his job		950	0.538
ACL	217	How satisfied respondent is with her/her family's/his/his family's present financial situation		2,193	0.637
ACL	238	How satisfied respondent is with her/his marriage/relationship with partner		1,140	0.756
NES50	5018	How satisfied respondent is with her/his present financial situation		1,090	0.500
SAFMO	8027	How happy respondent thinks her marriage is, taking things all together		626	0.769

Source	Variable Number	Question Content	Note	Sample Size	Reliability[a]
3. Employment					
ACL	152	Would respondent have liked to work more on her/his job or jobs over the past year		951	0.580
ACL	153	Would respondent have liked to work less on her/his job or jobs over the past year		597	0.637
ACL	246	Would respondent have liked to work more on her/his job or jobs over the past year	E,10	952	0.658
NES90	9144	Given how many hours respondent works on her/his job in the average week, is that more hours than respondent wants to work, fewer hours than respondent wants to work, or generally about right		347	0.638
NES90	9145	How worried respondent is about losing her/his job in the near future (R working/temp laid off)	E,4	345	0.818
NES90	9163	How worried respondent is about losing her/his job /not being able to find a job in the near future (all Rs)		429	0.815
4. Family					
SAFCH	8134	If respondent could start life over again, and have just the number of children she/he would like, what number of children would respondent want to have when her/his family is completed	S,2	866	0.786
SAFCH	8138	How much it would bother respondent if things turn out so that respondent does not marry		198	0.550
SAFMO	8080	If respondent could start life over again, knowing that things would turn out just about the way they have, what number of children would respondent want to have when her family is completed		870	0.793

II. Respondent Proxy-reports
A. Facts
1. Personal Characteristics

NES50	5031	Age of head of household, bracketed	D,2	459	1.016

2. Education

NES70	7006	Head of household's years of schooling	E,4	427	0.922
NES70	7010	Wife of family head's years of schooling	E,4	364	0.945

Source	Variable Number	Question Content	Note	Sample Size	Reliability[a]
NES70	7230	Spouse of R's years of schooling		805	0.930
NES70	8059	Husband's education		782	0.900

3. Employment

NES70	7029	Hours per week head of household works/worked on job	E,4	333	0.713
NES70	7034	Hours per week wife of family head works/worked on job	E,4	112	0.802
NES70	7232	Spouse of R's hours per week works/ worked on job		451	0.730
NES90	9171	Hours per week spouse or partner works/ worked on job		192	0.543

4. Occupation

NES70	7027	Family head's present/former Duncan SEI	E,4	415	0.766
NES70	7032	Wife of head's present/former Duncan SEI	E,4	132	0.931
NES70	7231	Spouse of R's present/former Duncan SEI		558	0.798
NES70	9167	Spouse's or partner's occupation— prestige score		240	0.781

III. Interviewer Reports

A. Facts

1. Household Characteristics

NES50	5029	Number of adults in the household		1,088	0.842
NES70	7016	Number of individuals in the household		1,283	0.967
NES70	7017	Number of politically eligible adults 18 years and older in the household		1,284	0.907

2. Place of Residence

NES70	7044	Number of stories in place of residence		1,203	0.925

B. Beliefs

1. Respondent Characteristics

ACL	214	How skilled would the interviewer say the respondent is in handling or dealing with other people		2,194	0.537
ACL	223	How was the respondent's understanding of the questions		2,122	0.604
ACL	226	How was the respondent's ability to express herself/himself		2,197	0.590
ACL	232	How much difficulty did the respondent have hearing the interviewer		2,201	0.723

Source	Variable Number	Question Content	Note	Sample Size	Reliability[a]
ACL	233	How much difficulty did the respondent have remembering things that the interviewer asked her/him about		2,195	0.377
ACL	234	How comfortable would the interviewer say the respondent is in dealing with other people		2,195	0.529
ACL	235	How self-confident did the respondent seem to be		2,197	0.456
ACL	236	How depressed did the respondent seem to be		2,194	0.400
NES70	7160	How high was the respondent's general level of information about politics and public affairs		1,252	0.697
NES70	7161	Rating of the respondent's apparent intelligence		1,258	0.726
NES90	9189	How high was the respondent's general level of information about politics and public affairs		580	0.713
NES90	9190	Rating of the respondent's apparent intelligence		577	0.719

2. Respondent Reaction to the Interview

Source	Variable Number	Question Content	Note	Sample Size	Reliability[a]
ACL	229	How was the respondent's cooperation during the interview		2,148	0.308
ACL	230	How tiring did the interview seem to be to the respondent		2,201	0.315
ACL	231	How much did the respondent seem to enjoy the interview		2,196	0.279
NES70	7159	How was the respondent's cooperation		1,258	0.667
NES70	7163	How great was the respondent's interest in the interview		1,265	0.602
NES70	7164	How sincere did the respondent seem to be in her/his answers		1,242	0.507
NES90	9192	How great was the respondent's interest in the interview		580	0.710
NES90	9193	How sincere did the respondent seem to be in her/his answers		580	0.870
NES90	9398	How was the respondent's cooperation (Pre Post Pre)		335	0.455
NES90	9399	How was the respondent's cooperation (Post Post Post)		335	0.413

IV. Organization Reports
A. Facts
1. Place of Residence

Source	Variable Number	Question Content	Note	Sample Size	Reliability[a]
NES50	5057	Size of place code for the respondent's place of residence		1,132	1.000

Source	Variable Number	Question Content	Note	Sample Size	Reliability[a]
NES70	7038	Actual population of the interview place		1,293	1.010[b]
NES70	7039	Belt code, or urbanicity code, for the respondent's place of residence		1,186	1.000
NES70	7040	Distance from the interview place to the center of the central city of the nearest SMSA, Standard Metropolitan Statistical Area		1,292	0.996
NES70	7041	Distance from the interview place to the center of the central city of the nearest SMSA, Standard Metropolitan Statistical Area, which has at least one city of 350,000 persons or more		1,292	0.978
NES70	7207	Degree of urbanization of the PSU, the Primary Sampling Unit		938	1.000
NES70	7208	Size of the interview place, master code	S,3	1,292	0.987
NES90	9021	Size of the place of interview	S,3	597	1.000
NES90	9022	Actual population of the interview location		478	1.000

[a]For variables having 2–15 response categories, the reliability estimate is based on polychoric correlations computed with threshold levels unconstrained and the assumption of equal reliabilities. For variables having 16 or more response categories, the reliability estimate is based on Pearson correlations under the assumption of equal reliabilities (see Heise, 1969).
[b]Reliability estimate set to 1.000 in data analysis.

Explanation of Note Codes:
 1. Variable based on questions already part of another variable
 2. Variable based on multiple questions not part of another variable
 3. Variable based on questions already part of another variable, but which adds new information
 4. Variable based on questions already part of another variable representing only a subset of respondents
 5. Variable based on questions already part of another variable, but other variable did not meet criteria for inclusion
 6. Variable duplicates questions already part of another variable for a large subset of cases
 7. Variable considered redundant because it shares two or more waves of data with another variable
 8. Variable not considered redundant because it shares only one wave of data with another variable
 9. Variable disaggregated from information contained in other variables
10. Variable synthesized from actual questions in interview

D. Transformed variable derived by NES from interview question
S. Summary variable created by NES or SAF from multiple questions
E. Variable omitted from analysis

References

Abelson, R.P. (1972). Are attitudes necessary? In B.T. King and E. McGinniew (Eds.), *Attitudes, conflict and social change* (pp. 19–32). New York: Academic Press.

Achen, C.H. (1975). Mass political attitudes and the survey response. *American Political Science Review*, 69, 1218–1231.

Allison, P.D. (1987). Estimation of linear models with incomplete data. In C.C. Clogg (Ed.), *Sociological methodology 1987* (pp. 71–103). Washington, DC: American Sociological Association.

Allport, G. (1968). The historical background of modern social psychology. In G. Lindzey and E. Aronson (Eds.), *The handbook of social psychology* (pp. 1–80). Reading, MA: Addison-Wesley.

Althauser, R.P., and Heberlein, T.A. (1970). A causal assessment of validity and the multitrait-multimethod matrix. In E.F. Borgatta and G.W. Bohrnstedt (Eds.), *Sociological methodology 1970* (pp. 151–169). San Francisco: Jossey-Bass.

Althauser, R.P., Heberlein, T.A., and Scott, R.A. (1971). A causal assessment of validity: The augmented multitrait-multimethod matrix. In H.M. Blalock, Jr. (Ed.), *Causal models in the social sciences*. Chicago: Aldine-Atherton.

Alwin, D.F. (1973). Making inferences from attitude-behavior correlations. *Sociometry*, 36, 253–278.

Alwin, D.F. (1974). Approaches to the interpretation of relationships in the multitrait-multimethod matrix. In H.L. Costner (Ed.), *Sociological methodology 1973–74* (pp. 79–105). San Francisco: Jossey-Bass.

Alwin, D.F. (1977). Making errors in surveys: An overview. *Sociological Methods and Research*, 6, 131–150.

Alwin, D.F. (1988). Structural equation models in research on human development and aging. In K.W. Schaie, R.T. Campbell, W. Meredith, and S.C. Rawlings (Eds.), *Methodological issues in aging research* (pp. 71–170). New York: Springer.

Alwin, D.F. (1989). Problems in the estimation and interpretation of the reliability of survey data. *Quality and Quantity*, 23, 277–331.

Alwin, D.F. (1991a). Family of origin and cohort differences in verbal ability. *American Sociological Review*, 56, 625–638.

Alwin, D.F. (1991b). Research on survey quality. *Sociological Methods and Research*, 20, 3–29.

Alwin, D.F. (1992). Information transmission in the survey interview: Number of response categories and the reliability of attitude measurement. In P.V. Marsden (Ed.), *Sociological methodology 1992* (pp. 83–118). Washington, DC: American Sociological Association.

Alwin, D.F. (1994). Aging, personality and social change: The stability of individual differences over the adult life-span. In D.L. Featherman, R.M. Lerner, and M. Permutter (Eds.), *Life-span development and behavior* (pp. 136–185), Vol. 12. New York: Lawrence Erlbaum.

Alwin, D.F. (1995). The reliability of survey data. Paper presented at the International Conference on Measurement Error and Process Quality. Bristol, England. April.

Alwin, D.F. (1997). Feeling thermometers vs. seven-point scales: Which are better? *Sociological Methods and Research*, 25, 318–340.

Alwin, D.F. (1999). Aging and errors of measurement: Implications for the study of life-span development. In N. Schwarz, D. Park, B. Knäuper, and S. Sudman (Eds.), *Cognition, aging, and self-reports* (pp. 365–385). Philadelphia, PA: Psychology Press.

Alwin, D.F. (2000). Factor analysis. In E.F. Borgatta and R. Montgomery (Eds.), *Encylopedia of sociology*. New York: Macmillan.

Alwin, D.F. (2005). Reliability. In K. Kempf-Leonard and others (Eds.), *Encyclopedia of social measurement* (pp. 351–359). New York: Academic Press.

Alwin, D.F., and Jackson, D.J. (1979). Measurement models for response errors in surveys: Issues and applications. In K.F. Schuessler (Ed.), *Sociological methodology 1980* (pp. 68–119). San Francisco: Jossey-Bass.

Alwin, D.F., and Krosnick, J.A. (1985). The measurement of values in surveys: A comparison of ratings and rankings. *Public Opinion Quarterly*, 49, 535–552.

Alwin, D.F., and Krosnick, J.A. (1991a). Aging, cohorts, and the stability of sociopolitical orientations over the life span. *American Journal of Sociology*, 97, 169–195.

Alwin, D.F., and Krosnick, J.A. (1991b). The reliability of survey attitude measurement: The influence of question and respondent attributes. *Sociological Methods and Research*, 20, 139–181.

Alwin, D.F., and Thornton, A. (1984). Family origins and the schooling process: Early vs. late influence of parental characteristics. *American Sociological Review*, 49, 784–802.

Alwin, D.F., Cohen, R.L., and Newcomb, T.M. (1991). *Political attitudes over the life span: The Bennington women after fifty years*. Madison, WI: University of Wisconsin Press.

Alwin, D.F., McCammon, R.J., and Rodgers, W.L. (2006). The reliability of measures in the Health and Retirement Study. Paper prepared for presentation at the 59th annual scientific meeting of the Gerontological Society of America, Dallas TX, November.

Alwin, D.F., McCammon, R.J., Wray, L.A., and Rodgers, W.L. (2007). *The aging mind in social and historical context*. Unpublished manuscript. Population Research Institute, Pennsylvania State University, University Park PA 16802.

Anderson, T.W. (1960). Some stochastic process models for intelligence test scores. In K.J. Arrow, S. Karlin, and P. Suppes (Eds.), *Mathematical models in the social sciences*. Stanford, CA: Stanford University Press.

Andrews, F.M. (1984). Construct validity and error components of survey measures: A structural modeling approach. *Public Opinion Quarterly*, 46, 409–42. Reprinted in W.E. Saris and A. van Meurs. (1990). *Evaluation of measurement instruments by meta-analysis of multitrait multimethod studies*. Amsterdam: North-Holland.

Andrews, F.M., and Herzog, A.R. (1986). The quality of survey data as related to age of respondent. *Journal of the American Statistical Association*, 81, 403–410.

APA (American Psychological Association). (2000). *Standards for educational and psychological testing.* Washington DC: American Psychological Association.

Arbuckle, J.L., and Wothke, W. (1999). *AMOS users guide. Version 4.0.* Chicago IL: Smallwaters Corporation.

Arbuckle, T.Y., Gold, D., and Andres, D. (1986). Cognitive functioning of older people in relation to social and personality variables. *Psychology and Aging*, 1, 55–62.

Bartholomew, D.J., and Schuessler, K.F. (1991). Reliability of attitude scores based on a latent-trait model. In P.V. Marsden (Ed.), *Sociological methodology 1991* (pp. 97–123). Oxford: Basil Blackwell.

Bassi, F., Hagenaars, J.A., Croon, M.A., and Vermunt, J.K. (2000). Estimating true changes when categorical panel data are affected by uncorrelated and correlated classification errors: An application to unemployment data. *Sociological Methods and Research*, 29, 230–268.

Beatty, P.C. (2003). *Answerable questions: Advances in the methods for identifying and resolving questionnaire problems in survey research.* Unpublished Ph.D. dissertation, Department of Sociology, University of Michigan—Ann Arbor.

Belson, W.A. (1981). *The design and understanding of survey questions.* Aldershot, England: Gower.

Bentler, P.M., and Bonett, D.G. (1980). Significance tests and goodness of fit in the analysis of covariance structures. *Psychological Bulletin*, 88, 588–606.

Bielby, W.T., and Hauser, R.M. (1977). Response error in earnings functions for nonblack males. *Sociological Methods and Research*, 6, 241–80.

Bielby, W.T., Hauser, R.M., and Featherman, D.L. (1977a). Response errors for nonblack males in models of the stratification process. *Journal of the American Statistical Association*, 72, 723–735.

Bielby, W.T., Hauser, R.M., and Featherman, D.L. (1977b). Response errors of black and nonblack males in models of status inheritance and mobility. *American Journal of Sociology*, 82, 1242–1288.

Biemer, P.P., and Stokes, S.L. (1991). Approaches to the modeling of measurement error. In P.P. Biemer et al. (Eds.), *Measurement Errors in Surveys* (pp. 487–516). New York: Wiley.

Biemer, P.P., and Trewin, D. (1997). A review of measurement error effects on the analysis of survey data. In L. Lyberg et al. (Eds.), *Survey measurement and process quality* (pp. 603–632). New York: Wiley.

Biemer, P.P., Groves, R.M., Lyberg, L.E., Mathiowetz, N.A., and Sudman, S. (Eds.). (1991). *Measurement errors in surveys.* New York: Wiley.

Bishop, G.F., Oldendick, R.W., and Tuchfarber, A.J. (1983). Effects of filter questions in public opinion surveys. *Public Opinion Quarterly*, 47, 528–546.

Blair, J., Menon, G., and Bickart, B. (1991). Measurement effects in self and proxy responses to survey questions: An information-processing perspective. In P.B. Biemer et al. (Eds.), *Measurement errors in surveys* (pp. 145–166). New York: Wiley-Interscience.

Blalock, H.M., Jr. (1965). Some implications of random measurement error for causal inferences. *American Journal of Sociology*, 71, 37–47.

Blalock, H.M., Jr. (1972). *Social statistics*. New York: McGraw-Hill.

Blau, P.M., and Duncan, O.D. (1967). *The American occupational structure*. New York: John Wiley and Sons.

Blumer, H. (1956). Sociological analysis and the "variable." *American Sociological Review*, 22, 683–690.

Bohrnstedt, G.W. (1969). A quick method for determining the reliability and validity of multiple item scales. *American Sociological Review*, 34, 542–548.

Bohrnstedt, G.W. (1983). Measurement. In P.H. Rossi, J.D. Wright, and A.B. Anderson (Eds.), *Handbook of survey research* (pp. 70–121). New York: Academic Press.

Bohrnstedt, G.W., Mohler, P.P, and Müller, W. (1987). An empirical study of the reliability and stability of survey research items. *Sociological Methods and Research*, 15, 171–176.

Bollen, K.A. (1989). *Structural equations with latent variables*. New York: Wiley.

Bound, J., Brown, C., Duncan, G.J., and Rodgers, W.L. (1990). Measurement error in cross-sectional and longitudinal labor market surveys: Validation study evidence. In J. Hartog, G. Ridder, and J. Theeuwes (Eds.), *Panel data and labor market studies*. Elsevier Science Publishers.

Bradburn, N.M., and Danis, C. (1984). Potential contributions of cognitive research to questionnaire design. In T. Jabine et al. (Eds.), *Cognitive aspects of survey methodology: Building a bridge between disciplines*. Washington, DC: National Academy Press.

Bradburn, N.M., and Miles, C. (1979). Vague quantifiers. *Public Opinion Quarterly*, 43, 92–101.

Bradburn, N.M, Sudman, S. (and associates). (1979). *Improving interviewing methods and questionnaire design: Response effects to threatening questions in survey research*. San Francisco: Jossey-Bass.

Bradburn, N.M., Rips, L.J., and Shevell, S.K. (1987). Answering autobiographical questions: The impact of memory and inference on surveys. *Science*, 236, 157–161.

Brislin, R.W. (1986). The wording and translation of research instruments. In W.J. Lonner and J.W. Berry (Eds.), *Field methods in cross-cultural research*. Newbury Park, CA: Sage.

Browne, M.W. (1984). The Decomposition of Multitrait-multimethod Matrices. *British Journal of Mathematical and Statistical Psychology*, 37, 1–21.

Browne, M.W., and Cudeck, R. (1993). Alternative ways of assessing model fit. In K. Bollen and J.S. Long (Eds.), *Testing structural equation models* (pp. 311–359). New York: Plenum Press.

Bryk, A.S., and Raudenbush, S.W. (1992). *Hierarchical linear models: Applications and data analysis methods*. Newbury Park, CA: Sage.

Campbell, A., and Converse, P.E. (1980). *The Quality of American Life: 1978 codebook*. Ann Arbor, MI: University of Michigan, Inter-University Consortium for Political and Social Research.

Campbell, A., Converse, P., Miller, W., and Stokes, D. (1971). *American Election Panel Study: 1956, 1958, 1960*. Ann Arbor, MI: Inter-University Consortium for Political and Social Research.

Campbell, D.T., and Fiske, D.W. 1959. Convergent and discriminant validation by the multitrait-multimethod matrix. *Psychological Bulletin*, 6, 81–105.

Cannell, C.F., Marquis, K.H., and Laurent, A. (1977). A summary of studies of interviewing methodology. *Vital and Health Statistics.* Series 2, No. 69, March.

Cannell, C.F., Miller, P.V., and Oksenberg, L. (1981). Research on interviewing techniques. In S. Lienhardt (Ed.), *Sociological methodology 1980* (pp. 389–437). San Francisco: Jossey-Bass.

Cantril, H., and Fried, E. (1944). The meaning of questions. In H. Cantril, *Gauging public opinion.* Princeton, NJ: Princeton University Press.

Castelloe, J.M. (2000). Sample size computations and power analysis with the SAS system. Paper 265-25. Paper presented at the 25th annual SAS Users Group International Conference, Indianapolis, IN. April 9–12.

Center for Political Studies. (1994). *Continuity guide to the American National Election Studies.* Institute for Social Research. Ann Arbor, MI: University of Michigan. March.

Clark, H.H. (1985). Language use and language users. In G. Lindzey and E. Aronson (Eds.), *The handbook of social psychology,* Vol. 2 (pp. 179–232). New York: Random House.

Cleary, T.A., Linn, R.L., and Walster, G.W. (1970). Effect of reliability and validity on power of statistical tests. In E.F. Borgatta and G.W. Bohrnstedt (Eds.), *Sociological methodology 1970* (pp. 30–38). San Francisco: Jossey-Bass.

Clogg, C.C. (1995). Latent class models. In G. Arminger, C.C. Clogg, and M.E. Sobel (Eds.), *Handbook of statistical modeling for the social and behavioral sciences* (pp. 311–359). New York: Plenum.

Clogg, C.C., and Goodman, L.A. (1984). Latent structure analysis of a set of multidimensional contingency tables. *Journal of the American Statistical Association,* 79, 762–71.

Clogg, C.C., and Manning, W.D. (1996). Assessing reliability of categorical measurements using latent class models. In A. von Eye and C.C. Clogg (Eds.), *Categorical variables in developmental research—methods of analysis* (pp. 169–182). New York: Academic Press.

Coenders, G., Saris, W.E., Batista-Foguet, J.M., and A. Andreenkova, A. (1999). Stability of three-wave simplex estimates of reliability. *Structural Equation Models,* 6, 135–157.

Coleman, J.S. (1964). *Models of change and response uncertainty.* Englewood Cliffs, NJ: Prentice-Hall.

Coleman, J.S. (1968). The mathematical study of change. In H.M. Blalock, Jr. and A.B. Blalock (Eds.) *Methodology in social research* (pp. 428–478). New York: McGraw-Hill.

Collins, L.M. (2001). Reliability for static and dynamic categorical latent variables: Developing measurement instruments based on a model of growth processes. In L.M. Collins and A.G. Sayer (Eds.), *New methods for the analysis of change* (pp. 271–288). Washington, DC: American Psychological Association.

Collins, L.M., and Flaherty, B.P. (2002). Latent class models for longitudinal data. In J.A. Hagenaars and A.L. McCutcheon (Eds.), *Applied latent class analysis* (pp. 287–303). New York: Cambridge University Press.

Collins, L.M., Flaherty, B.P, Hyatt, S.L., and Schafer, J.L.(1999). *WinLTA User's Guide.* Version 2.0. The Methodology Center. Pennsylvania State University.

Collins, L.M., Fidler, P.L., and Wugalter, S.E. (1996). Some practical issues related to the estimation of latent class and latent transition parameters. In A. von Eye and C.C. Clogg (Eds.), *Categorical variables in developmental research—methods of analysis* (pp. 133–146). New York: Academic Press.

Conrad, F.G. (1999). Customizing survey procedures to reduce measurement error. In M.G. Sirken, et al. (Eds.), *Cognition and survey research* (pp. 301–317). New York: Wiley.

Converse, J.M. (1976–77). Predicting "No Opinion" in the polls. *Public Opinion Quarterly*, 40, 515–530.

Converse, J.M. (1987). *Survey research in the United States: Roots and emergence, 1890–1960*. Berkeley, CA: University of California Press.

Converse, J.M., and Presser, S. (1986). *Survey questions: Handcrafting the standardized questionnaire*. Beverly Hills, CA: Sage.

Converse, P.E. (1964). The nature of belief systems in the mass public. In D.E. Apter (Ed.), *Ideology and discontent* (pp. 206–261). New York: Free Press.

Converse, P.E. (1970). Attitudes and non-attitudes: Continuation of a dialogue. In E.R. Tufte (Ed.), *The quantitative analysis of social problems* (pp. 168–189). Reading, MA: Addison-Wesley.

Converse. P.E. (1974). The status of non-attitudes. *American Political Science Review*, 68, 650–660.

Corcoran, M. (1980). Sex differences in measurement error in status attainment models. *Sociological Methods and Research*, 9, 199–217.

Costner, H.L. (1965). Criteria for measures of association. *American Sociological Review*, 30, 341–353.

Costner, H.L. (1969). Theory, deduction and rules of correspondence. *American Journal of Sociology*, 75, 245–263.

Cox, E.P. (1980). The optimal number of response alternatives for a scale: A review. *Journal of Marketing Research*, 17, 407–22.

Cronbach, L.J. (1951). Coefficient alpha and the internal structure of tests. *Psychometrika*, 16, 297–334.

Cronbach, L.J., and Meehl, P.E. (1955). Construct validity in psychological tests. *Psychological Bulletin*, 52, 281–302.

Crouse, J., and Trusheim, D. (1988). *The case against the SAT*. Chicago, IL: University of Chicago Press.

Dawes, R.M., and Smith, T.L. (1985). Attitude and opinion measurement. In G. Lindzey and E. Aronson (Eds.), *Handbook of social psychology* (pp. 509–66). New York: Random House.

Dex, S. (1991). The reliability of recall data: A literature review. Working Paper No. 11. ESRC Research Centre on Micro-Social Change. Colchester: Essex, England.

Diggle, P.J., Liang, K.Y., and Zeger, S.L. (1994). *Analysis of longitudinal data*. New York: Oxford University Press.

Dillman, D.A. (1978). *Mail and telephone surveys: The total design method*. New York: Wiley.

Duncan, G.J., Mathiowetz, N.A., and others. (1985). *A validation study of economic survey data*. Ann Arbor, MI: Survey Research Center, Institute for Social Research, University of Michigan.

Duncan, O.D. (1984a). *Notes on social measurement*. New York: Academic Press.

Duncan, O.D. (1984b). The latent trait approach in survey research: The Rasch measurement model; Rasch measurement: Further examples and discussion. In C.F. Turner and

E. Martin (Eds.), *Surveying subjective phenomena*. Vols. 1 and 2 (pp. 210–229; 367–440). New York: Russell Sage Foundation.

Elias, P. (1991). Methodological, statistical and practical issues arising from the collection and analysis of work history information by survey techniques. *Bulletin de Methodologie Sociologique*, 31, 3–31.

Esser, H. (1986). Können Befragte lügen? Zum Konzept des "wahren" Wertes im Rahmen der handlungstheoretischen Interpretation des Befragtenverhaltens. *Kölner Zeitschrift für Soziologie und Sozialpsychologie*, 38, 314–336.

Esser, H. (1993). Response set: Habit, frame or rational choice? In D. Krebs and P. Schmidt (Eds.), *New directions in attitude measurement* (pp. 293–314). New York: Walter de Gruyter.

Fazio, R.H., Herr, P.M., and Olney, T.J. (1984). Attitude accessibility following a self-perception process. *Journal of Personality and Social Psychology*, 47, 277–286.

Featherman, D.L., and Hauser, R.M. (1975). Design for a replicate study of social mobility in the United States. In K.C. Land and S. Spilerman (Eds.), *Social indicators models* (pp. 219–252). New York: Russell Sage Foundation.

Featherman, D.L., and Hauser, R.M. (1978). *Opportunity and change*. New York: Academic Press.

Ferber, R. (1966). Item nonresponse in a consumer survey. *Public Opinion Quarterly*, 30, 399–415.

Flaherty, B.P. (2002). Assessing reliability of categorical substance use measures with latent class analysis. *Drug and Alcohol Dependence*, 68, S7-S20.

Fowler, F.J. (1992). How unclear terms affect survey data. *Public Opinion Quarterly* , 56, 218–231.

Francis, J.D., and Busch, L. (1975). What we know about "I don't knows." *Public Opinion Quarterly*, 39, 207–218.

Freedman, D., Thornton, A., and Camburn, D. (1980). Maintaining response rates in longitudinal studies. *Sociological Methods and Research*, 9, 87–98.

Freedman, D., Thornton, A., Camburn, D., Alwin, D.F., and Young-DeMarco, L. (1988). The life history calendar: A technique for collecting retrospective data. In C.C. Clogg (Ed.), *Sociological methodology 1988* (pp. 37–68). Washington D.C.: American Sociological Association.

Garner, W.R. (1960). Rating scales, discriminability, and information transmission. *Psychological Review*, 67, 343–52.

Garner, W.R., and Hake, H.W. (1951). The amount of information in absolute judgments. *Psychological Review*, 58, 446–59.

Gergen, K., and Back, K. (1966). Communication in the interview and the disengaged respondent. *Public Opinion Quarterly*, 30, 385–398.

Goldstein, H. (1995). *Multilevel statistical models*, 2nd edition. London: Arnold.

Goodman, L. (2002). Latent class analysis: The empirical study of latent types, latent variables and latent structures. In J.A. Hagenaars and A.L. McCutcheon (Eds.), *Applied latent class analysis* (pp. 3–55). New York: Cambridge University Press.

Greene, V.L., and Carmines, E.G. (1979). Assessing the reliability of linear composites. In K.F. Schuessler (Ed.), *Sociological methodology 1980* (pp. 160–175). San Francisco: Jossey-Bass.

Groves, R.M. (1989). *Survey errors and survey costs*. New York: Wiley.

Groves, R.M. (1991). Measurement error across the disciplines. In P.P. Biemer et al. (Eds.), *Measurement errors in surveys* (pp. 1–25). New York: Wiley.

Groves, R.M., and Couper, M.P. (1998). *Nonresponse in household interview surveys.* New York: Wiley.

Groves, R.M., Fultz, N.H., and Martin, E. (1992). Direct questioning about comprehension in a survey setting. In J. Tanur (Ed.), *Questions about questions: Inquiries into the cognitive bases of surveys* (pp. 49–61). New York: Russell Sage Foundation.

Groves, R.M., Dillman, D.A., Eltinge, J.L., and Little, R.J.A. (Eds.). (2002). *Survey nonresponse.* New York: Wiley.

Guttman, L. (1954). A new approach to factor analysis: The radex. In P.F. Lazarsfeld (Ed.), *Mathematical thinking in the social sciences.* New York: Columbia University Press.

Hadaway, C.K., Marler, P.L., and Chaves, M. (1993). What the polls don't show: A closer look at U.S. church attendance. *American Sociological Review,* 58, 741–752.

Hagenaars, J.A., and McCutcheon, A.L. (Eds.). (2002). *Applied latent class analysis.* New York: Cambridge University Press.

Hansen, M.H., Hurwitz, W.N., and Bershad, M.A. (1961). Measurement errors in censuses and in surveys. *Bulletin of the International Statistical Institute,* 38, 359–374.

Hansen, M.H., Hurwitz, W.N., and Madow, W. (1953). *Sample survey methods and theory.* New York: John Wiley.

Hauser, R.M., and Featherman, D.L. (1977). *The process of stratification: Trends and analyses.* New York: Academic Press.

Hauser, R.M., and Goldberger, A.S. (1971). The treatment of unobservable variables in path analysis. In H.L. Costner (Ed.), *Sociological methodology 1971* (pp. 81–117). San Francisco: Jossey-Bass.

Hauser, R.M., and Warren, J.R. (1997). Socioeconomic indexes for occupations: A review, update and critique. In A.E. Raftery (Ed.), *Sociological methodology 1997* (pp. 177–298). Cambridge, MA: Basil Blackwell.

Hauser, R.M., Tsai, S-L., and Sewell, W.H. (1983). A model of stratification with response error in social and psychological variables. *Sociology of Education,* 56, 20–46.

Hedecker, D., and Gibbons, R.D. (1997). Application of random-effects pattern-mixture models for missing data in longitudinal studies. *Psychological Methods,* 2, 64–78.

Heise, D.R. (1969). Separating reliability and stability in test-retest correlation. *American Sociological Review,* 34, 93–191.

Heise, D.R., and Bohrnstedt, G.W. (1970). Validity, invalidity, and reliability. In E.F. Borgatta and G.W. Bohrnstedt (Eds.), *Sociological methodology 1970* (pp. 104–129). San Francisco: Jossey-Bass.

Herzog, A.R., and Rodgers W.R. (1989). Age differences in memory performance and memory ratings as measured in a sample survey. *Psychology and Aging,* 4, 173–182.

Hippler, H.J., and Schwarz, N. (1989). No opinion filters: A cognitive perspective. *International Journal of Public Opinion Research,* 1, 77–87.

Hippler, H.-J., Schwarz, N., and Sudman, S. (1987). *Social information processing and survey methodology.* New York: Springer-Verlag.

Holt, D., McDonald, J.W., and Skinner, C.J. (1991). The effect of measurement error on event history analysis. In P.P. Biemer et al. (Eds.), *Measurement errors in surveys* (pp. 665–685). New York: Wiley.

House, J. (No date.) *Study guide: Americans' changing lives.* Survey Research Center, Institute for Social Research. Ann Arbor, MI: University of Michigan.

Humphreys, L.G. (1960). Investigations of the simplex. *Psychometrika*, 25, 313–323.

Hunter, J.E., and Schmidt, F.L. (1990). Methods of meta-analysis: Correcting error and bias in research findings. Newbury Park, CA: Sage Publications.

Hyman, H.H. (and associates). (1975). *Interviewing in social research.* (Original publication, 1954). Chicago: University of Chicago Press.

Jabine, T.B., Straf, M.L., Tanur, J.M., and Tourangeau, R. (1984). *Cognitive aspects of survey methodology: Building a bridge between disciplines. Report of the advanced research seminar on cognitive aspects of survey methodology.* Washington, DC: National Academy of Sciences Press.

Jackson, D.N. (1969). Multimethod factor analysis in the evaluation of convergent and discriminant validity. *Psychological Bulletin*, 72, 30–49.

Jöreskog, K.G. (1970). Estimating and testing of simplex models. *British Journal of Mathematical and Statistical Psychology*, 23, 121–145.

Jöreskog, K.G. (1971a). Statistical analysis of sets of congeneric tests. *Psychometrika*, 36, 109–133.

Jöreskog, K.G. (1971b). Simultaneous factor analysis in several populations. *Psychometrika*, 36, 409–426.

Jöreskog, K.G. (1974). Analyzing psychological data by structural analysis of covariance matrices. In D.H. Dranz, R.C. Atkinson, R.D. Luce, and P. Suppes (Eds.), *Measurement, psychophysics and neural information processing.* San Francisco: Freeman.

Jöreskog, K.G. (1978). Structural analysis of covariance and correlation matrices. *Psychometrika*, 43, 443–477.

Jöreskog, K.G. (1990). New developments in LISREL: Analysis of ordinal variables using polychoric correlations and weighted least squares. *Quality and Quantity*, 24, 387–404.

Jöreskog, K.G. (1994). On the estimation of polychoric correlations and their asymptotic covariance matrix. *Psychometrika*, 59, 381–389.

Jöreskog, K.G. and D. Sörbom. (1996a). *PRELIS2: User's reference guide—a program for multivariate data screening and data summarization; a preprocesser for LISREL*, 2nd edition. Chicago: Scientific Software.

Jöreskog, K.G. and D. Sörbom. (1996b). *LISREL8: User's reference guide*, 2nd edition. Chicago: Scientific Software.

Juster, F.T. and Suzman, R. (1995). An overview of the Health and Retirement Study. *Journal of Human Resources*, 30, S7-S56.

Kaase, M. (Ed.). (1999). *Quality criteria for survey research.* Berlin: Akademie Verlag GmbH.

Kalton, G., and Schuman, H. (1982). The effect of the question on survey responses: A review. *Journal of the Royal Statistical Association*, 145, 42–73.

Kaplan, A. (1964). *The conduct of inquiry.* San Francisco: Chandler.

Kessler, R.C., Mroczek, D.K., and Belli, R.F. (1994). Retrospective adult assessment of childhood psychopathology. In D. Shaffer and J. Richters (Eds.), *Assessment in child and adolescent psychopathology.* New York: Guilford.

Kipling, R. (1899). *From sea to sea and otehr sketches.* New York: Doubleday & McClure.

Kish, L. (1965). *Survey sampling.* New York: Wiley.

Knäuper, B., Belli, R.F., Hill, D.H. and Herzog, A.R. (1997). Question difficulty and respondents' cognitive ability: The impact of on data quality. Unpublished paper. Survey Methods Program, Institute for Social Research. Ann Arbor, MI: University of Michigan.

Krosnick, J.A. (1999). Survey research. *Annual Review of Psychology*, 50, 537–567.

Krosnick, J.A. (2002). The causes of no-opinion responses to attitude measures in surveys: They are rarely what they appear to be. In R.M. Groves, D.A. Dillman, J.L. Eltinge, and R.J.A. Little (Eds.), *Survey nonresponse* (pp. 87–100). New York: Wiley.

Krosnick, J.A., and Alwin, D.F. (1987). An evaluation of a cognitive theory of response order effects in survey measurement. *Public Opinion Quarterly*, 51, 201–219.

Krosnick, J.A., and Alwin, D.F. (1988). A test of the form-resistant correlation hypothesis: Ratings, rankings, and the measurement of values. *Public Opinion Quarterly*, 52, 526–538.

Krosnick, J.A., and Alwin, D.F. (1989). Response strategies for coping with the cognitive demands of attitude measures in surveys. Unpublished paper. Institute for Social Research. Ann Arbor, MI: University of Michigan. (Also published as Krosnick, J.A. (1991). Response strategies for coping with the cognitive demands of attitude measures in surveys. *Applied Cognitive Psychology*, 5, 213–236.)

Krosnick, J.A., and Fabrigar, L.R. (1997). Designing rating scales for effective measurement in surveys. In L. Lyberg et al. (Eds.), *Survey measurement and process quality* (pp. 141–164). New York: Wiley-Interscience.

Kuhn, T.S. (1961). The function of measurement in modern physical science. *Isis*, 52, 161–193.

Langeheine, R., and van de Pol., F.J.R. (1990). A unifying framework for Markov modeling in discrete space and discrete time. *Sociological Methods and Research*, 18, 416–441.

Langeheine, R., and van de Pol, F.J.R. (2002). Latent Markov chains. In J.A. Hagenaars and A.L. McCutcheon (Eds.), *Applied latent class analysis* (pp. 304–341). New York: Cambridge University Press.

Lawley, D.N., and Maxwell, A.E. (1971). *Factor analysis as a statistical model*. New York: American Elsevier.

Lazarsfeld, P.F., and Henry, N.W. (1968). *Latent structure analysis*. Boston: Houghton Mifflin.

Lee, S.-Y., Poon, W.-Y., and Bentler, P.M. (1990). A three-stage estimation procedure for structural equation models with polytomous variables. *Psychometrika*, 55, 45–51.

Likert, R. (1932). A technique for the measurement of attitudes. *Archives of General Psychology*, 140, 5–55.

Little, R.J.A. and Rubin, D.B. (1987). *Statistical analysis with missing data*. New York: Wiley.

Little, R.J.A., and Rubin, D.B. (1989). The analysis of social science data with missing values. *Sociological Methods and Research*, 18, 292–326.

Little, R.J.A., and Schenker, N. (1995). Missing data. In G. Arminger, C.C. Clogg, and M.E. Sobel (Eds.), *Handbook of statistical modeling for the social and behavioral sciences* (pp. 39–75). New York: Plenum.

Lord, F.M., and Novick, M.L. (1968). *Statistical theories of mental test scores*. Reading, MA: Addison-Wesley.

Lyberg, L., Biemer, P., Collins, M., de Leeuw, E., Dippo, C., Schwarz, N., and Trewin, D. (1997). *Survey measurement and process quality*. New York: Wiley.

Mare, R.D., and Mason, W.M. (1980). Children's reports of parental socioeconomic status: A multiple group measurement model. *Sociological Methods and Research*, 9, 178–198.

Marquis, K.H. (1978). *Record check validity of survey responses: A reassessment of bias in reports of hospitalizations.* Santa Monica, CA: The Rand Corporation.

Marquis, K.H., Cannell, C.F., and Laurent, A. (1972). Reporting health events in household interviews: Effects of reinforcement, question length, and reinterviews. *Vital and Health Statistics.* Series 2, No. 45.

Marquis, M.S. and Marquis, K.H. (1977). *Survey measurement design and evaluation using reliability theory.* Santa Monica, CA: The Rand Corporation.

Mason, W.M., Hauser, R.M., Kerckhoff, A.C., Poss, S.S., and Manton, K. (1976). Models of response error in student reports of parental socioecnomic characteristics. In W.H. Sewell, R.M. Hauser, and D.L. Featherman (Eds.), *Schooling and achievement in American society* (pp. 443–494). New York: Academic Press.

Maxwell, A.E. (1971). Estimating true scores and their reliabilities in the case of composite psychological tests. *British Journal of Mathematical and Statistical Psychology,* 24, 195–204.

McArdle, J.J., and Bell, R.Q. (2000). An introduction to latent growth models for developmental data analysis. In T.D. Little, K.U. Schnabel, and J. Baumert (Eds.), *Modeling longitudinal and multilevel data: Practical issues, applied approaches and specific examples* (pp. 69–107). Mahwah, NJ: Lawrence Erlbaum Associates, Publishers.

McArdle, J.J., and Hamagami, F. (1992). Modeling incomplete longitudinal and cross-sectional data using latent growth structural models. *Experimental Aging Research,* 18, 145–166.

McArdle, J.J., and Nesselroade, J.R. (1994). Using multivariate data to structure developmental change. In S.H. Cohen and H.W. Reese (Eds.), *Life-span developmental psychology.* Hillsdale NJ: Lawrence Erlbaum Associates, Publishers.

McClendon, J.J. (1991). Acquiescence and recency response-order effects in interview surveys. *Sociological Methods and Research,* 20, 60–103.

McClendon, M.J., and Alwin, D.F. (1993). No-opinion filters and attitude measurement reliability. *Sociological Methods and Research,* 21, 438–464.

McGuire, W.R. (1969). Suspiciousness of experimenter's intent. In R. Rosenthal and R.L. Rosnow (Eds.), *Artifact in behavioral research.* New York: Academic Press.

McKeon, R. (Ed.). (1992). *Introduction to Aristotle.* New York: Modern Library.

McNemar, Q. (1946). Opinion-attitude methodology. *Psychological Bulletin,* 43, 289–374.

Merton, R., and Lazarsfeld, P.F. (1950). *Continuities in social research: Studies in the scope and method of "The American Soldier."* Glencoe, IL: Free Press.

Miller, G.A. (1956). The magical number seven, plus or minus two: Some limits on our capacity for processing information. *The Psychological Review,* 63, 81–97.

Miller, W.E. and NES (National Election Study) staff. (1993). *American National Election Studies: Cumulative data file 1952–1992.* Center for Political Studies. Institute for Social Research. Ann Arbor, MI: University of Michigan.

Moore, J.C. (1988). Self/proxy report status and survey response quality. *Journal of Official Statistics,* 4, 155–172.

Moser, C.A., and Kalton, G. (1972). *Survey methods in social investigation,* 2nd edition. New York: Basic Books.

Murphy, K.R. and Myors, B. (2004). *Statistical power analysis: A simple and general model for traditional and modern hypothesis tests,* 2nd edition. Mahwah, NJ: Lawrence Erlbaum Associates.

Muthén, B.O. (1984). A general structural equation model with dichotomous, ordered categorical, and continuous latent variable indicators. *Psychometrika,* 49, 115–132.

Muthén, B.O. (1998–2004). M*plus—technical appendices*. Los Angeles: Muthén and Muthén.

Muthén, L.K., and Muthén, B.O. (2001–2004). M*plus—the comprehensive modeling program for applied researchers. User's guide*. Version 2.0. Los Angeles: Muthén and Muthén.

Oksenberg, L., and Cannell, C.F. (1977). Some factors underlying the validity of response in self report. *Proceedings of the 41st Session of the International Statistical Institute*, New Delhi, India.

O'Muircheartaigh, C.A. (1997). Measurement error in surveys: A historical perspective. In L. Lyberg, et al. (Eds.), *Survey measurement and process quality* (pp. 1–25). New York: Wiley-Interscience.

O'Muircheartaigh, C.A., Gaskell, G.D., and Wright, D.B. (1993). Intensifiers in behavioral frequency questions. *Public Opinion Quarterly*, 57, 552–565.

Palmquist, B., and Green, D.P. (1992). Estimation of models with correlated errors from panel data. In P.V. Marsden (Ed.), *Sociological methodology 1992* (pp. 119–146). Oxford: Basil Blackwell.

Park, D.C., Smith, A.D., Lautenschlager, G., Earles, J.L., Frieske, D., Zwahr, M., and Gaines, C.L. (1996). Mediators of long-term memory performance across the life span. *Psychology and Aging*, 11, 621–637.

Payne, S.L. (1951). *The art of asking questions*. Princeton, NJ: Princeton University Press.

Perlmutter, M. (1978). What is memory aging the aging of? *Developmental Psychology*, 14, 330–345.

Petersen, T. (1993). Recent advances in longitudinal methodology. *Annual Review of Sociology*, 19, 25–54.

Pirsig, R.M. (1974). *Zen and the art of motorcycle maintenance: An inquiry into values*. New York: Perennial Classics. [2nd edition, 2000]

Presser, S. (1990). Can change in context reduce overreporting in surveys? *Public Opinion Quarterly*, 54, 586–93.

Presser, S., and Traugott, M. (1992). Little white lies and social science models. *Public Opinion Quarterly*, 56, 77–86.

Rasch, G. (1961). On general laws and the meaning of measurement in psychology. In J. Neyman (Ed.), *Proceedings of the 4th Berkeley Symposium on Mathematical Statistics and Probability*, Vol. 4. Berkeley, CA: University of California Press.

Rasch, G. (1966a). An individualistic approach to item analysis. In P.F. Lazarsfeld and N.W. Henry (Eds.), *Readings in mathematical social science*. Chicago: Science Research Associates.

Rasch, G. (1966b). An item analysis which takes individual differences into account. *British Journal of Mathemetical and Statistical Psychology*, 19, 49–57.

Rasch, G. (1960). *Probabalistic models for some intelligence and attainment tests*. Copenhagen: Danmarks Paedogogishe Institut.

Riley, M.W. (1973). Aging and cohort succession: Interpretations and misinterpretations. *Public Opinion Quarterly*, 37, 35–47.

Rodgers, W.L. (1989). Measurement properties of health and income ratings in a panel study. Presented at the 42nd Annual Scientific Meeting of the Gerontological Society of America, Minneapolis, November.

Rodgers, W.L., and Herzog, A.R. (1987a). Interviewing older adults: The accuracy of factual information. *Journal of Gerontology*, 42, 387–394.

Rodgers, W.L., and Herzog, A.R. (1987b). Covariances of measurement errors in survey responses. *Journal of Official Statistics*, 3, 403–418.

Rodgers, W.L., and Herzog, A.R. (1992). Collecting data about the oldest old: Problems and procedures. In R.M. Suzman, D.P. Willis, and K.G. Manton (Eds.), *The oldest old* (pp. 135–156). New York and Oxford: Oxford University Press.

Rodgers, W.L., Andrews, F.M., and Herzog, A.R. (1992). Quality of survey measures: A structural modeling approach. *Journal of Official Statistics*, 3, 251–275.

Rodgers, W.L., Herzog, A.R., and Andrews, F.M. (1988). Interviewing older adults: Validity of self-reports of satisfaction. *Psychology and Aging*, 3, 264–272.

Rodgers, W.L., McCammon, R.J., and Alwin, D.F. (2004). Assessing mode effects through analysis of covariances across four waves of data. Health and Retirement Study Conference, Traverse City, MI. August.

Rokeach, M. (1970). *Beliefs, attitudes and values*. San Francisco: Jossey-Bass.

Ross, M. (1988). The relation of implicit theories to the construction of personal histories. *Psychological Review*, 96, 341–357.

Ruckmick, C.A. (1930). The uses and abuses of the questionnaire procedure. *Journal of Applied Psychology*, 14, 32–41

Rugg, D., and Cantril, H. (1944). The wording of questions. In H. Cantril (Ed.), *Gauging public opinion* (pp. 23–50). Princeton, NJ: Princeton University Press.

Salthouse, T.A. (1991). *Theoretical perspectives on cognitive aging*. Hillsdale, NJ: Erlbaum.

Salthouse, T.A. (1996). The processing-speed theory of adult age differences in cognition. *Psychological Review*, 103, 403–428.

Saris, W.E. (1988). *Variation in response functions: A Source of measurement error in attitude research*. Amsterdam, the Netherlands: Sociometric Research Foundation.

Saris, W.E., and Andrews, F.M. (1991). Evaluation of measurement instruments using a structural modeling approach. In P.B. Biemer et al. (Eds.), *Measurement errors in surveys* (pp. 575–597). New York: Wiley.

Saris, W.E., and van den Putte, B. (1988). True score or factor models: A secondary analysis of the ALLBUS test-retest data. *Sociological Methods and Research*, 17, 123–157.

Saris, W.E., and van Meurs, A. (1990). *Evaluation of measurement instruments by meta-analysis of multitrait multimethod studies*. Amsterdam: North-Holland.

Satorra, A., and Bentler, P.M. (2001). A scaled difference chi-square test statistic for moment structure analysis. *Psychometrika*, 66, 507–514.

Schaeffer, N.C. (1991a). Hardly ever or constantly? Group comparisons using vague quantifiers. *Public Opinion Quarterly*, 55, 395–423.

Schaeffer, N.C. (1991b). Conversation with a purpose—or conversation? Interaction in the standardized interview. In P.B. Biemer et al. (Eds.), *Measurement error in surveys* (pp. 367–391). New York: Wiley.

Schaeffer, N.C., and Presser, S. (2003). The science of asking questions. *Annual Review of Sociology*, 29, 65–88.

Schaie, K.W. (1996). *Intellectual development in adulthood: The Seattle Longitudinal Study*. Cambridge UK: Cambridge University Press.

Schaie, K.W. (2005). *Developmental influences on adult intelligence: The Seattle Longitudinal Study*. New York: Oxford University Press.

Scherpenzeel, A.C. (1995). *A question of quality: Evaluating survey questions by multitrait-multimethod studies*. Ph.D. thesis. University of Amsterdam.

Scherpenzeel, A.C., and Saris, W.E. (1997). The validity and reliability of survey questions: A meta-analysis of MTMM studies. *Sociological Methods and Research*, 25, 341–383.

Schuman, H., and Kalton, G. (1985). Survey methods. In G. Lindzey and E. Aronson (Eds.), *The handbook of social psychology* (pp. 634–697), 3rd edition. New York: Random House.

Schuman, H., and Presser, S. (1981). *Questions and answers: Experiments in question wording, form and context*. New York: Academic Press.

Schwarz, N. (1999a). Cognitive research into survey measurement: Its influence on survey methodology and cognitive theory. In M.G. Sirken, et al. (Eds.), *Cognition and survey research* (pp. 65–75). New York: Wiley.

Schwarz, N. (1999b). Self-reports: How the questions shape the answer. *American Psychologist*, 54, 93–105.

Schwarz, N., and Sudman, S. (1994). *Autobiographical memory and the validity of retrospective reports*. New York: Springer-Verlag Inc.

Scott, J., and Alwin, D.F. (1997). Retrospective vs. prospective measurement of life histories in longitudinal research. In J.Z. Giele and G.H. Elder, Jr. (Eds.), *Methods of life course research: Qualitative and quantitative approaches* (pp. 98–127). Newbury Park, CA: Sage Publications.

Sears, D.O. (1983). The persistence of early political predispositions: The roles of attitude object and life stage. *Review of Personality and Social Psychology*, 4, 79–116.

Sewell, W.H., Hauser, R.M., Springer, K.W., and Hauser, T.S. (2003). As we age: The Wisconsin Longitudinal Study, 1957–2001. In K. Leicht (Ed.), *Social stratification and mobility*, Vol. 20 (pp. 3–111). London: Elsevier.

Shannon, C., and Weaver, W. (1949). *The mathematical theory of communication*. Urbana, IL: University of Illinois Press.

Sherif, M., and Hovland, C.I. (1961). *Social judgement: Assimiliation and contrast effects in communication and attitude change*. New Haven, CT: Yale University Press.

Sherif, C.W., Sherif, M., and Nebergall, R.E. (1965). *Attitude and attitude change*. Philadelphia, PA: Saunders.

Siegel, P.M., and Hodge, R.W. (1968). A causal approach to the study of measurement error. In H.M. Blalock, Jr. and A.B. Blalock (Eds.), *Methodology in social research* (pp. 28–59). New York: McGraw-Hill.

Simon, H.A. (1957). *Models of man*. New York: Wiley.

Simon, H.A., and Stedry, A.C. (1968). Psychology and economics. In G. Lindzey and E. Aronson (Eds.), *Handbook of social psychology* (pp. 269–314), Vol. 5, 2nd edition. Reading, MA: Addison-Wesley.

Sirken, M.G., Herrmann, D.J., Schechter, S., Schwarz, N., Tanur, J.M., and Tourangeau, R. (1999). *Cognition and survey research*. New York: Wiley.

Skrondal, A., and Rabe-Hesketh, S. (2004). *Generalized latent variable modeling: Multilevel, longitudinal, and structural equation models*. New York: Chapman and Hall/CRC.

Sobel, D. (1999). *Galileo's daughter: A historical memoir of science, faith, and love*. New York: Penguin Books.

Soldo, B.J., Hurd, M.D., Rodgers, W.L., and Wallace, R.B. (1997). Asset and health dynamics among the oldest old: An overview of the AHEAD study. *Journal of Gerontology: Psychological Sciences*, 52B, 1–20.

Steiger, J.H., and Lind, J.M. (1980). Statistically based tests for the number of common factors. Paper presented at the annual meeting of the Psychometric Society, Iowa City, Iowa.

Strack, F., and Martin, L. (1987). Thinking, judging and communicating: A process account of context effects in attitude surveys. In H.-J. Hippler et al. (Eds.), *Social information processing and survey methodology* (pp. 123–148). New York: Springer-Verlag.

Sudman, S., and Bradburn, N.M. (1974). *Response effects in surveys*. Chicago, IL: Aldine.

Sudman, S., and Bradburn, N.M. (1982). *Asking questions: A practical guide to questionnaire design*. San Francisco: Jossey-Bass.

Sudman, S., Bradburn, N.M., and Schwarz, N. (1996). *Thinking about answers: The application of cognitive processes to survey methodology*. San Francisco: Jossey-Bass.

Thomson, W. (Lord Kelvin). (1883). *Electrical units of measurement*. Cambridge University, England.

Thornton, A., Freedman, D., and Camburn, D. (1982). Obtaining respondent cooperation in family panel studies. *Sociological Methods and Research*, 11, 33–51.

Thornton, A., and Binstock, G. (2001). Reliability of measurement and cross-time stability of individual and family variables. *Journal of Marriage and Family*, 63, 881–894.

Tourangeau, R. (1984). Cognitive science and survey methods. In T. Jabine et al. (Eds.), *Cognitive aspects of survey methodology: Building a bridge between disciplines*. Washington, DC: National Academy Press.

Tourangeau, R. (1987). Attitude measurement: A cognitive perspective. In H.-J. Hippler et al. (Eds.), *Social information processing and survey methodology* (pp. 149–162). New York: Springer-Verlag.

Tourangeau, R. (1999). Context effects to answers to attitude questions. In M.G. Sirken, et al. (Eds.), *Cognition and survey research* (pp. 111–131). New York: Wiley.

Tourangeau, R., and Rasinski, K. (1988). Cognitive processes underlying context effects in attitude measurement. *Psychological Bulletin*, 103, 299–314.

Tourangeau, R., Rips, L.J., and Rasinski., K. (2000). *The psychology of survey response*. Cambridge: Cambridge University Press.

Traugott, M., and Katosh, J.P. (1979). Response validity in surveys of voting behavior. *Public Opinion Quarterly*, 43, 359–77.

Traugott, M., and Katosh, J.P. (1981). The Consequences of validated and self-reported voting measures. *Public Opinion Quarterly*, 45, 519–35.

Tuma, N.B., and Hannan, M.T. (1984). *Social dynamics: Models and methods*. Orlando, FL: Academic.

Turner, C.F., and Martin, E. (1984). *Surveying subjective phenomena*. New York: Russell Sage.

van de Pol, F., and Langeheine, R. (1990). Mixed Markov latent class models. In C.C. Clogg (Ed.), *Sociological methodology 1990* (pp. 213–247). Oxford: Blackwell.

van de Pol, F., and de Leeuw, J. (1986). A latent Markov model to correct for measurement error. *Sociological Methods and Research*, 15, 118–141.

van de Pol, F., Langeheine, R., and de Jong, W. (1991). *PANMARK user manual: PANel Analysis using MARKov Chains*. Voorburg: Netherlands Central Bureau of Statistics.

van der Zouwen, J. (1999). An assessment of the difficulty of questions used in the ISSP questionnaires, the clarity of their wording, and the comparability of the responses. *ZA-Information*, 46, 96–114.

van Meurs, A., and Saris, W.E. (1990). Memory effects in MTMM studies. In W.E. Saris and A. van Meurs (Eds.), *Evaluation of measurement instruments by meta-analysis of multitrait multimethod matrices* (pp. 52–80). Amsterdam: North-Holland.

Vermunt, J.K., Langeheine, R., and Böckenholt, U. (1999). Discrete-time discrete-state latent markov models with time-constant and time-varying covariates. *Journal of Educational and Behavioral Statistics*, 24, 179–207.

Weisberg, H., and Miller, A.H. (No date). Evaluation of the feeling thermometer: A report to the National Election Study Board based on data from the 1979 pilot survey. Ann Arbor, MI: University of Michigan, Center for Political Studies, Institute for Social Research.

Werts, C.E., and Linn, R.L. (1970). Path analysis: Psychological examples. *Psychological Bulletin*, 74,194–212.

Werts, C.E., Jöreskog, K.G., and Linn, R.L. (1971). Comment on "The estimation of measurement error in panel data." *American Sociological Review*, 36, 110–112.

Werts, C.E., Linn, R.L., and Jöreskog, K.G. (1977). A simplex model for analyzing academic growth. *Educational and Psychological Measurement*, 37, 745–756.

Werts, C.E., Linn, R.L., and Jöreskog, K.G. (1978). The reliability of college grades from longitudinal data. *Educational and Psychological Measurement*, 38, 89–95.

Werts, C.E., Breland, H.M., Grandy, J., and Rock, D.R. (1980). Using longitudinal data to estimate reliability in the presence of correlated measurement errors. *Educational and Psychological Measurement*, 40, 19–29.

Wicker, A.W. (1969). Attitudes versus actions: The relationship of verbal and overt behavioral responses to attitude objects. *Journal of Social Issues*, 25, 41–78.

Wiggins, L.M. (1973). *Panel analysis: Latent probability models for attitude and behavior processes*. New York: Elsevier Scientific Publishing Company.

Wiley, D.E., and Wiley, J.A. (1970). The estimation of measurement error in panel data. *American Sociological Review*, 35, 112–117.

Wothke, W. (2000). Longitudinal and multigroup modeling with missing data. In T.D. Little, K.U. Schnabel, and J. Baumert (Eds.), *Modeling longitudinal and multilevel data: Practical issues, applied approaches and specific examples* (pp. 219–240). Mahwah, NJ: Lawrence Erlbaum Associates.

Index

Activities, 126, 156–158, 329–330
Americans' Changing Lives (ACL) panel
    study
  age and reliability, 218–220, 223–227
  attrition, 135–146, 155
  measures, 126–127
  question position, 174–175
  reliability, 155–162, 175–179, 181–210,
    217–227, 303
  sample, 119, 121
  sample weights, 130
  schooling and reliability, 221–227, 247,
    261
  use of Don't Know option, 199
AMOS software package, 84, 140–144, 229,
    298, 369
Assets and Health Dynamics (AHEAD)
    study, 231–261, *see* Health and
    Retirement Study
Accessibility of information, xi, 12, 22,
    25–26, 213, 218
Age, 13, 44, 157, *see also* reliability and age
Agree-disagree questions, *see* Likert-type
    questions
Architecture of questionnaires, 14, 165–171,
    179–180, 319–320
Assets, 184, 329
Attenuation, 9, 71, 100–101, 192, 298, 299,
    307, 310, 315, 323
Attitudes, 1, 11, 12–13
  definition, 122–124
  Don't Know option, 196–200
  measurement, 44, 118, 120, 126–127, 186,
    188–190

Attitudes *(continued)*
  nonattitudes, *see* nonattitudes
  reliability estimates, 131–136, 153–162,
    168–171, 177–179, 346–351
  unipolar vs. bipolar, 195–196
Attrition, 13–14, 116, 119, 121, 135–148

Batteries, 13, 45, 57, 76, 79, 163, 234–235,
    see also series
  characteristics, 168–169
  definition, 167–168
  introduction length, 177–179
  location, 172–175
  position in batteries, 175–177
  reliability, 171–172, 175–179, 184–191,
    203–211, 318–321, 324
Beliefs, 11, 12
  definition, 122–124
  Don't Know option, 196–200
  measurement, 120, 126–127, 186,
    188–190, 268–271, 280–286
  reliability estimates, 131–136, 153–162,
    168–171, 177–179, 336–346
  unipolar vs. bipolar, 195–196
Bennington College study, 28–29
Bias
  due to attrition, 135–146
  in reliability estimation, 60, 101, 109,
    128, 272, 305, 307
  measurement bias, 18, 20, 27–29, 46,
    48–49
  survey bias, 5–6
  statistical bias, 32–33, 91, 233, 292–299,
    302, 315

Bipolar questions, 183, 185, 186–187, 188, 190–191, 195–197, 199, 211, 321–323
Birth cohorts, 121, 223, 233, 235
Body weight, 8, 156–157, 234–238, 241, 258, 318, 327
Bureau of Census data, 18, 150, 302–303

Categorical variables, 14, 28, 44, 45, 57, 116, 129, 257, 263–287, 317
Church attendance, 20, 156–157, 330
Classical test theory, *see* classical true score theory
Classical true-score theory (CTST), 35–42
  composite score theory, 51–53, 308–315, *see* Cronbach's alpha
  CTST and latent class models, 263–265, 271–272, 286–287
  CTST and longitudinal design, 101–116
  CTST and MTMM, 67–79, 80–82, 91–93, 229–230, *see* multitrait-multimethod
  multiple indicators, 65–67
  nonrandom error, 53–55
  reliability, *see* reliability of measurement
  replicate measures, 61–65
  sampling, 55–57
  scaling of variables, 43–45
  test-retest method, 96–101
  validity, 46–50
Closed questions, *see* forced-choice questions, *see also* open questions vs. closed questions
Coefficient alpha (α), *see* Cronbach's alpha (α)
Cohort effects, 14, 215–216, 220–221, 223, 225–226, 262
Common-factor model, 6, 10, 36, 42–43, 51–53, 60–61, 65–70, 71–76, 77, 81–82, 92, 111–113, 272–273, 307
Composite variables, 10, 13, 51–53, 59–60, 117, 308–315
Comprehension, xi, 9, 12, 17, 22, 23–25, 163, 203, 212, 213, 218, 315
Congeneric measures, 41, 42, 43, 52, 62, 64, 66–67, 111, 216, 315
Construct validity, *see* validity of measurement
Constructs, 6–7, 11, 23, 54, 95, 113, 117, 190, 301, 310, *see also* construct validity

Content validity, *see* validity of measurement
Context effects, 24–25, 50, 167–171, 320, *see also* question context
Convergent validity, *see* validity of measurement
Correction for attenuation, 9, 71, 299, 307, 310, *see also* attenuation
Coverage error, 3–4, 117, 289
Criterion validity, *see* validity of measurement
Cronbach's alpha (α), 10, 36, 51–52, 59, 60, 90, 113, 118, 309–315
Cross-cultural differences, 17, 24, 182
Cross-national differences, 6, 24, 32, 182, *see* cross-cultural differences

Delighted-terrible scale, 83–88, 90, 91, 229–232
Detroit Area Study, 121
Discriminant validity, *see* validity of measurement
Don't know response, 17, 25–26, 198, 218
Don't know option, 20, 25, 79, 163, 166, 178, 180, 183, 186, 196–200, 201, 210–211, 321, 323, 324

Earnings, 18–19, 302–304, *see also* income
Economy, 120, 126, 161, 337–338
Education, 1, 6, 7, 10, 11, 13, 14, 27–28, 57, 116, 125, 126, 152, 156–157, 213, 215, 216–217, 221–231, 237–253, 262, 302–304, 318, 328
Employment, 18–19, 27–28, 122, 126, 156–157, 161, 232, 277–278, 328
Essential tau-equivalence, 50, 97–99
Event history models (EHM), 10–11
Expectations, 25, 29, 122, 126–127, 218, 234–253, 257–260

Factor model, *see* common factor model
Face-to-face interviews, 2, 79, 120, 121, 122, 201–202
Facts, 11, 123, 126–127, 130, *see* reliability—content of question
Family, 127, 161, 327, *see also* quality of life
Feeling thermometer, 62, 83–91, 120, 131, 178, 185, 187–190, 193, 195, 201, 229–231, 322

Filter questions, 17, 20, 26, 50, 198–200
Forced-choice questions, 184, 185, 187, 188–191, 193, 321–322
Full-information maximum likelihood (FIML), 84–85, 140–144, 147, 229
Functional status, 1, 160–161, 353

Generational differences, 227

Health and health behavior, 126, 157, 330–331, *see* self-reported health *and* quality of life
Health and Retirement Study (HRS)
  age and reliability, 237, 242–253, 254–261
  measures, 233–235
  reliability, 235–261
  samples, 232–233
  schooling and reliability, 237–253
Hierarchical linear models (HLM), 10–11
Household characteristics, 327
Housing, 126, 219, *see also* place of residence and quality of life

Income, 1, 6, 19, 30, 36, 44, 123, 126, 156–157, 185, 302–304, 305–308, 328–329, *see also* quality of life
Incomplete data, *see* missing data
Index of reliability, 40, 47, 48, 49, 54, 82, 89, 108, 267–268, 287, 291–292, 310
Indicator validity, *see* validity of measurement
Indicators, 1, 6–9, 11–12, 22–23, 54, 111, *see also* multiple indicators
Instrument error, 3–6, 7, 106, 213
Internal consistency reliability (ICR), 36, 37, 51–53, 57, 59–61, 71, 77, 93, 118, 308–315
Interviewer error, 5–6
Interviewer reports, 13, 125–126, 150–152, 363–364, *see also* reliability of interviewer observations
Introductions to questions, 79, 203–207, 210, 211, *see also* reliability—length of questions
Introductions to units (series and batteries), 79, 166, 167–170, 177–180, 189–190, 198, *see also* reliability and length of introductions

Invalidity of measurement, 80–82, 85–91, 228–232, *see* validity of measurement

Latent class models, 57, 129, 263–287, 317
Latent classes, *see* latent class models
Latent Markov model, 129, 277–287
Latent variable models, *see* latent variables
Latent variables, 13–14, 129, 317
  categorical, 129, 263–287
  continuous, 35–57, 59–94, 95–116, 127–130, 137–146
Length of introductions, *see* introductions to questions, and *see* introductions to units
Length of questions, *see* question length
Length of unit, *see* unit length
Life history calendar, 27
Likert-type questions, 62, 183–184, 185, 186, 188, 190–191, 193, 322
LISREL computer program, 45, 128, 142, 298, 375
LISREL-type models, 9, 11, 117
Local independence, *see* measurement independence
Location in questionnaire, 166, 172–175, *see* reliability of meaurement and location of question
Longitudinal design, 13–14, 16, 45–46, 60, 67, 95–116, 117–148, 263–265, 272–287, 299–301, 304, 305, 317

Mail questionnaires, 317, *see* self-administered questionnaries
Marital status, 125
Maternal employment, 277–278
Maximum-likelihood, 65, 84–86, 142, 229–232
Measurement, 8–9
  survey measurement, *see* survey measurement error
Measurement bias, *see* bias
Measurement error, *see* classical true-score theory, CTST
  consequences, 8–10, 12, 32–33, 292–299
  defined, 3, 12–13
Measurement independence, 38, 57, 64, 115, 266, 287
Measurement quality, 1, 11, 16, 79, 166, 210–212, 290–292, 301, 316, 318–325

Measurement standardization, 8–9
Memory, 14, 15, 18, 21, 25, 46, 57, 74,
    91–93, 103, 110
  age and memory, 218–221
  reconstructive memory, 158
  retrieval, 26–29, see retrieval
  test-retest design, 96–101
Method variance, 33, 53, 67–94, 228–231
Middle alternatives, 20, 31, 188, 191
Missing data, 83, 130, 137–140, 144, 229
  missing at random (MAR), 84, 137–139,
    229
  missing completely at random (MCAR),
    137–139, 142–143
Mode of administration, 5–6, 317
Motivation, xi, 2, 12, 19–22, 29–30, 74, 95,
    113, 166, 172, 175, 192, 213, 219, 301
Mplus, 45, 128, 142, 242, 257, 269, 279,
    298, 378
Multiple indicators, 6, 10, 11, 35–36, 46,
    60–62, 65–67, 72, 91, 93, 95, 102,
    111–113, 114, 117–118, 265–272, 281,
    285, 301, 306, 307
Multiple measures, 10, 11, 35–36, 46, 50,
    61–64, 111, 114, 272–277, 306, 310
Multitrait-multimethod (MTMM), xiv, 16,
    23, 60–61, 65–67, 68–94, 95, 111, 216,
    217, 228–231, 301, 305
  common factor model of MTMM, 71–73
  CTST and MTMM, 77–79
  decomposition of variance, 74–76, 80–82
  MTMM studies, 79
  multitrait-multimethod matrix, 68–71

National Election Studies (NES)
  age and reliability, 218–220, 223–227
  latent class models, 268–286
  measures, 126–127
  panel attrition, 135–146, 155
  proxy reports, 152–153
  question position, 174–175
  reliability, 155–162, 175–179, 181–210,
    218–227, 303
  samples, 119–121
  sample weights, 130
  schooling and reliability, 221–227, 247, 261
  thermometers, see feeling thermometers
  use of Don't Know options, 199

No-opinion filters, see filter questions
Nonattitudes, 25–26, 154, 218, see also
    attitudes
Nonrandom measurement errors, 10, 32–33,
    41–42, 53–54, 66–68, 79, 291
Nonresponse errors, 5–6, see survey
    nonresponse
Number of response categories, 13, 191–195,
    see also reliability

Occupation, 6, 17, 18, 27, 125, 126, 153,
    156, 157, 184, 302–307, 328
Open questions vs. closed questions, 30,
    183–185
Optimizing, 19–20
Ordinal measurement, 43–45, 57, 84,
    128–129, 134, 148, 229, 242, 257,
    263–264, 317
Organization reports, 125–126, 150–152,
    319, 364–365
Oversampling, 121, 233

Panel Study of Income Dynamics (PSID),
    305
Parallel measures, 41, 43, 49, 99, 111
Parental nativity, 277–278
Parental socioeconomic characteristics,
    17–18, 305–306, 308
Pearson correlations, 43–45, 84–87, 89,
    128–129, 131–136, 142, 146–147, 148,
    153, 157, 192–193, 235–236, 298–299
Personal characteristics, 126–127, 327
Place of residence, 126, 157, 327–328, see
    also quality of life
Political behavior and attitudes, 28–29,
    120, 123, 127, 153, 157, 160–162, 178,
    188–189, 205–207, 330, see also voting
Polychoric correlations, 44–45, 84–89,
    128–129, 130–137, 146–148, 150, 157,
    189, 192–194, 229, 235–236, 241–257
Population coverage, 3–6, 117, 289, see
    coverage error
Predictive validity, see validity of
    measurement
Proxy reports, 13, 26, 125–127, 131–136,
    152–153, 162, 318–319, 324, 362–363
Psychological well being, 126, 161–162, 352,
    see also quality of life

Quality of data, 1–14, 15–33, 53, 79, 119, 163, 165–166, 172–173, 182–183, 200, 289, 315–325

Quality of life, 83–91, 126–127, 161, 228–231, 353, 361

Quality of Life (QoL) survey, 83, 228

Quasi-simplex model, 13, 16, 57, 60, 67, 140–141, 217, 255–257, *see* quasi-simplex Markov model

Quasi-simplex Markov model, 102–110

Question adequacy, xi, 12, *see* content validity

Question batteries, *see* batteries

Question content, 12–13, 153–163

Question context, 13, 163, 165–180, 210, 321

Question length, 202–210

Question series, *see* series

Questionnaire design, 9, 14, 79, 93, 154, 163, 165–167, *see also* architecture of questionnaires

Race, 1, 125, 263, 277–278

Rational choice theory, 29

Record-check studies, 20, 23, 48–49

Region, 263

Reinterview design, 18, 27–28, 46, 61, 96–116, 119–122, 216–217, 232–233, 301

Reliability of measurement
age and reliability, 157, 215–262
composite scores, *see* composite variables
content of question and reliability, 149–150, 153–163
context of question and reliability, 165–180
definition of, 36–41
designs for estimation, 45–46, 93–94
Don't Know options and reliability, 196–200
education and reliability, 213–262
estimation of, 61–64, 95–116, 130–148
internal consistency (distinguished from), 60, 308–315
interviewer observations and reliability, 150–152
length of introduction and reliability, 177–180
length of question and reliability, 202–210

Reliability of measurement *(continued)*
location of question and reliability, 172–175
question form and reliability, 183–191
MTMM and reliability, 71
nonrandom measurement error and reliability, 41–42, 53–55
number of response categories and reliability, 191–200
position in series and batteries, 175–177
properties of questions and reliability, 13, 181–212, 318–325
proxy reports and reliability, 152–153
question form and reliability, 183–191
source of information and reliability, 150–153
unit length and reliability, 175–177
validity and reliability, 46–47, 291–292, *see* MTMM
verbal labeling and reliability, 200–202

Religious activities, 157, 330, *see also* church attendance

Religious preference, 125

Repeated measures, *see* test-retest design

Retrieval, xi, 2, 12, 21, 22, 25, 26–29, 149, 213, 218

Retrospective measurement, 17–19, 25–29, 156, 158

Root mean square error of approximation (RMSEA), 85, 145–146

Sample bias, 5–6

Sample weights, 130, 142–144, 146

Samples, 4–5, 83, 119–122, 232–233

Sampling, 35, 55–57, 130

Sampling error, 3–6, 9, 117, 289

Sampling frame, *see* sample coverage

Satisficing, 19–20

Satisfied-dissatisfied scale, 83–88, 90, 91, 229–232

Scale construction, 10, 36, 59, 308–315

Scaling of variables, 43–45, 128–129, 130–136

School attendance, 28, *see also* education

Schooling, *see* education

Self-administered questionnaires, 2, 202, 317

Self-appraisals, 11, 153, *see* self-assessments

Self-assessments, 12
  definition, 123
  measurement, 126–127, 233–234
  reliability, 83–90, 131–136, 158–162,
    168–171, 196–197, 360–362
  unipolar vs. bipolar, 195–196
Self-descriptions, 123, 124, 125–127, 153
Self-evaluations, 123, 158, 235, *see also*
  self-assessments
Self-image, 26, 218
Self-perceptions
  definition, 123
  measurement, 126–127, 233–234
  reliability, 131–136, 158–162, 168–171,
    196–197, 233–236, 351–360
  unipolar vs. bipolar, 195–196
Self-ratings of health, *see* self-reported
  health
Self-reported health, 126, 157, 233–261
Self reports, 13, 18, 19, 20, 26, 48–49, 123,
    125–127, 142, 149–150, 152–153, 156–
    162, 168–169, 170, 172–175, 233–236,
    316–319, 324, 327–362
Series, 13, 57, 163
  characteristics, 168–169
  definition, 167–168
  introduction length, 177–179
  location, 172–175
  position in series, 175–177
  reliability, 171–172, 175–179, 184–191,
    203–211, 318–321, 324–325
Sex, 263, 277–278
Simplex models, xvi, 112, 255, 302–304, *see*
  also quasi-simplex models *and* quasi-
  simplex Markov models
Social activities, *see* activities
Social and political trust, 120, 127,
    268–272, 280–286
Social desirability, 20
Social interactions, 127, 157, 329
Social relationships, 127, 336–337
Social sciences, xi, xvi, 1–2, 291
Source of information, 149–153
Split-ballot experiments, 50
Stand-alone questions, 167–180, 204, 207,
    210, 211, 320, 321, 324, 325
Standards of measurement, 8
Stata statistical program, 298

Statistical Analysis Software (SAS), 142, 371
Statistical bias, *see* bias
Statistical sampling, 291
Structural equation models (SEM), 9, 11,
    36, 46, 57, 61, 70, 84, 102–110, 128,
    140–146, 217, 229, 242, 257, 264
Study of American Families (SAF)
  attrition, 135–146, 155
  measures, 126–127
  proxy reports, 152–153
  question position, 174–175
  reliability, 153–162, 217–228, 303
  samples, 119, 121–122, 130
  schooling and reliability, 221–227, 247,
    261
  use of Don't Know options, 199
Survey bias, *see* bias
Survey errors, 3–6, 117, 289
  nonobservational errors, 5
  observational errors, 5
Survey measurement error, 1–3, 6–8, 15–33,
    see also measurement error
Survey nonresponse, 3–6, 289
Survey research, 1–2, 6, 8–9, 10, 11–14, 117,
    152, 165, 167, 172, 181–185, 210–212,
    261–262, 289–290, 315–325

Tau-equivalent measures, 40–41, 43, 52, 62,
    111, 309, 315
Telephone interviews, 2, 79, 121, 202, 302
Test-retest design, 46, 96–102, 110, 216–217,
    301–304
Test-retest reliability, *see* test-retest design
Tetrachoric correlations, 44–45, 128, 135
True-score validity, *see* validity of
  measurement
True scores
  classical true scores, *see* classical true
    score theory (CTST)
  platonic true scores, 20, 23, 48

Unipolar questions, 183, 185, 186–187, 188,
    190–191, 195–197, 199, 211, 321–323
Unit length, 175–177
Univocal measures, 26, 36–38, 43, 53, 54,
    57, 62–67, 74, 76, 95, 110–111, 154,
    216, 299, 307
Univocity, *see* univocal measures

Validity of measurement, xi, xv, 3, 15–16, 18, 21–24, 46–50, 59, 67–70, 80–91, 114–115, 117, 289–290, 291–292, 304, 307, 316
    content validity, 22, 47, 213, *see also,* question adequacy
    construct validity, 22–23, 47, 49, 60, 70, 80, 90, 292
    convergent validity, 68, 70–71
    criterion validity, 22, 47, 49, 54, 64, 80, 292
    discriminant validity, 68, 70–71
    indicator validity, 82, 86–91, 228–232
    predictive validity, *see* criterion validity
    record validity, 48
    true-score validity, 82, 86–91, 228–232
Validity coefficient, 47, 80–91, 228–232
Validity values, *see* convergent validity
Values
    definition, 122–124
    don't know option, 196–200
    measurement, 126–127

Values *(continued)*
    reliability, 131–136, 153–162, 168–171, 186, 190, 196–197, 331–335
    unipolar vs. bipolar, 195–196
Variance
    components of reliability, *see* classical true-score theory (CTST)
    survey errors, 5–6
    measurement error variance, *see* CTST
    decomposition of MTMM variance, 80–82
    true-score variance, *see* CTST

Verbal labeling, 13, 30, 83, 84, 91, 163, 183, 190, 200–202, 211, 318, 321, 323
Voting, 20, 48, 156–157, 219, 273–275, *see also* political behavior and attitudes

Weighted least squares, 45, 242
Wisconsin Longitudinal Study (WLS), 303–308

WILEY SERIES IN SURVEY METHODOLOGY
Established in Part by WALTER A. SHEWHART AND SAMUEL S. WILKS

Editors: *Robert M. Groves, Graham Kalton, J. N. K. Rao, Norbert Schwarz, Christopher Skinner*

The *Wiley Series in Survey Methodology* covers topics of current research and practical interests in survey methodology and sampling. While the emphasis is on application, theoretical discussion is encouraged when it supports a broader understanding of the subject matter.

The authors are leading academics and researchers in survey methodology and sampling. The readership includes professionals in, and students of, the fields of applied statistics, biostatistics, public policy, and government and corporate enterprises.

ALWIN · Margins of Error: A Study of Reliability in Survey Measurement

*BIEMER, GROVES, LYBERG, MATHIOWETZ, and SUDMAN · Measurement Errors in Surveys

BIEMER and LYBERG · Introduction to Survey Quality

BRADBURN, SUDMAN, and WANSINK ·Asking Questions: The Definitive Guide to Questionnaire Design—For Market Research, Political Polls, and Social Health Questionnaires, *Revised Edition*

BRAVERMAN and SLATER · Advances in Survey Research: New Directions for Evaluation, No. 70

CHAMBERS and SKINNER (editors · Analysis of Survey Data

COCHRAN · Sampling Techniques, *Third Edition*

COUPER, BAKER, BETHLEHEM, CLARK, MARTIN, NICHOLLS, and O'REILLY (editors) · Computer Assisted Survey Information Collection

COX, BINDER, CHINNAPPA, CHRISTIANSON, COLLEDGE, and KOTT (editors) · Business Survey Methods

*DEMING · Sample Design in Business Research

DILLMAN · Mail and Internet Surveys: The Tailored Design Method

GROVES and COUPER · Nonresponse in Household Interview Surveys

GROVES · Survey Errors and Survey Costs

GROVES, DILLMAN, ELTINGE, and LITTLE · Survey Nonresponse

GROVES, BIEMER, LYBERG, MASSEY, NICHOLLS, and WAKSBERG · Telephone Survey Methodology

GROVES, FOWLER, COUPER, LEPKOWSKI, SINGER, and TOURANGEAU · Survey Methodology

*HANSEN, HURWITZ, and MADOW · Sample Survey Methods and Theory, Volume 1: Methods and Applications

*HANSEN, HURWITZ, and MADOW · Sample Survey Methods and Theory, Volume II: Theory

HARKNESS, VAN DE VIJVER, and MOHLER · Cross-Cultural Survey Methods

KALTON and HEERINGA · Leslie Kish Selected Papers

KISH · Statistical Design for Research

*KISH · Survey Sampling

KORN and GRAUBARD · Analysis of Health Surveys

LESSLER and KALSBEEK · Nonsampling Error in Surveys

LEVY and LEMESHOW · Sampling of Populations: Methods and Applications, *Third Edition*

LYBERG, BIEMER, COLLINS, de LEEUW, DIPPO, SCHWARZ, TREWIN (editors) · Survey Measurement and Process Quality

*Now available in a lower priced paperback edition in the Wiley Classics Library.

MAYNARD, HOUTKOOP-STEENSTRA, SCHAEFFER, VAN DER ZOUWEN ·
  Standardization and Tacit Knowledge: Interaction and Practice in the Survey Interview
PORTER (editor) · Overcoming Survey Research Problems: New Directions for
  Institutional Research, No. 121
PRESSER, ROTHGEB, COUPER, LESSLER, MARTIN, MARTIN, and SINGER
  (editors) · Methods for Testing and Evaluating Survey Questionnaires
RAO · Small Area Estimation
REA and PARKER · Designing and Conducting Survey Research: A Comprehensive
  Guide, *Third Edition*
SÄRNDAL and LUNDSTRÖM · Estimation in Surveys with Nonresponse
SCHWARZ and SUDMAN (editors) · Answering Questions: Methodology for
  Determining Cognitive and Communicative Processes in Survey Research
SIRKEN, HERRMANN, SCHECHTER, SCHWARZ, TANUR, and TOURANGEAU
  (editors) · Cognition and Survey Research
SUDMAN, BRADBURN, and SCHWARZ · Thinking about Answers: The Application
  of Cognitive Processes to Survey Methodology
UMBACH (editor) · Survey Research Emerging Issues: New Directions for Institutional
  Research No. 127
VALLIANT, DORFMAN, and ROYALL · Finite Population Sampling and Inference: A
  Prediction Approach